ADVANCES IN GENETICS

VOLUME 28

Genomic Responses to Environmental Stress

Contributors to This Volume

Christopher A. Cullis
Richard A. Dixon
Maria J. Harrison
Luis Herrera-Estrella
Arthur J. Hilliker
Joe L. Key
Janice A. Kimpel
Ronald T. Nagao
Nancy S. Petersen
John P. Phillips
John G. Scandalios
June Simpson
Michael F. Thomashow

ADVANCES IN GENETICS

Edited by

JOHN G. SCANDALIOS

Department of Genetics
North Carolina State University
Raleigh, North Carolina

THEODORE R. F. WRIGHT

Department of Biology
University of Virginia
Charlottesville, Virginia

VOLUME 28

Genomic Responses to Environmental Stress

Edited by

JOHN G. SCANDALIOS

Department of Genetics
North Carolina State University
Raleigh, North Carolina

ACADEMIC PRESS, INC.
Harcourt Brace Jovanovich, Publishers

San Diego   New York   Boston
London   Sydney   Tokyo   Toronto

Academic Press, Inc.
San Diego, California 92101

United Kingdom Edition published by
Academic Press Limited
24-28 Oval Road, London NW1 7DX

Library of Congress Catalog Card Number: 47-30313

ISBN   0-12-017628-9   (alk. paper)

Printed in the United States of America
90 91 92 93   9 8 7 6 5 4 3 2 1

CONTENTS

Response of Plant Antioxidant Defense Genes to Environmental Stress

JOHN G. SCANDALIOS

Genetic Analysis of Oxygen Defense Mechanisms in *Drosophila melanogaster*

JOHN P. PHILLIPS AND ARTHUR J. HILLIKER

CONTENTS

Molecular and Cellular Biology of the Heat-Shock Response

RONALD T. NAGAO, JANICE A. KIMPEL, AND JOE L. KEY

Effects of Heat and Chemical Stress on Development

NANCY S. PETERSEN

CONTRIBUTORS TO VOLUME 28

Numbers in parentheses indicate the pages on which the authors' contributions begin.

CHRISTOPHER A. CULLIS (73), *Department of Biology, Case Western Reserve University, Cleveland, Ohio 44106*

RICHARD A. DIXON (165), *Plant Biology Division, The Samuel Roberts Noble Foundation, Inc., Ardmore, Oklahoma 73402*

MARIA J. HARRISON (165), *Plant Biology Division, The Samuel Roberts Noble Foundation, Inc., Ardmore, Oklahoma 73402*

LUIS HERRERA-ESTRELLA (133), *Centro de Investigación y Estudios, Avanzados del I. P. N., 36500 Irapuato, Guanajuato, Mexico*

ARTHUR J. HILLIKER (43), *Department of Molecular Biology and Genetics, University of Guelph, Guelph, Ontario, Canada N1G 2W1*

JOE L. KEY (235), *Botany Department and Office of the Vice President for Research, University of Georgia, Athens, Georgia 30602*

JANICE A. KIMPEL (235), *Office of the Vice President for Research, University of Georgia, Athens, Georgia 30602*

RONALD T. NAGAO (235), *Botany Department, University of Georgia, Athens, Georgia 30602*

NANCY S. PETERSEN (275), *Department of Molecular Biology, University of Wyoming, Laramie, Wyoming 82071*

JOHN P. PHILLIPS (43), *Department of Molecular Biology and Genetics, University of Guelph, Guelph, Ontario, Canada N1G 2W1*

JOHN G. SCANDALIOS (1), *Department of Genetics, North Carolina State University, Raleigh, North Carolina 27695*

JUNE SIMPSON (133), *Centro de Investigación y Estudios, Avanzados del I. P. N., 36500 Irapuato, Guanajuato, Mexico*

MICHAEL F. THOMASHOW (99), *Department of Crop and Soil Sciences and Department of Microbiology and Public Health, Michigan State University, East Lansing, Michigan 48824*

PREFACE

Every living organism is affected by its environment. Since the environment, internal or external to the organism, is changing continuously, the organism must adapt in order to survive. Although numerous studies in the past have demonstrated clearly "cause–effect" relationships upon exposure of a given organism or cell to a particular environmental factor or stressor, it was only recently that certain environmental stresses were shown to elicit specific responses at the DNA level. Some examples of environmental stresses that have been shown to elicit specific genomic responses in a variety of organisms are pathogenicity, temperature shifts, light, anaerobiosis, and oxidative stress.

With the dramatic advances in molecular technologies during the last few years, it has become possible to investigate the underlying mechanisms utilized by organisms to cope with environmental insult. An in-depth investigation of genomic responses to challenge is beginning to shed some light on unique DNA sequences capable of perceiving stress signals, thus allowing the organism to mobilize its defenses in response to challenge. Such investigations are enhancing our knowledge base on gene structure, regulation, and expression and are providing the bases for ultimately engineering organisms to better adapt to hostile environments, whether natural or anthropogenic.

The above advances are the topics discussed by active investigators in this, the fourth "thematic" volume in the *Advances in Genetics* series. As with the previous volumes, it is hoped that the material covered in the various articles will prove of lasting value in view of the rapid developments in this field.

It is with gratitude that I acknowledge the cooperation of each of my fellow contributors to this volume. I thank several colleagues who aided me in the review process and the Academic Press staff for their continued and gracious cooperation.

JOHN G. SCANDALIOS

RESPONSE OF PLANT ANTIOXIDANT DEFENSE GENES TO ENVIRONMENTAL STRESS

John G. Scandalios

Department of Genetics, North Carolina State University,
Raleigh, North Carolina 27695

1

I. Introduction

The genome constantly faces a variety of "shocks" or "stresses" to which it must respond in a programmed manner for any organism to function properly and survive. Some examples are temperature shock, anaerobiosis, pathogenicity, and hyperoxia responses in eukaryotes, and the "SOS" responses in microorganisms.

A terminally differentiated cell expresses an array of genes required for its stable functioning and metabolic roles. However, when severe changes in environmental conditions occur, a genome can respond in a rapid and specific manner by selectively increasing or decreasing the expression of specific genes. Genes whose expression is increased during times of stress presumably are critical to the survival of the organism under adverse conditions, and examination of such "stress-responsive" genes has biological, health, and agricultural significance. However, in addition to their role in aiding the organism under stress, genetic systems that are modified by stress can be used to study the molecular events that occur during periods of increased or decreased gene expression. The stress-induced changes in expression of the *Drosophila* heat-shock genes are a well-characterized example of the usefulness of stress in the study of gene regulation (Craig, 1985). Similarly, the mechanisms by which a plant recognizes a signal to alter gene expression and responds to fill that need are not only important physiologically, but permit examination of gene regulation.

Investigators also have employed stress-induced changes in gene expression to look for similarities in the mechanism of induction or stress-induced factors common to various stress responses. Heat shock, anaerobiosis, and pathogenic infections are examples of well-characterized environmentally induced stresses that affect gene expression. Oxidative stress, heavy metals, chilling, and water stress, although less well characterized, also induce specific changes in gene expression. The broader aspects of genomic responses to challenge have been reviewed by numerous investigators (McClintock, 1984; Cullis, 1986; Matters and Scandalios, 1986a; Bienz and Pelham, 1987).

Among higher eukaryotes, plants have evolved diverse ways of responding to their environment. It has frequently been suggested that the absence of motility among higher plants has resulted in their acquiring unique sets of responses to environments from which they cannot escape. Plants have incorporated a variety of environmental signals into their developmental pathways that have provided for their wide range of adaptive capacities over time. An example of such an environmental signal is light, which, in addition to driving photosynthesis, serves as a trigger and modulator of complex regulatory and developmental mechanisms (Tobin and Silverthorne, 1985). An example of the latter will be discussed later with respect to catalase gene expression in maize.

Herein, I discuss genomic responses to oxidative stress, particularly in maize, and the mechanisms plants and other aerobes have evolved for protection against the toxic effects of reduced oxygen species. Particular emphasis is placed on discussing the molecular dissection of two antioxidant gene families, catalase and superoxide dismutase, in an effort to unravel the mechanisms by which the genome perceives and responds to oxidative stress signals.

II. Oxygen and Oxygenation of the Earth

Of all the planets in our solar system, only Earth is known to contain molecular oxygen in its atmosphere and to support aerobic life. However, when Earth was formed about 4.5 billion years ago, its atmosphere was unlike the present, being primarily reducing and essentially free of oxygen. Most likely, the earliest living organisms were anaerobic heterotrophs living in the primitive ocean depths, shielded from the damaging ionizing radiation of the sun. The earliest relatively low levels of oxygen were likely the result of photolytic dissociation of water by the sun's ionizing radiation. The bulk of Earth's present oxygen concentration (i.e., 21% O_2) is derived from the photosynthetic activities of cyanobacteria and plants.

It has been estimated that Earth contains approximately 410×10^3 Erda moles (1 Emol = 10^8 mol) of oxygen, and of this, 38.4×10^3 Emol is in the hydrosphere as water (Elstner, 1987). Molecular oxygen is present in the atmosphere (37 Emol) and in the hydrosphere (0.4 Emol) and undergoes continuous turnover, with the total oxygen exchange estimated at about 15×10^3 Emol/10^6 years (Gilbert, 1981). Aerobic life is responsible for the major part of the oxygen turnover, with photosynthesis being the main input into the oxygen reservoir, and respiration the main output. The two processes are in approximate equilibrium, and fossil fuel combustion is the major source of oxygen loss from the reservoir (Elstner, 1987).

The accumulation of dioxygen in Earth's atmosphere allowed for the evolution of aerobic organisms that use dioxygen (O_2) as the electron acceptor, thus providing a higher yield of energy compared with fermentation and anaerobic respiration.

III. Active Oxygen and Oxygen Toxicity

In its ground state, molecular oxygen is surprisingly unreactive, yet is capable of giving rise to frightfully reactive excited states, as free radicals and derivatives (Taub, 1965). Its complete four-electron reduction to water proceeds most readily via a stepwise pathway which generates partially reduced intermediates (Fig. 1). The reactive species of reduced dioxygen include the superoxide radical ($O_2^{\cdot-}$), hydrogen peroxide (H_2O_2), and the hydroxyl radical (OH·). These and the physically energized form of dioxygen, singlet oxygen ($^1O_2$), are the biologically most important oxygen species. An activation energy of ≈ 22 kcal/mol is required to raise molecular oxygen from its ground state to its first singlet state. In higher plants, this energy is readily obtained from light quanta via such transfer molecules as chlorophyll (Foote, 1976). All of these oxygen species are extremely reactive and cytotoxic in all organisms. Thus, oxygen provides a paradox, in that it is essential for aerobic life, yet in its reduced form is one of the most toxic substances with which life on Earth must cope.

Oxygen activation in biological systems is known to be effected by physical and chemical (enzymatic and nonenzymatic) means. Physical activation of oxygen does not change the electron number but it changes the spin state, from parallel to antiparallel. The light-dependent, pigment-catalyzed production of singlet oxygen ($^1O_2$) with an energy consumption of 22 kcal/mol is of great importance in biological systems (Elstner, 1987). The chemical activation of oxygen yields the superoxide anion ($O_2^{\cdot-}$) after a one-electron reduction. The superoxide anion can also be produced by the univalent oxidation of hydrogen peroxide. The $O_2^{\cdot-}$ is the conjugate base of a weak acid, the perhydroxyl radical (HO_2), whose pK_a is 4.69 ± 0.08 (Bielski, 1978). Thus, under acidic conditions, the very reactive perhydroxyl radical may predominate following a one-electron reduction of dioxygen, while at higher pH values the $O_2^{\cdot-}$ is predominant. A number of excellent detailed reviews

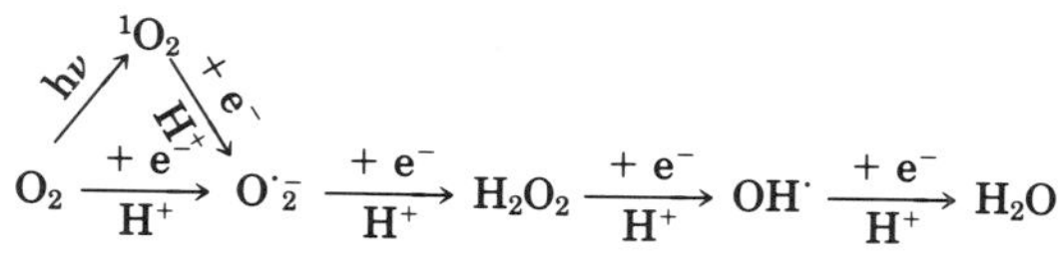

$$O_2 \xrightarrow[H^+]{+ e^-} O_2^{\cdot-} \xrightarrow[H^+]{+ e^-} H_2O_2 \xrightarrow[H^+]{+ e^-} OH^{\cdot} \xrightarrow[H^+]{+ e^-} H_2O$$

(Overall reaction $O_2 + 4e^- + 4H^+ \longrightarrow 2H_2O$)

FIG. 1 Pathways in the reduction of oxygen to water leading to the formation of various intermediate reactive oxygen species.

on the chemistry of active oxygen species have been published (Halliwell and Gutteridge, 1985; Asada and Takahashi, 1987; Elstner, 1987).

As mentioned above, both enzymatic and nonenzymatic production of active oxygen species occur in living organisms (Fridovich, 1986). Furthermore, $O_2^{\cdot-}$ and H_2O_2 participate in a metal-catalyzed Haber–Weiss reaction, producing the $OH\cdot$ radical, one of the most potent oxidizing agents known and implicated as the major factor in oxidative stress. The sum of the reactions is shown below.

$$O_2^{\cdot-} + M^{3+} \rightarrow M^{2+} + O_2 \tag{1}$$

$$\frac{M^{2+} + H_2O_2 \rightarrow M^{3+} + OH\cdot + OH^-}{O_2^{\cdot-} + H_2O_2 \rightarrow O_2 + OH^- + OH\cdot} \tag{2}$$

In the above reaction "M" depicts a transition metal. Note that the sum of reactions 1 and 2 is production of the powerful hydroxyl radical. Due to its high reactivity, immediate indiscriminate substrate attack occurs.

IV. Biology of Oxy Intermediates

In plants, the superoxide radical and singlet oxygen are commonly produced in illuminated chloroplasts by the occasional transfer of an electron from an excited chlorophyll molecule to molecular oxygen, or from photosystem I components under conditions of high NADPH/NADP ratios. In addition, various enzymes, such as xanthine oxidase, aldehyde oxidase, dihydrorotate dehydrogenase, and other flavin dehydrogenases, can generate superoxide as a catalytic product (Fridovich, 1978). The dismutation reaction of two superoxide anions produces hydrogen peroxide, which is also a product of the microbody-associated β-oxidation of fatty acids and glyoxylate cycle, and peroxisomal photorespiration reactions (Beevers, 1979; Tolbert, 1982). The most highly reactive and destructive oxygen free radical, the hydroxyl radical ($OH\cdot$) is generated by the transfer of an electron from $O_2^{\cdot-}$ to H_2O_2, or less frequently in a Fenton-type reaction of H_2O_2 with Fe^{2+} or reduced ferredoxin (Fridovich, 1978; Elstner, 1982). These highly reactive oxygen species can react with unsaturated fatty acids to cause peroxidation of essential membrane lipids in the plasmalemma or in intracellular organelles (Fridovich, 1976). Peroxidation damage to the plasmalemma leads to leakage of cellular contents, rapid dessication, and cell death. Intracellular membrane damage can affect respiratory activity in mitochondria, or cause pigment breakdown and loss of carbon-fixing ability in chloroplasts. Several Calvin-cycle enzymes within plant chloroplasts are very sensitive to hydrogen peroxide, and high levels of H_2O_2 can directly inhibit CO_2 fixation (Kaiser, 1979; Robinson et al., 1980; Charles and Halliwell, 1981). In addition, hydrogen peroxide has been clearly shown to be active with mixed-function oxidases in marking proteins, in particular, several types of enzymes, for proteolytic degradation (Levine et al., 1981; Fucci et al., 1983). In mammalian cells, superoxide appears to be the toxic agent in phagocytic elimination of bacterial cells, possibly through its membrane peroxidative effect (Halliwell, 1982). Most importantly, the hydroxyl radical can attack lipids and proteins indiscriminately and causes lesions in the DNA (Lesko et al., 1980). Some of the damaging effects in biological systems attributed to oxygen free radicals and derivatives are listed in Table 1. The accrual of free-radical damage has been attributed as the major factor in the aging process of both animals and plants (Sohal, 1987; Leshem, 1988). However, the evidence has been largely correlative even though one can readily reason that the cumulative effects of oxy radicals in biological systems are likely contributors to the various disease states frequently associated with the aging process. Perhaps the utility of longevity mutants in various organisms amenable to genetic and molecular analyses will contribute to the resolution of the aging and senescence connection to active oxygen species.

V. Oxy-Intermediate Defense Systems

As a challenge to the toxic and potentially lethal effects of active oxygen, aerobic organisms evolved protective scavenging or antioxidant defense systems, both enzymatic and nonenzymatic (Halliwell and Gutteridge, 1985). Among the latter are included low-molecular-weight nonprotein sulfhydryls such as glutathione, cysteine, and cysteinylglycine. Other substances shown to possess antioxidant activity are hydroquinones, allopurinol, mucin, manitol, dimethysulfoxide (DMSO), vitamins C and E, urates, and β-carotene (Ames, 1983; Cerutti, 1985).

The enzymatic protective defenses include enzymes capable of removing, neutralizing, or scavenging free radicals and oxy intermediates. Peroxidases, catalases, and superoxide dismutases have been shown to play a major role in protecting cells from oxidative damage. Catalases (EC 1.11.1.6; $H_2O_2 : H_2O_2$ oxidoreductase; CAT) and peroxidases (EC 1.11.1.7) are classes of enzymes that remove H_2O_2 very efficiently, as depicted below.

TABLE 1
Some Sources and Effects of Activated Oxygen Species[a]

Oxygen species	Sources	Biological effects
Molecular oxygen (O_2)	Atmospheric oxygen, photosystem II ($2H_2O \rightarrow 4H^+ + O_2 + 4e^- 0$, and various enzymes (catalase and superoxide dismutase)	Inhibition of photosynthesis (competes with CO_2 for RUBP carboxylase) and random production of free radicals
Superoxide radical ($O_2^{\cdot-}$)	Illuminated chloroplasts, herbicides (paraquat and nitrofen), and enzymatic reactions (uricase, aldehyde oxidase, xanthine oxidase, flavin dehydrogenases, and oxidation of ferredoxin$_{red}$)	Membrane lipid peroxidation, enzyme inactivation, depolymerization of polysaccharides, reaction with H_2O_2 to form $OH\cdot$, aging, and autoimmune diseases
Hydrogen peroxide (H_2O_2)	Glycolate oxidase in glyoxysomes, illuminated chloroplasts, and enzymatic reactions (superoxide dismutase, urate oxidase, and various peroxisomal enzymes)	Inhibition of CO_2 fixation, marking of some proteins (e.g., glutamine synthetase) for proteolytic degradation, oxidation of flavonols and sulfhydryls, mutagenesis, and inactivation of Calvin cycle enzymes
Hydroxyl radical ($OH\cdot$)	Haber–Weiss reaction $H_2O_2 + O_2^{\cdot-} \rightarrow OH^- + OH\cdot + O_2$ Fenton reaction $H_2O_2 + Fe^{2+} \rightarrow OH^- + OH\cdot + Fe^{3+}$	Most potent oxidant known! Causes DNA lesions, protein degradation, peroxidation of membrane lipids, ethylene production, and is implicated in rheumatoid arthritis
Singlet oxygen ($^1O_2$)	Photoexcited chlorophyll molecules in triplet state, air pollutants (NO_2, O_3, etc.), and pathogenic fungi (cercosporin toxin)	Mutagenesis, membrane lipid peroxidation, and photooxidation of amino acids

[a] From Scandalios (1987).

$$H_2O_2 + H_2O_2 \rightarrow 2H_2O + O_2 \quad (K_1 = 1.7 \times 10^7 \, M^{-1} \, sec^{-1}) \qquad \text{Catalase}$$
$$H_2O_2 + R(OH)_2 \rightarrow 2H_2O + R(O)_2 \quad (K_4 = 0.2\text{--}1 \times 10^3 \, M^{-1} \, sec^{-1}) \qquad \text{Peroxidases}$$

Superoxide dismutase (EC 1.15.1.1; $O_2^{\cdot-} : O_2^{\cdot-}$ oxidoreductase; SOD) comprises a class of metal-combining proteins that are highly efficient at scavenging the superoxide radical ($O_2^{\cdot-}$). The enzyme catalyzes a disproportionation reaction at a rate very near that of diffusion. To accomplish this reaction, the mechanism employs an alternating reduction/oxidation of the respective metal associated with the enzyme.

$$O_2^{\cdot-} + M^{2+} \rightarrow O_2 + M^+$$
$$\underline{O_2^{\cdot-} + M^+ + 2H^+ \rightarrow H_2O_2 + M^{2+}}$$
$$2O_2^{\cdot-} + 2H^+ \xrightarrow{SOD} H_2O_2 + O_2 \quad (K_2 = 2.4 \times 10^9 \, M^{-1} \, sec^{-1}) \qquad \text{Overall}$$

$$O_2 + e^- \longrightarrow O_2^{\cdot -} \qquad\qquad (1)$$

$$O_2^{\cdot -} + O_2^{\cdot -} + H^+ \xrightarrow{\text{SOD}} H_2O_2 + O_2 \qquad\qquad (2)$$

$$H_2O_2 + O_2^{\cdot -} \longrightarrow OH^- + OH^{\cdot} + O_2 \qquad\qquad (3)$$

$$H_2O_2 + H_2O_2 \xrightarrow{\text{CAT}} 2H_2O + O_2 \qquad\qquad (4)$$

FIG. 2. Sequence of reactions showing the cooperative action of superoxide dismutase (SOD) and catalase (CAT) in converting the toxic superoxide radical and hydrogen peroxide to water and oxygen, thus abating the formation of the potent oxidant, the hydroxyl radical.

The combined action of catalase and superoxide dismutase converts the potentially toxic superoxide radical and hydrogen peroxide to water and molecular oxygen (Fig. 2), and in the process abate the formation of the most toxic and highly reactive oxidant, the hydroxyl radical (OH·), thus preventing cellular damage. It is perhaps this series of cooperative interactions that plays a more crucial role in the detoxification of reactive oxygen species rather than the elimination of $O_2^{\cdot -}$ or H_2O_2 per se. It is for these reasons that superoxide dismutases and catalases are deemed the most important of the "antioxidant enzymes." Although there are no known direct scavengers of singlet oxygen ($^1O_2$) or the hydroxyl radical (OH·), SOD is believed to function in their elimination by chemical reaction (Matheson et al., 1975).

VI. Oxidative Stress

The term "oxidative stress" is used to describe a cellular situation characterized by an elevation in the steady-state concentration of reactive oxygen species (Cadenas, 1985; Matters and Scandalios, 1986a). Oxidative stress occurs when the balance between the mechanisms triggering oxidative conditions (e.g., hyperoxia, radiation, Fenton reactions, impairment of the cytochrome electron-transport chain, peroxisome proliferators, xenobiotic agents, and pollutants such as O_3 and SO_2) and the cellular antioxidant defenses (e.g., superoxide dismutase, catalase, peroxidases, and vitamins E, C, and A) is impaired (Cadenas, 1985; Cerutti, 1985).

Among the effects associated with oxidative stress are aging, carcinogenesis, atherosclerosis, arthritis, and in some instances cell death (Halliwell and Gutteridge, 1985). When bacteria are exposed to conditions resulting in oxidative stress, SOD and CAT activities are elevated (Fridovich, 1986). In eukaryotes, exposure to oxidative stress also correlates with increased SOD and CAT activities (Halliwell, 1974; Foster and Hess, 1980; Matters and Scandalios, 1986a). In addition, prior exposure of cells to agents that induce SOD and CAT activity conferred a partial protection against oxidative stress (Gregory and Fridovich, 1973). Resistance to environmentally induced oxidative stress has been shown in several instances to be associated with high levels of both CAT and SOD (Harper and Harvey, 1978; Hassan and Fridovich, 1979). The redox-active herbicide paraquat is known to produce oxidative stress in bacteria, animals, and plants (Harbour and Bolton, 1975; Hassan and Fridovich, 1979; Matters and Scandalios, 1986b). This compound acts by accepting an electron to form a radical cation, which then donates the electron to O_2 to form O_2^- (Farrington et al., 1973). In green plant tissues, the primary site of reduction of paraquat is in the chloroplast at photosystem I (Farrington et al., 1973). Paraquat is a potent oxidant and desiccant in plants and is widely used as a preemergence herbicide and harvest aid. Paraquat-resistant varieties of ryegrass, tobacco, and horseweed exhibit higher levels of SOD and sometimes catalase activities than do nonresistant varieties (Harper and Harvey, 1978; Miller and Hughes, 1980; Youngman and Dodge, 1980).

VII. Intracellular Concentration of Oxygen and Maize Photosynthesis

In plants, the primary reaction of carbon fixation is the reaction of CO_2 with ribulose bisphosphate (RuBP), catalyzed by the enzyme ribulose bisphosphate carboxylase/oxygenase (Rubisco) (Lorimer, 1981). Rubisco may use O_2 as an alternate substrate for CO_2, leading to the formation of a three-carbon acid and a two-carbon acid (phosphoglycollate). In maize, there is a spatial separation of the primary fixation of CO_2 in the mesophyll cells and its subsequent release and refixation in the bundle sheath cells via Rubisco (Chollet and Ogren, 1975). This functions to increase the local concentration of CO_2 at the Rubisco active site, decreasing the reaction of the enzyme with O_2 (Hatch and Slack, 1969). In addition, differences in the distribution of the light reactions of photosynthesis in maize leaf cell types further increase photosynthetic efficiency by decreasing the O_2 concentration in the bundle sheath cells where Rubisco is localized (Woo et al., 1970). The maize mesophyll cell contains both photosystems I and II (PSI and PSII). The O_2-evolving complex of photosynthesis is contained in PSII and, in conjunction with PSI, gives the light-dependent production of NADPH and ATP. The maize bundle sheath cells are deficient in PSII; therefore, little O_2 is evolved in these cells. The anatomy of the maize leaf, in which the bundle sheath cells are surrounded by mesophyll cells, limits inward diffusion of atmospheric oxygen. Thus, the concentration of oxygen in bundle sheath cells is substantially lower than that in mesophyll cells and, because paraquat acts by donating an electron to O_2, the level of paraquat-induced oxidative stress might be higher in mesophyll cells.

VIII. Gene Expression and Response of Plants to Environmentally Imposed Stress

The response of plants to several environmentally imposed stresses has been well characterized (Matters and Scandalios, 1986a). When root tissue is exposed to anaerobiosis, through flooding or by being placed in an argon–nitrogen environment, normal protein synthesis is repressed and the synthesis of a characteristic set of 20 proteins is induced (Sachs and Ho, 1986). When plants are exposed to a 10°C upward shift in ambient temperature, normal transcriptional and translational activity is shifted to a characteristic set of proteins known as heat-shock proteins (HSPs) (Kimpel and Key, 1985). However, rather than being degraded, the "normal" cellular mRNAs are sequestered within the cell. Unfortunately, the full identity and function of HSPs are not known. Pathogenic infection by fungi, bacteria, or viruses can cause the induction of pathogenesis-related proteins (PRPs) and phytoalexins. The PRPs are low-M_r, protease-resistant proteins that are secreted into the plant's intracellular spaces (van Loon, 1985). Although the identity and function of PRPs are unknown, they are distinct from other stress-related proteins. When plants are treated with elicitors (glucan fragments from bacterial cell walls), de novo synthesis of the enzymes phenylalanine ammonia lyase, chalcone synthase, chalcone isomerase, and 4-coumarate : CoA ligase arises from increased transcription of the genes involved in phytoalexin production (Edwards et al., 1985; Cramer et al., 1986). Thus, changes in gene expression as a consequence of environmentally imposed stress may occur at both the transcriptional and translational levels in plants.

IX. Translational and Transcriptional Regulation of Gene Expression

As indicated above, *de novo* synthesis of a protein may arise via increased transcription of a gene and/or increased translation of the mRNA transcript. In turn, transcription and translation may be affected in several different ways. For example, increases in protein synthesis in the absence of a change in RNA concentration (i.e., translational control) have been identified under several conditions, although the mechanisms involved have been defined in only a few cases. In etiolated pea plants, the mRNAs coding for the chlorophyll *a/b*-binding protein are present, as measured by hybridization and *in vitro* translation of polysomal RNA, but the polypeptides for these proteins do not accumulate. The apparent increase in translation of the chloropyll *a/b*-binding proteins in the light is due to a decreased degradation of newly synthesized protein (Ellis, 1986). A similar phenomenon occurs in *Volvox*, where *in vitro* translation products of RNA isolated from light- and dark-grown cultures are virtually indistinguishable. However, there is a distinct shift in the pattern of protein synthesis in cultures shifted from dark to light, indicating that translational control of protein synthesis is occurring *in vivo* (Kirk and Kirk, 1985). The polyadenylation of mRNA may also lead to differences in the ability of the RNA to be translated into protein. It has been shown that a decrease in the translation of a certain mRNA in *Dictyostelium* correlated with the mean reduction in poly(A) tail length of this RNA (Palatnik *et al.*, 1984). In addition, mRNA may be bound to polysomes, but not translated under certain conditions. For example, the rat liver sterol carrier protein, has a daily 7-fold fluctuation in concentration, but the concentration of mRNA coding for this protein does not change. The RNA is always bound to large polysomes, yet it is translated only at night (McGuire *et al.*, 1985). Also, the availability of prosthetic groups may influence the translation of mRNA. Yeasts which cannot synthesize the catalase prosthetic group, heme, make catalase mRNA, but do not accumulate the apoprotein. Addition of heme or heme precursors to the culture stimulates the synthesis of the enzyme. However, mRNA from heme-sufficient and -deficient cultures could be translated with equal efficiency *in vitro*, but only ribosomes isolated from a heme-sufficient culture could be used for translation of catalase mRNA (Hamilton *et al.*, 1982).

Increased transcription of genes resulting in the increased concentration of a specific mRNA and *de novo* protein synthesis can occur in response to various environmental changes. When etiolated plants are exposed to light there is an increase in the transcription of the Rubisco small subunit and light-harvesting complex genes (Gallagher and Ellis, 1982; Timko *et al.*, 1985). The induction of metallothionine by metals, hormones, and lipopolysaccharides and the induction of galactose-metabolizing enzymes by galactose are also cases of enzyme induction via increases in gene transcription (Stuart *et al.*, 1984).

The activation of nuclear gene transcription has been shown to be controlled by nontranscribed DNA sequences within the gene. Two types of sequences are involved in the activation of transcription: a promoter region and an enhancer region(s). Promoters are present in all transcribed eukaryotic genes and are immediately 5' to the transcription initiation site. Protein structural gene promoters (TATA and CAAT boxes) are involved in the initiation of transcription under different conditions. That is, transcription is initiated from different promoters under different conditions (Minty and Newark, 1980). Enhancer sequences are important in the activation of gene transcription under certain environmental or developmental conditions, such as the transcription of the Rubisco small subunit and light-harvesting complex genes in light (Serfling *et al.*, 1985; Ellis, 1986). These sequences activate transcription of a linked gene from the correct initiation site. Their ability to activate transcription is independent of orientation, and they can function from large distances [>1000 base pairs (bp)] from the initiation site and give preferential stimulation of the most proximal of two tandem promoters (Serfling *et al.*, 1985).

From the preceding discussion, it becomes apparent that in order to understand the underlying mechanisms by which plants and other eukaryotes respond to oxidative stress, or to other environmental signals, we must identify the responsive genes and understand their structure and mode of action. Only through such information can one hope to be able to manipulate the genome for greater tolerance to stress. Even though we have ample information on the physiological consequences of oxidative stress in many organisms, our understanding of the underlying factors involved is just beginning to emerge. As mentioned earlier, the two most effective enzymatic defenses against oxidative damage are the roles played by the enzymes catalase and superoxide dismutase. These two gene families are discussed in the following sections within the framework of the preceding introduction.

X. Maize Catalase Gene—Enzyme System

A. BIOCHEMISTRY

The maize catalase (CAT) system has been extensively characterized (Scandalios, 1965, 1968, 1979, 1983). Three distinct catalase isozymes (CAT-1, CAT-2, and CAT-3) have been identified in various tissue extracts, at different developmental stages of the maize life cycle, by conventional separation procedures. Each isozyme has been purified to homogeneity (Chandlee et al., 1983) and characterized biochemically (Table 2). Both biochemical and genetic data indicate that the maize catalases, like all eukaryotic catalases, are tetrameric heme proteins of ≈240 kDa, with each subunit being ≈56–60 kDa. CAT-1 and CAT-2 subunits, when expressed in the same cell, hybridize to form heterotetramers *in vivo* (Scandalios, 1979). CAT-3 subunits do not hybridize with CAT-1 or CAT-2 *in vivo* (Scandalios et al., 1980a). However, the three kinds of subunits hybridize readily *in vitro* in dissociation–reassociation experiments. When utilized as an antigen, each of the purified isozyme proteins elicits monospecific polyclonal antibodies (Chandlee et al., 1983).

The three isozymes have similar pH optima, with maximum activity from pH 7–9. CAT-2 has a slightly broader pH range with activity from pH 4–13. The isozymes have similar thermostabilities. CAT-1 is the least stable at 40°C with $T_{1/2} = 12.5$ minutes (Table 2). At 50°C CAT-2 has a $T_{1/2} = 141$ minutes, whereas CAT-1 and CAT-3 are rapidly denatured. Each isozyme exhibits varying degrees of sensitivity to different inhibitors (i.e., aminotriazole, KCN, NaN_3, dithiothreitol, and 2-mercaptoethanol) tested. CAT-3 proved less sensitive than CAT-1 and CAT-2 to all the inhibitors (Table 2).

B. GENETICS

CAT-1, CAT-2, and CAT-3 are respectively encoded by the three, unlinked, structural genes, *Cat1, Cat2,* and *Cat3* (Scandalios, 1968; Scandalios et al., 1972, 1980a). *Cat1* is located on the short arm of chromosome 5, *Cat2* is on the short arm of chromosome 1, and *Cat3* is on the long arm of chromosome 1 (Roupakias et al., 1980). Six alleles have been identified at the *Cat1* locus, and three each at the *Cat2* and *Cat3* loci. Null alleles for *Cat2* and *Cat3* have been isolated from screening large numbers of maize lines. The *Cat2* null allele has been investigated in some detail biochemically (Tsaftaris and Scandalios, 1981), and more recently was shown to encode a truncated *Cat2* transcript

TABLE 2
Biochemical, Genetic, and Molecular Properties of the
Maize Catalase Gene–Enzyme System[a]

Property	CAT-1	CAT-2	CAT-3
Subunit number	4	4	4
Native M_r	240,000	240,000	240,000
Specific activity (units · mg protein)	561	2343	688
K_m	0.143 M	0.040 M	0.062 M
pH optima	7–9	7–9	7–9
Turnover rate[b]	47 hours	22 hours	22–24 hours
Thermostability[c]	12.5 minutes	Stable	18 minutes
Aminotriazole inhibition[d]	93.1%	98.2%	31.7%
Subcellular localization	Peroxisomes/ cytosol	Peroxisomes/ cytosol	Mitochondria- /cytosol
Gene locus	Cat1	Cat2	Cat3
Chromosomal location	5 S	1 S	1 L
Number of alleles	6	3	3
Transcript size[e]	2000 nt	1800 nt	1800 nt
Percentage similiarity[f]	—	67%	62%

[a] Data from Scandalios (1979), Chandlee et al. (1983), and Redinbaugh et al. (1988).

[b] Turnover rate, the time required to reach 50% saturation (50% of enzyme molecules are fully labeled) under our density labeling conditions.

[c] Thermostability was measured at 45°C in 100 mM Tris–HCl (pH 7.8).

[d] Inhibition (%) by 10 mM aminotriazole.

[e] Transcript size (nt, nucleotides) determined by RNA blot analysis.

[f] Percentage similarity was determined by pair-wise comparisons of the nucleic acid sequences of cDNA clones to the sequence of the Cat1 cDNA sequence.

(Bethards and Scandalios, 1988). Plants expressing the normal CAT-2 phenotype have a Cat2 transcript of ≈1800 nucleotides (nt), whereas the Cat2 null variant has an aberrant transcript of ≈1400 nt. This shortened transcript lacks regions of the Cat2 gene that encode the carboxy-terminal domain of the CAT-2 isozyme. Molecular analysis of the Cat3 null variant using run-off transcription and RNA blots showed that the Cat3 gene is not transcribed, and Cat3 mRNA does not accumulate in this line (IDS28). Genomic Southern blots suggest that the molecular basis for the Cat3 null variant is a deletion in the 5′ end of the Cat3 gene (Wadsworth and Scandalios, 1990).

C. Subcellular Localization

The catalases of maize are differentially compartmentalized within the cell (Scandalios, 1974; Scandalios *et al.*, 1980a). CAT-1 and CAT-2 are localized in peroxisomes, glyoxysomes, and cytosol. CAT-3, however, copurifies and sediments with mitochondria on sucrose gradients. In this respect, CAT-3 is unique among eukaryotic catalases, including the other maize catalases. This alternate subcellular localization might also account for the lack of *in vivo* hybridization between CAT-3 and the other maize catalases to form heterotetramers at times when CAT-3 is coexpressed with either CAT-2 or CAT-1 (Scandalios *et al.*, 1980a).

D. Developmental Expression

Each of the maize catalases exhibits a unique and complex pattern of developmental expression throughout the maize life cycle (Scandalios *et al.*, 1984). In this respect, this system has served as a model to study the differential expression of genes in higher eukaryotes (Scandalios, 1987). CAT-3 is detected in the pericarp (a maternal tissue) during ovule development, and during very early postpollination kernel development, but declines rapidly shortly after pollination. CAT-1 is also detected, but at extremely low levels, in the pericarp. By days 12–15 postpollination, CAT-1 becomes the predominant form, and is in fact the only catalase detected in the kernel at this time. CAT-1 is the only catalase detected in the milky endosperm of developing kernels. The aleurone, a specialized kernel tissue derived from endosperm cells, expresses both CAT-1 and CAT-2 during the latter stages of kernel development. At days 9–20 postpollination, the scutellum expresses only the CAT-1 isozyme. However, at later stages, as the seed matures, the *Cat2* gene is induced to moderate levels until the seed desiccates and enters developmental arrest. In the scutellum isolated from dry "dormant" seed, CAT-1 appears to be the only isozyme expressed. Upon imbibition and germination, CAT-1 is initially the only form expressed in the scutellum, but as the embryo grows, *Cat2* is rapidly induced and by day 5 postimbibition CAT-2 becomes the predominant isozyme. By day 10 postimbibition, catalase expression has shifted entirely to *Cat2* and no *Cat1* expression can be detected at the protein level. In the coleoptile, CAT-3 is the only isozyme detected on zymograms. Both CAT-1 and CAT-3 are detected in dark-grown leaf tissue. CAT-2 is only expressed with greening of the leaf tissue in the light. The root has very low levels of catalase and is difficult to analyze. However, in young roots both CAT-1 and CAT-3 can be detected. As roots mature and grow, catalase activity decreases and rapidly drops below assay sensitivity. Only CAT-3 is detected in the stem of the mature plant, and only CAT-1 is detected in the pollen after anthesis (Acevedo and Scandalios, 1990).

The overall developmental program of catalase gene expression presented herein has been corroborated by zymogram analyses, activity profiles, immunochemical measurements, and transcript levels for each of the isozymes (Scandalios, 1987; Redinbaugh *et al.*, 1988; Wadsworth and Scandalios, 1989).

E. Expression in Early Sporophytic Development

Catalase activity in the scutellum increases rapidly after germination, peaking at 4 days postgermination and thereafter declining to moderate levels. A time-course zymogram analysis reflects the temporal shift in expression from *Cat1* to *Cat2* during this time period (Scandalios, 1983). This shift in expression of the two *Cat* genes was shown to be largely due to changes in the rates of synthesis and degradation of the CAT-1 and CAT-2 isozymes during this period (Quail and Scandalios, 1971).

That this shift in expression of the two *Cat* genes in the scutellum is under genetic control has been verified by use of several maize lines shown to exhibit altered patterns of catalase expression during the same developmental period (Scandalios *et al.*, 1980b; Chandlee and Scandalios, 1984a). For example, line R6-67 exhibits a continuous increase in catalase activity after 4 days postimbibition compared to the usual decline seen in most lines (e.g., W64A). The continuous increase in catalase activity is concurrent with an increase in CAT-2 protein synthesis and accumulation. By genetic analysis, a genetic regulatory element, *Car1*, was identified and shown to be responsible for the overexpression of the *Cat2* gene in R6-67 (Scandalios *et al.*, 1980b). *Car1* maps 37 map units from the *Cat2* structural gene on the short arm of chromosome 1. The product of *Car1* and the mechanism by which it increases *Cat2* expression are unclear. However, R6-67 has significantly more *Cat2* mRNA than do the normal lines, and it is suggested that *Car1* may encode a trans-acting factor affecting the transcription of *Cat2* (Kopczynski and Scandalios, 1986).

A second regulatory element, *Car2*, was identified and shown to function in decreasing the rate of CAT-1 synthesis in the scutellum during early sporophytic development in line TX303 (Chandlee and Scandalios, 1984b). The decline in CAT-1 synthesis is associated with a decline in *Cat1* mRNA. *Car2* acts independently of *Car1*, and is either closely linked or contiguous with the *Cat1* structural gene.

Catalase expression in the scutellum is apparently regulated by "signals" from tissues that spatially interact with the scutellum. Histoimmunological assays have shown that the postgermination induction of *Cat2* occurs as a consequence of a temporal and spatial gradient. CAT-2 is first detected in cells that interact with the aleurone. As development proceeds, CAT-2 accumulates in a spatial gradient, occurring first in cells nearest the aleurone, then in cells further and further away from the aleurone, until all scutellar cells express the CAT-2 protein (Tsaftaris and Scandalios, 1986). This suggested that the aleurone may be releasing a diffusible molecular signal triggering the induction of *Cat2* in the scutellum.

It was also demonstrated (Skadsen and Scandalios, 1989) that the embryonic axis exerts a specific effect on the accumulation of glyoxysomal proteins, including catalase. Upon excising the embryonic axis from the scutellum prior to imbibition, the developmental accumulation of all glyoxysomal proteins is drastically reduced. Whereas the developmental patterns of nonglyoxysomal proteins are unaffected; this suggested that an axis-specific factor modulates the level of expression of the glyoxysomal proteins, including catalase. The exact nature of this "factor" is yet unclear. However, it is apparent that possibly two molecular signals may be involved in regulating the expression of *Cat2* in the scutellum during early sporophytic development; one emanating from the aleurone may act to "turn on" the gene, and the other emanating from the embryonic axis may modulate the level of expression of *Cat2*. Both aspects are being investigated.

XI. Maize Superoxide Dismutase Gene–Enzyme System

A. BIOCHEMISTRY

In maize, depending on the inbred lines examined, four or five distinct superoxide dismutase (SOD) isozymes are resolved by conventional separation procedures; the four-isozyme phenotype is the most common (Baum and Scandalios, 1979, 1981). The isozymes SOD-1, SOD-2, SOD-4, and SOD-5 are dimeric, Cu- and Zn-containing proteins with subunits ranging from ≈ 14 to 17 kDa. SOD-3 is a Mn-containing tetrameric protein with a subunit of 24 kDa. The maize SODs have characteristics similar to Cu–ZnSODs and MnSODs from other organisms (Kitagawa *et al.*, 1986; Salin, 1988). SOD-3 is unique in its properties in that it is insensitive to cyanide (1 mM), hydrogen peroxide (5 mM), diethyldithiocarbamate (1 mM), and temperature (55°C). The Cu–ZnSOD isozymes exhibit 10-fold greater activities at pH 10 than at pH 7.5, whereas the MnSOD-3 isozyme is relatively unaffected by pH. The Cu–ZnSODs have very low UV absorbance, presumably due to their low abundance of tryptophan; SOD-3 exhibits an absorbance maximum at 260–270 nm (Baum and Scandalios, 1981; Baum *et al.*, 1983).

Turnover studies of SOD during seedling growth demonstrated that SOD is synthesized *de novo* following seed imbibition, and accumulates to roughly 1% of the total soluble protein in scutella between days 1 and 8 post-imbibition (Baum and Scandalios, 1982a). All of the maize SOD isozymes are encoded by nuclear genes, synthesized on cytosolic ribosomes, and translocated to various intracellular compartments.

Monospecific polyclonal antibodies were generated for each of the purified SOD isozymes (Baum and Scandalios, 1981) and were used to quantitate the levels of SOD proteins synthesized during seedling development (Baum and Scandalios, 1982a) and after exposing plants to various environmental stresses (Matters and Scandalios, 1986b). Some of the biochemical properties of the maize SODs are summarized in Table 3.

TABLE 3

Biochemical Properties of Maize Superoxide Dismutases

Isozyme	SOD-1	SOD-2	SOD-3	SOD-4
Subcellular compartment	Chloroplast	Cytosol	Mitochondria	Cytosol
Metal content	2Cu–2Zn	2Cu–2Zn	4Mn	2Cu–2Zn
Mass (kDa)				
Subunit	14.5	17.0	24.0	15.9
Holoenzyme	31–33	31–33	90	31–33
Quaternary structure	Dimeric	Dimeric	Tetrameric	Dimeric
$T_{1/2}$ (minutes)				
55°C	Stable	7.9	Stable	7.1
0.5 mM H$_2$O$_2$	80	43	Stable	37
1 mM DDC	Stable	8.4	Stable	9.6
Relative activities				
pH 7.8	1.00	1.00	1.00	1.00
1 mM KCN, pH 10.0	0.058	0.093	1.095	0.067
pH 10.0	3.13	11.34	1.160	10.56
Carbohydrate (%)	<1.0	4.5	<1.0	2.0
Monospecific antiserum	SOD-1	SOD-2/4	SOD-3	SOD-2/4

B. Intracellular Localization

Unlike catalase, superoxide dismutase is not highly regulated temporally or spatially, but it does exhibit a great degree of intracellular compartmentalization. Cu–ZnSOD-1 is associated with chloroplasts and etioplasts, and MnSOD-3 is associated with mitochondria (Baum and Scandalios, 1979). The other Cu–Zn isozymes, SOD-2, SOD-4, and SOD-5, are located in the cytosolic fraction of maize cells. Although SOD-1 is structurally similar to the other Cu–ZnSODs, antibodies to SOD-1 will not cross-react with the other Cu–Zn isozymes.

The intracellular compartmentalization of SOD is believed to ensure a critical defense against oxygen toxicity in organelles in which O_2^- is generated during electron transport, photorespiration, and other metabolic processes. Because O_2^- is a charged molecule, it must be eliminated *in situ,* as it cannot traverse membranes. Understanding the processes by which nuclear-encoded gene products find their way into specific compartments to perform specialized functions is crucial if one wishes eventually to be able to engineer cells for greater capacities to resist the effects of oxidative stress. To this end, the importation of SOD-3 into maize mitochondria was examined, and it was the first such study to be reported in higher plants (White and Scandalios, 1989). Deletion mutagenesis coupled with *in vitro* transcription, translation, and importation into isolated mitochondria provided information about the requirements for higher plant mitochondrial import. The data showed that a transit peptide, 31 amino acids in length, was required for efficient import. The relative import efficiency was dependent on the extent of the deletion within the transit peptide region. Significant findings in the study were that *in vitro*-synthesized SOD-3 proteins were not only imported and processed, but were also assembled into tetrameric SOD-3 holoenzymes in the mitochondrial matrix (White and Scandalios, 1987, 1989).

C. Genetics

Genetic analysis of electrophoretic SOD variants demonstrated that the maize isozymes SOD-1, SOD-2, SOD-3, SOD-4, and SOD-5 are, respectively, encoded by the nuclear, unlinked structural genes *Sod1, Sod2, Sod3, Sod4,* and *Sod5* (Baum and Scandalios, 1982b). Furthermore, in the process of isolating cDNAs for the various *Sod* genes, we recently identified yet another cytosolic SOD isozyme similar to SOD-4. Two cDNAs encoding two proteins differing by only 3 out of 153 amino acids were isolated. One of the differing amino acids was located near the N-terminus at residue number 12, where GAG (Glu) was changed to GAT (Asp). Upon deciphering this, we purified the SOD-4 protein, N-terminally sequenced it, and found that at residue 12, both Asp and Glu were present. The 5' and 3' cDNA sequence differences allowed for gene-specific genomic DNA and RNA blots. Data from this work confirmed the existence and expression of two genes (*Sod4* and *Sod4A*) that encode two virtually identical SOD-4-like proteins (Cannon and Scandalios, 1989).

XII. Catalase and Superoxide Dismutase Gene Expression in Response to Environmental Stress

There have been numerous reports that, in both prokaryotes and eukaryotes, oxidative stress enhances or induces the activity of superoxide dismutase and catalase. The formation of superoxide and other active oxygen species can be accelerated as a consequence of various stress conditions, including ultraviolet radiation, high light intensity, low CO_2 concentration, and treatment with herbicides that serve as preferred terminal electron acceptors at the reducing site of photosystem I (e.g., diquat and paraquat) or that are known to block electron transport (e.g., atrazine and diuron). Increases in antioxidant enzyme activities have been reported in response to heat and light conditions that cause sunscald in vegetables, fruits, and flowers (Rabinowitch and Sklan, 1980). The fungal toxin cercosporin, produced by the pathogenic fungi *Cercospora,* which cause damaging leaf spot diseases on various economically important crops, acts by generating increased levels of singlet oxygen (Daub and Hangarter, 1983). Increases in SOD and CAT have been observed in response to ozone (Lee and Bennett, 1982) and SO_2 (Tanaka and Sugahara, 1980; Alscher *et al.,* 1987) levels in the environment. However, the mechanisms for the observed increases in antioxidant enzymes in response to oxidative stress have yet to be resolved. Changes in individual *Sod* and/or *Cat* genes in response to environmental stresses have not been previously examined in detail, nor have the responses to different stress factors within a single SOD or CAT multienzyme system been studied. The maize systems described above have provided an opportunity to study the response of specific *Sod* and *Cat* genes to imposed environmental stresses in an effort to unravel the mechanisms regulating such responses.

A. Photoresponse of *Cat2*
and *Cat3* Genes in Maize Leaves

One of the most fundamental and important processes in nature is the utilization of light by plant cells as the basic energy source to drive biological reactions. Photodynamic reactions are known to occur in green plant tissues in the presence of light and oxygen, and are capable of cell injury and tissue death (Rabinowitch *et al.*, 1982).

Green plants have more possibilities to produce activated oxygen species than do animal cells. Consequently, the challenge to overcome oxidative stress is greater in plants than other eukaryotes because plants both consume O_2 during respiration and generate O_2 during photosynthesis. The question then arises as to the role(s) light might play in modulating, directly or indirectly, the antioxidant defense systems of green plants.

In dark-grown maize leaves, the catalase isozyme CAT-2 is absent, but it appears soon after exposure of etiolated leaves to white light (Scandalios, 1979). With continuous white light, CAT-2 protein levels increase, due to *de novo* synthesis, and plateau after 24 hours. When total poly(A)$^+$ RNA (mRNA), polysomes, or isolated polysomal mRNA from light- and dark-grown leaves were translated *in vitro,* CAT-2 protein was detected only among the light-grown leaf products (Skadsen and Scandalios, 1987). RNA blot analyses, using a gene-specific *Cat2* cDNA probe, showed that *Cat2* mRNA was present in approximately equal quantities in total mRNA and polysomal mRNA derived from both light- and dark-grown maize leaves. Furthermore, *Cat2* mRNA was equally distributed in identical high-molecular-weight fractions in polysomes from light- and dark-grown leaves, indicating that the *Cat2* mRNA is not sequestered in ribonucleoprotein particles in dark-grown leaves (Skadsen and Scandalios, 1987). Thus, the control of *Cat2* expression in maize leaves in response to light appears to involve a unique form of translational inhibition in dark-grown leaves, preventing translation of the isolated *Cat2* mRNA. The mRNA is rendered translatable only after the leaves are exposed to white light.

There is evidence that, like maize *Cat2*, the mRNA for chlorophyll *a/b*-binding protein is present in dark-grown peas (Giles *et al.*, 1977) and in *Lemna gibba* (Slovin and Tobin, 1982) and may be activated by a translational control mechanism in the light. The most dramatic translational control by light has been found in *Volvox* (Kirk and Kirk, 1985).

It is tempting to speculate that a physiological role should attend this CAT-2 specific light response. A plausible explanation is that the cell reserves CAT-2 synthesis until it is needed to destroy H_2O_2 generated during photorespiration (Zelitch, 1971). Photorespiration in maize, a C_4 plant, occurs to a lesser degree as compared to C_3 plants. C_3 plants often form large catalase crystals in their peroxisomes (Huang *et al.*, 1983), which may reflect the degree to which catalase is vital. This point is underscored in a catalase underexpression mutant of barley, which does not survive under photorespiratory conditions (Kendall *et al.*, 1983). However, the existence of normally growing *Cat2* null mutants in maize (Tsaftaris and Scandalios, 1981; Bethards and Scandalios, 1988) suggests that other mechanisms may accommodate the C_4 level of photorespiration.

Light-response studies of the *Cat3* gene have revealed two major forms of expression. In etiolated seedlings, CAT-3 protein levels are relatively high. Upon exposure to light, CAT-3 levels (and its mRNA) decline gradually. By the end of 7 days of growth in the light (14 days from germination), *Cat3* mRNA is no longer detectable. Thus, *Cat2* and *Cat3* respond oppositely to light. An intriguing possibility is that although *Cat2* and *Cat3* probably originated from a prototypical *Cat* gene, they likely evolved separate and opposite light-response elements (LREs, denoting DNA sequences controlling a gene's response to light) (Fluhr *et al.*, 1986).

Following an initial exposure to light (up to 48 hours), and upon transferring seedlings to a regular photoperiod of 12 hours of light and 12 hours of dark, another type of light–dark response is observed. The *Cat3* gene exhibits a cyclic expression. The CAT-3 protein and mRNA levels fall after light exposure and rise after the end of the light period.

The accumulation of *Cat3* mRNA in green maize leaves clearly exhibits a novel circadian rhythm (M. G. Redinbaugh and J. G. Scandalios, unpublished observations). The steady-state transcript level of *Cat3* varies with a 24-hour period, with high levels of the mRNA present late in the photoperiod (Fig. 3). This diurnal variation in transcript level does not occur with the *Cat1* or *Cat2* transcripts. The periodicity ob-

FIG. 3. An S_1 nuclease protection analysis of maize RNA using a *Cat3* gene-specific probe. The level of expression of the *Cat3* gene, but not *Cat1* or *Cat2*, is regulated by both a diurnal and a circadian rhythm in maize. Seedlings transferred from the dark–light cycle to either continuous darkness or light retain the diurnal pattern of *Cat3* mRNA; the rhythm is independent of phytochrome. Numbers at the top indicate the hour on a 24-hour clock.

served in *Cat3* mRNA levels persists when the seedlings are transferred to continuous darkness or light, and it is not phytochrome regulated. The rate of *Cat3* transcription was shown to parallel the changes in steady-state mRNA levels (M. G. Redinbaugh and J. G. Scandalios, unpublished observations).

The circadian rhythm in *Cat3* transcript levels is unique among the three maize catalases. In addition, light affects the accumulation of the three catalase transcripts in different ways (Skadsen and Scandalios, 1987; Redinbaugh *et al.*, 1990), suggesting that each isozyme may fulfill a different physiological role. The CAT-3 isozyme, which in leaves is expressed only in mesophyll cells (Tsaftaris *et al.*, 1983), has considerably more peroxidatic than catalatic activity (Havir and McHale, 1989) as compared to CAT-2, which is localized in the bundle sheath cells of maize leaves. Because in maize there is a differential localization of the O_2-evolving and CO_2-fixing reactions of photosynthesis in the mesophyll and bundle sheath cells, respectively, the different catalases may indeed perform distinct metabolic functions.

Studies with other plant genes have fortified the assumption that light-responsive sequences are located in the 5'-flanking regions of structural genes (Herrera-Estrella *et al.*, 1984; Nagy *et al.*, 1987). In fact, light-grown CAT-2 null plants [due to a deletion of the 3' half of the *Cat2* transcript (Bethards and Scandalios, 1988)] produce polyadenylated *Cat2* mRNA, which is recruited onto leaf polysomes *in vivo* (R. W. Skadsen and J. G. Scandalios, unpublished observations). This provides preliminary evidence that the light control of *Cat2* gene transcription lies at the 5' end of the structural gene. A reasonable assumption is that LREs may be found in the 5'-flanking region of *Cat2* and *Cat3*. If so, *Cat2*-promoter constructs should not express CAT-2 protein in transformed dark-grown callus, whereas they will express CAT-2 in green callus; the opposite should hold for CAT-3. Such results would allow the manipulation of these antioxidant genes under different light regimens and should lead to a better understanding of their response to light, and perhaps to other environmental signals.

B. Superoxide Dismutase and Catalase Gene Responses to Paraquat, Temperature, Hyperoxia, and Pollutants

1. Response to Paraquat

The herbicidal action of paraquat and similar compounds has been attributed to increased production of oxygen-derived species (Halliwell, 1984). Paraquat treatments (10^{-5} *M*) of 10-day-old maize leaves resulted in a 40% increase in superoxide dismutase activity and a smaller increase in catalase activity (Matters and Scandalios, 1986b). The increase in total superoxide dismutase activity correlated with higher levels of specific SOD isozymes. The chloroplast (SOD-1) and cytosolic (SOD-2 and SOD-4) forms were increased significantly, whereas the mitochondrial form (SOD-3) was increased only slightly. Higher levels of SOD-4 and SOD-3 after paraquat exposure were the result of increased synthesis of these proteins, as determined by labeling *in vivo* with [^{35}S]methionine. Isolation and *in vitro* translation of polysomes from 10^{-5} *M* paraquat-treated leaves indicated paraquat increased the amount of polysomal mRNA that codes for SOD-4 and SOD-3. Superoxide dismutase induction does not appear to be a response specific to paraquat, because another superoxide-generating compound, juglone, caused a similar increase in total superoxide dismutase activity. However, treatment with the paraquat structural analog, benzyl viologen, which does not generate oxy radicals, had no effect on SOD or CAT (Table 4). Therefore, the effect of these compounds on the expression of the maize *Sod* and *Cat* genes is via their ability to generate superoxide.

Elevated levels of total SOD activity were dependent on paraquat dosage and were accompanied by increases in the activities of maize catalase and glutathione reductase, which are involved in the detoxification of an SOD end product, H_2O_2 (Table 5). Other enzymes, such as

TABLE 4

Comparison of the Effects of the Redox-Active Agent Juglone and Paraquat Analog Benzyl Viologen on Superoxide Dismutase and Catalase in Maize Leaves[a]

Enzyme activity		Treatment		
	Buffer	Paraquat (10^- *M*)	Benzyl viologen (10^{-5} *M*)	Juglone (10^{-3} *M*)
Superoxide dismutase				
U/gFW	100	140	101	152
U/mg protein	100	141	105	163
Catalase				
U/gFW	100	128	101	95
U/mg protein	100	115	100	105

[a] Changes in total activity of superoxide dismutase and catalase in extracts of 10-day-old leaves exposed for 12 hours to 10^{-5} *M* benzyl viologen or for 1 hour to 10^{-3} *M* juglone. The change in enzyme activity, expressed on a per gram fresh weight (U/gFW) or on a per milligram of total protein (U/mg protein) basis, is the percentage activity in paraquat-, benzyl viologen-, or juglone-treated tissue as compared to activity in tissue treated only with buffer.

TABLE 5
Activity of Oxygen Free-Radical-Scavenging Enzymes in Paraquat-Treated Maize Leaves[a]

	Catalase		Superoxide dismutase		Glutathione reductase	
	U/gFW	U/mg protein	U/gFW	U/mg protein	U/gFW	U/mg protein
Control	12.00 ± 0.82	0.232 ± 0.013	1084 ± 117	20.50 ± 1.36	0.060 ± 0.004	3.25 ± 0.33
Paraquat	15.33 ± 1.23	0.266 ± 0.013	1522 ± 99	28.87 ± 1.26	0.072 ± 0.005	4.32 ± 0.45
Increase (%)	28%	15%	41%	40%	20%	33%

[a] Total catalase, superoxide dismutase, and glutathione reductase activity in extracts of leaves exposed to 10^{-5} M paraquat or buffer (control) for 12 hours. Units of activity ± standard errors of the mean are expressed as units per gram fresh weight of leaf tissue (U/gFW) or as units per milligram of total protein (U/mg protein).

triose phosphate isomerase, hydroxypyruvate reductase, and malate dehydrogenase, were not increased by paraquat treatments and, at high concentrations of the herbicide, were significantly lower, reflecting the damaging effects of paraquat-generated O_2^- on leaf tissue (Matters and Scandalios, 1986b). The increases observed in polysome-bound $Sod3$ and $Sod4$ mRNAs suggest that the response to paraquat of these isozymes, and perhaps the others, might be due to enhanced transcription of these genes.

The link between paraquat's presence in the cell and increased levels of SOD could involve several intermediary factors, for it is unlikely that the herbicide molecule directly alters Sod gene expression. The ability of a redox-active compound such as juglone, and other compounds that are structurally different from paraquat, to increase SOD levels in maize leaves and in $Escherichia\ coli$ (Hassan and Fridovich, 1979) indicates the change in SOD is more likely a response of the cell to paraquat-generated superoxide radicals.

2. Response to Ozone, Sulfur Dioxide, and Oxygen

The activities of superoxide dismutase and catalase were determined in maize leaves treated with O_3 or SO_2 for 8 hours or elevated levels of oxygen for up to 96 hours. Neither O_3 nor SO_2 significantly increased the levels of superoxide dismutase or catalase activity. However, after 72 hours of continuous, 90% oxygen treatment, an increase in superoxide dismutase activity was observed (Matters and Scandalios, 1987); SOD activity remained high throughout 96 hours of treatment. The activities of other enzymes, including catalase, ascorbate peroxidase, and malate dehydrogenase, were not increased during this period. Immunological analysis showed that amounts of the cytosolic Cu–Zn superoxide dismutase isozymes, SOD-2 and SOD-4, were increased by elevated oxygen but that the chloroplast (SOD-1) or mitochondrial (SOD-3) isozymes were not increased. Immunoprecipitation of translation products of 72-hour oxygen-treated or nontreated leaf polysomes indicated that the higher levels of SOD-2 and SOD-4 were due to increased amounts of polysome-bound mRNA coding for these proteins. The specific response of SOD-2 and SOD-4 to increased oxidative stress during 90% oxygen treatments contrasts with the increase in all SOD isozymes in maize leaves treated with the herbicide paraquat, and may represent an alternative type of stress response.

The variation in the response to different oxidative stresses may reflect the nature of the stress-inducing agent or level of stress induction. Because paraquat scavenges electrons from the chloroplast photosystem I and mitochondrial electron transport chain (Farrington $et\ al.$, 1973), a significant portion of paraquat-generated superoxide may occur within these organelles. The mechanisms by which oxygen can increase oxidative stress are less clear. Formation of oxygen-induced radicals may occur through numerous pathways, many of which may be cytosolic. Therefore, the need for superoxide dismutase activity within organelles may not be as great during oxygen-induced stress, and increases only in the cytosolic isozymes may be adequate. Whereas increased SOD activity was seen after 96 hours of 90% O_2, the amounts of SOD-2 and SOD-4 protein at this time are not increased. Although it has not been examined in more detail, it is possible that general protein turnover is occurring more rapidly in the O_2-stressed tissue.

The level at which the *Sod* genes respond to elevated oxygen treatments appears not to be posttranslational. Oxygen-treated leaves contain more polysome-bound *Sod2* and *Sod4* mRNA than do nontreated leaves, as shown by the higher amounts of SOD-2 and SOD-4 translation products produced from isolated leaf polysomes (Matters and Scandalios, 1987). The increases in *Sod2* and *Sod4* polysome-bound mRNA in oxygen-treated tissues could be the result of more effective initiation of *Sod* mRNA during translation, or increased amounts of *Sod* mRNA available to be translated. Oxygen is known to affect transcription of several yeast genes (Lowry and Zitomer, 1984). The induction of these genes by oxygen involves upstream DNA sequences (Guarente *et al.*, 1984), and several DNA-binding proteins (Arcangioli and Lescure, 1985). Similar types of regulatory factors may be in effect during oxidative stress-induced changes for the maize SOD genes. This study defines a system with which to continue to explore the mechanisms involved in oxidative stress responses, and characterizes more accurately the steps leading to environmentally induced changes in gene expression.

3. Catalase and Superoxide Dismutase Gene Responses to Temperature

The similarity of responses to elevated temperature (heat shock), various oxidative stresses in bacteria (Lee *et al.*, 1983; Bonner *et al.*, 1984) and *Drosophila* (Ashburner and Bonner, 1979), and the pleiotropic induction of some of the heat-shock proteins in higher plants in response to a diverse range of stresses (Czarnecka *et al.*, 1984; Heikkla *et al.*, 1984) suggests the possibility that increased temperature and oxidative stress may be related. If elevated temperature increases the oxidative stress placed on a cell, it may be possible to alter levels of oxygen free-radical detoxifying enzymes by growing the organisms at higher temperatures.

Acute, 40°C heat shock of W64A maize (normal CAT expression line) seedlings did not significantly change total SOD activity in the scutellum of treated seedlings, and total catalase activity was reduced. Exposure of seedlings to persistent 40°C temperatures over a period of 10 days also did not increase SOD or CAT activity. Total SOD activity, although lower on days 1 and 3 at 40°C, was not different from 25°C-grown seedling activity at the later stages of germination. At days 2 and 3 postimbibition, catalase activity was increased slightly in 35°C-grown seedlings and decreased slightly in 40°C-grown seedlings. However, at days 6–10, little change in catalase activity was observed at 25, 35, or 40°C. In contrast to W64A, the high-catalase-activity line R6-67 (overexpresses CAT-2) showed a dramatic decrease in catalase activity when grown at 35 or 40°C, to levels closely resembling the catalase activity profile in W64A (Fig. 4). Rocket immunoelectrophoresis of R6-67 scutellar extracts from 25°C-grown and 40°C-grown seedlings indicated that lower levels of catalase activity were due to reduced amounts of CAT-2 protein at 40°C. Use of the catalase synthesis inhibitor AIA showed the rate of catalase synthesis at 40°C was decreased in R6-67 to levels similar to those in W64A, whereas degradation rates were increased slightly. Therefore, elevated temperatures did not appear to increase CAT or SOD activity in developing maize seedlings. Chronic elevated temperature decreased catalase activity, levels of CAT-2 protein, and catalase synthesis rates in the high-activity variant R6-67 (Matters and Scandalios, 1986c). Because CAT-2 levels in R6-67 are known to be affected by the trans-acting locus *Car1* (Scandalios *et al.*, 1980b), it is possible that the *Car1* product, perhaps a protein, is inhibited at high temperature. Recently, experiments have demonstrated that the drop in CAT-2 in 40°C-grown R6-67 seedlings may be due to decreased levels of *Cat2* message. If the *Car1* product is heat sensitive, this may provide one criterion by which the *Car1* product can be identified and characterized.

4. Response of Catalase and Superoxide Dismutase Genes to Additional Factors

The maize catalases and superoxide dismutases have been shown to respond to numerous other exogenously applied signals. Among these are hydrogen peroxide, which in a concentration of $\approx 0.01\%$ in a nutrient medium results in more than doubling of catalase activity at day 4 postimbibition without perturbing the morphological pattern of seedling growth. Lower amounts of H_2O_2 (below 0.004%) had no effect, and levels higher than 0.01% caused tissue damage in young seedlings. The increased levels of catalase activity in response to H_2O_2 are due to

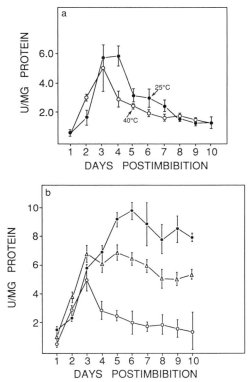

FIG. 4. Total catalase activity in scutella of developing maize seedlings from days 1–10 postimbibition. The inbred maize lines (a) W64A (normal-catalase-activity profile) and (b) R6-67 (high-catalase-activity profile variant) were germinated and grown at 25°C (●), 35°C (△), and 40°C (○). The bars represent the standard error of the mean of five independent experiments. Note that under normal growing conditions (25°C) R6-67 exhibits scutellar catalase activity levels three to four times higher than is seen in W64A, due to extended synthesis of CAT-2 in R6-67. Nevertheless, when grown at elevated temperatures (35 and 40°C), the high catalase activity is not observed in R6-67, its activity more closely resembling that observed for W64A. There is no apparent difference in catalase activity for W64A whether it is grown at 25 or 40°C.

catalase protein (CAT-2) accumulation, as determined by immunoelectrophoresis (Scandalios, unpublished).

Exposure of mature maize seed to the fungal toxin cercosporin in the presence of light resulted in an approximately 60% increase in total catalase activity in scutella after germination. Rocket immunoelectrophoresis showed a corresponding increase in CAT-2 protein of sufficient magnitude to account for all of the increased activity (J. D. Williamson and J. G. Scandalios, unpublished observations). Under identical conditions, no changes in SOD activity or protein levels were detected. However, preliminary results suggest that a measurable increase in *Sod3* transcript may occur in response to cercosporin. Cercosporin, which is produced by the fungus *Cercospora,* produces singlet oxygen and superoxide through a photosensitized reaction (Daub and Hangarter, 1983), and the active oxygen produced damages the infected plant.

Numerous other environmental conditions (e.g., flood, drought, chilling, and anoxia) have been shown to create oxidative stress, and to often result in increases in antioxidant defenses. Increases in SOD activity were observed in anoxia-tolerant *Iris pseudacorus* rhizomes, during and after anoxic stress (Monk *et al.,* 1987). Plants completely deprived of oxygen have been shown to survive a period of anoxia only to perish on reexposure to air, indicating oxidative damage occurred during the recovery period (Monk *et al.,* 1989).

Although many factors are known to increase oxidative stress and to mobilize the antioxidant defense systems in all aerobic organisms, we have to understand the underlying mechanisms involved. How genes encoding antioxidant proteins perceive and respond to oxidative stress signals is just beginning to be unraveled via the dissection of antioxidant gene structure, regulation, and expression.

XIII. Molecular Biology

In order to understand the underlying mechanisms by which the catalase and superoxide dismutase genes of maize respond to environmental signals, it is essential that their structure, regulation, and expression be better understood. To these ends, cDNA clones were isolated and characterized for each of the *Cat* and *Sod* genes.

A. CLONING AND CHARACTERIZATION OF CATALASE GENE cDNAs

Full-length cDNAs were isolated (Fig. 5) and characterized for the *Cat1, Cat2,* and *Cat3* transcripts encoding each of the maize catalases (Bethards *et al.,* 1987; Redinbaugh *et al.,* 1988). DNA sequence analyses confirmed that each cDNA encodes a unique catalase protein. The *Cat* transcripts have regions of extensive homology to each other and to *Cat* transcripts of other species. Due to the sequence diversity at the 3′ ends, we were able to produce gene-specific cDNA probes that are highly

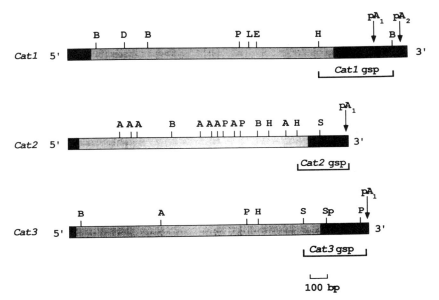

FIG. 5. Restriction endonuclease maps of *Cat1, Cat2,* and *Cat3* cDNAs. The polyadenylation sites pA$_1$ and pA$_2$ are indicated above the restriction maps. The gene-specific probe (gsp) regions are underlined below the maps. (▢) Coding regions and (▉) nontranslated 5′ and 3′ regions. A, *Ava*I; B, *Bam*HI; D, *Hind*III; H, *Hind*II; L, *Sal*I; P, *Pst*I; S, *Sac*I; Sp, *Sph*I.

selective under stringent hybridization conditions. RNA blot analyses, using gene-specific probes, confirmed a clear correlation between the *Cat1, Cat2,* and *Cat3* transcripts and the tissue-specific expression of the CAT-1, CAT-2, and CAT-3 isozymes (Redinbaugh *et al.,* 1988). The gene-specific probes hybridized with maize genomic DNA blots in simple, unique patterns (Fig. 6), indicating that there is one or very few copies of each catalase gene in different locations on the maize genome, in accordance with the earlier genetic and mapping data (Roupakias *et al.,* 1980).

The coding region of the *Cat3* cDNA comprises 66% G + C, which led to a strong codon usage bias in this gene. This codon bias was also observed with the *Cat2* transcripts, but not with those for *Cat1,* and might be involved in regulating the expression of these genes. A high degree of similarity was found between the maize catalase nucleic acid and deduced amino acid sequences and those of sweet potato and mammalian catalase (Redinbaugh *et al.,* 1988).

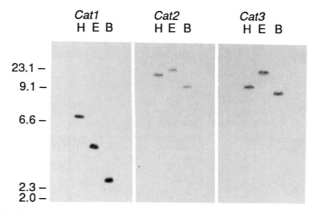

FIG. 6. Genomic DNA blot analysis. Maize genomic DNA was digested with *Hin*dIII (H), *Eco*RI (E), or *Bam*HI (B) and electrophoresed on a 0.7% agarose gel. Following transfer to nitrocellulose, the genomic DNA was hybridized with the indicated gene-specific probe under the conditions appropriate with homologous probes. The migration of the molecular weight standards (in kilobases) is indicated on the left.

B. CLONING AND CHARACTERIZATION OF SUPEROXIDE DISMUTASE GENE cDNAs

Full-length cDNAs for the maize Cu–Zn *Sod2, Sod4,* and *Sod4A* and for the Mn *Sod3* have been isolated, cloned, sequenced, and characterized (Cannon *et al.,* 1987; White and Scandalios, 1988; Cannon and Scandalios, 1989). Nucleic acid and amino acid sequences show significant homologies to the respective Cu–ZnSODs and MnSODs characterized and reported for other organisms (Perl-Treves *et al.,* 1988; Scioli and Zilinskas, 1988). The Cu–Zn *Sod* genes are highly conserved in the coding regions, leading to cross-hybridization problems when full-length cDNAs are used as probes. However, this can be alleviated by using gene-specific probes utilizing the 3′ untranslated region of each Cu–Zn *Sod* (Fig. 7). Genomic DNA blots indicate that, like the catalases, the *Sod* genes exist in single or very few copies, and RNA blots indicate that the tissue and temporal distribution of the *Sod* transcripts parallel the earlier isozyme profiles of the various maize tissues.

C. GENOMIC CLONES FOR CATALASE AND SUPEROXIDE DISMUTASE GENES

Genomic clones have been isolated for *Sod2, Sod3, Sod4,* and *Sod4A,* and for *Cat1* and *Cat3* (R. E. Cannon and J. G. Scandalios, unpublished

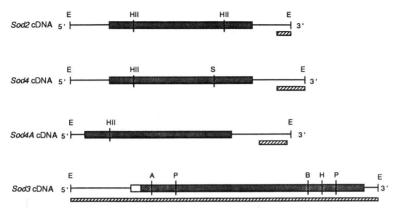

FIG. 7. Restriction endonuclease maps and regions used as sequence-specific, gene-specific probes of *Sod2, Sod4, Sod4A,* and *Sod3* cDNAs. (▨) Coding regions, (——) noncoding regions, and (▮) gene-specific probe regions. A, *Ava*I; B, *Bam*HI; E, *Eco*RI; HII, *Hinc*II; H, *Hind*III; P, *Pst*I; S, *Sal*I.

observations; G. J. Wadsworth and J. G. Scandalios, unpublished observations). These clones are being restriction mapped and sequenced to determine intron–exon boundaries, promoter-binding sites, light-responsive (or stress-responsive) elements, and cis-regulatory elements. Preliminary evidence suggests that the genomic clones for both *Cat* and *Sod* may have several kilobases of flanking DNA around the structural genes, which may contain cis-acting regions involved in the differential expression of the *Cat* and *Sod* genes.

D. Significance of Molecular Studies

Resolution of the molecular biology of these systems would render possible their investigation in transient assays *in vivo* to determine how the *Cat* and *Sod* genes of maize are regulated (i.e., cis regions and trans factors) and how they respond to specific stresses or signals. It would permit the direct investigation of the biological role(s) of catalase and superoxide dismutase by engineering transgenic plants with higher levels of any given form (isozyme) of CAT or SOD. Exposing such plants to oxidative stress should prove interesting and may clarify the precise biological role of each form of CAT and SOD. Lowering the existing endogenous levels of any SOD or CAT by using engineered plants that express the antisense of a given *Sod* or *Cat* transcript should also indicate the functional significance of each isozyme.

These studies, taken in concert, should help elucidate how the *Sod* and *Cat* genes in maize respond to oxidative stress, and should directly test the biological significance of these genes in higher plants, and perhaps all aerobic organisms.

XIV. Concluding Remarks

Activated oxygen species, generated both endogenously as by-products of normal metabolism, or exogenously as a consequence of various environmental factors, are highly reactive molecules capable of causing extensive damage to plant cells. The effects of oxidative stress may range from simple inhibition of enzyme function to the production of random lesions in proteins and nucleic acids, and the peroxidation of membrane lipids. Loss of membrane integrity due to peroxidation, together with direct damage to enzymatic and structural proteins and their respective genes, can result in decreased mitochondrial and chloroplast functions. Decreased organellar function in turn lowers the plant's ability to fix carbon and to properly utilize the resulting products. The resulting decrease in metabolic efficiency results in loss of yield. Extreme oxidative damage leads to cell and tissue death, and hence to an even greater loss of yield.

To cope with oxidative stress, aerobic organisms evolved protective antioxidant defenses, both nonenzymatic and enzymatic. However, aside from numerous correlative responses (i.e., increases in oxidative stress leading to increased levels of some antioxidant defenses), there currently exists very little information and understanding of the underlying molecular mechanisms for the mobilization of the antioxidant defenses in aerobic organisms. The maize CAT and SOD systems described herein provide an opportunity to examine how higher plant genomes may perceive, mobilize, and respond to oxidative insult.

The role of oxidative stress in effecting specific gene expression is intricate in nature when one considers that oxidative stress can affect specific organelles, or the entire plant cell or tissue. The plant has defense responses (gene expression mechanisms) to accommodate either or both situations. How the plant responds to a specific organellar stress versus a cellular or systemic oxidative stress, and what genetic elements may be involved in the defense response, can now be resolved utilizing state-of-the-art molecular technologies.

The advantages presented for such investigations by both the catalase and superoxide dismutase gene–enzyme systems of maize include their in-depth definition at the biochemical, developmental, genetic, and molecular levels. Having isolated and sequenced virtually all of the *Cat* and *Sod* cDNAs and having genomic clones at hand will permit us to examine "stress response elements" in these important oxy-defense genes. Once such information is available, and a better understanding of the defense response is attained, the engineering of higher plants for greater resistance to oxidative stress imposed under hostile environments (natural or manmade) will be possible.

ACKNOWLEDGMENTS

Research from the author's laboratory was supported by grants from the National Institutes of Health (GM33817) and from the United States Environmental Protection Agency (R-812404 and R-814013). I thank my colleagues, past and present, for their contributions to ongoing research in my laboratory. I especially thank Stephanie Ruzsa and Sheri P. Kernodle for their dedicated and expert technical assistance. The excellent typing of Suzanne Quick is appreciated.

REFERENCES

Acevedo, A., and Scandalios, J. G. (1990). Expression of the catalase and superoxide dismutase genes in mature pollen in maize. *Theor. Appl. Genet.* (in press).

Alscher, R., Franz, M., and Jeske, C. (1987). Sulfur dioxide and chloroplast metabolism. *In* "Phytochemical Effects of Environmental Compounds" (J. Saunders, L. Channing, and E. Conn, eds.), pp. 1–28. Plenum, New York.

Ames, B. (1983). Dietary carcinogens and anticarcinogens. *Science* 221, 1256–1264.

Arcangioli, B., and Lescure, B. (1985). Identification of proteins involved in the regulation of yeast iso-1-cytochrome C expression by oxygen. *EMBO J.* 4, 2627–2633.

Asada, K., and Takahashi, M. (1987). Production and scavenging of active oxygen in photosynthesis. *In* "Photoinhibition" (D. J. Kyle, C. B. Osmond, and C. J. Arntzen, eds.), pp. 227–287. Elsevier, Amsterdam.

Ashburner, M., and Bonner, J. J. (1979). The induction of gene activity in *Drosophila* by heat shock. *Cell* 17, 241–254.

Baum, J. A., and Scandalios, J. G. (1979). Developmental expression and intracellular localization of superoxide dismutases in maize. *Differentiation (Berlin)* 13, 133–140.

Baum, J. A., and Scandalios, J. G. (1981). Isolation and characterization of the cytosolic and mitochondrial superoxide dismutases of maize. *Arch. Biochem. Biophys.* 206, 249–264.

Baum, J.A., and Scandalios, J. G. (1982a). Expression of genetically distinct superoxide dismutases in the maize seedling during development. *Dev. Genet.* 3, 7–23.

Baum, J. A., and Scandalios, J. G. (1982b). Multiple genes controlling superoxide dismutase expression in maize. *J. Hered.* 73, 95–100.

Baum, J. A., Chandlee, J. M., and Scandalios, J. G. (1983). Purification and partial characterization of a genetically defined superoxide dismutase (SOD-1) associated with maize chloroplasts. *Plant Physiol.* 73, 31–35.

Beevers, H. (1979). Microbodies in higher plants. *Annu. Rev. Plant Physiol.* **30,** 159–193.

Bethards, L. A., and Scandalios, J. G. (1988). Molecular basis for the CAT-2 null phenotype in maize. *Genetics* **118,** 149–153.

Bethards, L. A., Skadsen, R. W., and Scandalios, J. G. (1987). Isolation and characterization of a cDNA clone for the *Cat2* gene in maize and its homology with other catalases. *Proc. Natl. Acad. Sci. U.S.A.* **84,** 6830–6834.

Bielski, B. H. J. (1978). Reevaluation of the spectral and kinetic properties of HO_2 and O_2^- free radicals. *Photochem. Photobiol.* **28,** 645–649.

Bienz, M., and Pelham, H. R. B. (1987). Mechanisms of heat-shock gene activation in higher eukaryotes. *Adv. Genet.* **24,** 31–72.

Bonner, B. R., Lee, P. C., Wilson, S. W., Cutter, C. W., and Ames, B. N. (1984). AppppA and related adenylated nucleotides are synthesized as a consequence of oxidative stress. *Cell* **37,** 225–232.

Cadenas, E. (1985). Oxidative stress and formation of excited species. *In* "Oxidative Stress" (H. Sies, ed.), pp. 311–330. Academic Press, New York.

Cannon, R. E., and Scandalios, J. G. (1989). Two cDNAs encode two nearly identical Cu/Zn superoxide dismutase proteins in maize. *Mol. Gen. Genet.* **219,** 1–8.

Cannon, R. E., White, J. A., and Scandalios, J. G. (1987). Cloning of cDNA for maize superoxide dismutase 2 (SOD-2). *Proc. Natl. Acad. Sci. U.S.A.* **84,** 179–183.

Cerutti, P. A. (1985). Prooxidant states and tumor production. *Science* **227,** 375–381.

Chandlee, J. M., and Scandalios, J. G. (1984a). Analysis of variants affecting the catalase developmental program in maize scutellum. *Theor. Appl. Genet.* **69,** 71–77.

Chandlee, J. M., and Scandalios, J. G. (1984b). Regulation of *Cat1* gene expression in the scutellum of maize during early sporophytic development. *Proc. Natl. Acad. Sci. U.S.A.* **81,** 4903–4907.

Chandlee, J. M., Tsaftaris, A. S., and Scandalios, J. G. (1983). Purification and partial characterization of three genetically defined catalases of maize. *Plant Sci. Lett.* **29,** 117–131.

Charles, S., and Halliwell, B. (1981). Light activation of fructose bisphosphatase in isolated spinach chloroplasts and deactivation by hydrogen peroxide—A physiological role for the thioredoxin system. *Planta* **151,** 242–246.

Chollet, R., and Ogren, W. (1975). Regulation of photorespiration in C_3 and C_4 species. *Bot. Rev.* **41,** 137–179.

Craig, E. A. (1985). The heat-shock response. *CRC Crit. Rev. Biochem.* **18,** 239–280.

Cramer, C. L., Ryder, T. B., Bell, J. N., and Lamb, C. J. (1986). Rapid switching of plant gene expression induced by fungal elicitor. *Science* **227,** 1240–1243.

Cullis, C. A. (1986). Phenotypic consequences of environmentally induced changes in plant DNA. *Trends Genet.* **2,** 307–309.

Czarnecka, E., Edelman, L., Schoffl, F., and Key, J. L. (1984). Comparative analysis of physical stress responses in soybean seedlings using cloned heat shock cDNAs. *Plant Mol. Biol.* **3,** 45–58.

Daub, M. E., and Hangarter, R. P. (1983). Production of singlet oxygen and superoxide by the fungal toxin, cercosporin. *Plant Physiol.* **73,** 855–857.

Edwards, K., Cramer, C. L., Bolwell, G. P., Dixon, R. A., Schuch, W., and Lamb, C. J. (1985). Rapid transient induction of phenylalanine ammonia-lyase mRNA in elicitor-treated bean cells. *Proc. Natl. Acad. Sci. U.S.A.* **82,** 6725–6731.

Ellis, R. J. (1986). Photoregulation of gene expression. *Biosci. Rep.* **6,** 127–136.

Elstner, E. F. (1982). Oxygen activation and oxygen toxicity. *Annu. Rev. Plant Physiol.* **33,** 73–96.

Elstner, E. F. (1987). Metabolism of activated oxygen species. *In* "The Biochemistry of Plants" (D. D. Davies, ed.), Vol. 2, pp. 253–314. Academic Press, Orlando, Florida.

Farrington, J., Ebert, M., Land, E., and Fletcher, K. (1973). Bipyridylium quaternary salts and related compounds. V. Pulse radiolysis studies of the reaction of paraquat radical with oxygen. Implications for the mode of action of bipyridyl herbicides. *Biochim. Biophys. Acta* **314**, 372–381.

Fluhr, R., Kuhlemeier, C., Nagy, F., and Chua, N.-H. (1986). Organ-specific and light-induced expression of plant genes. *Science* **232**, 1106–1112.

Foote, C. S. (1976). Photosensitized oxidation and singlet oxygen: Consequences in biological systems. *Free Radicals Biol.* **2**, 85–133.

Foster, J., and Hess, J. (1980). Responses of superoxide dismutase and glutathione reductase activities in cotton leaf tissue exposed to an atmosphere enriched in oxygen. *Plant Physiol.* **66**, 482–487.

Fridovich, I. (1976). Oxygen radicals, hydrogen peroxide and oxygen toxicity. *Free Radicals Biol.* **1**, 239–277.

Fridovich, I. (1978). Superoxide dismutases. *Annu. Rev. Biochem.* **44**, 147–152.

Fridovich, I. (1986). Superoxide dismutases. *Adv. Enzymol.* **58**, 61–85.

Fucci, L., Oliver, C., Coon, M., and Stadtman, E. (1983). Inactivation of key metabolic enzymes by mixed-function oxidation reactions: Possible implication in protein turnover and aging. *Proc. Natl. Acad. Sci. U.S.A.* **80**, 1521–1525.

Gallagher, T. F., and Ellis, R. J. (1982). Light-stimulated transcription of genes for two chloroplast polypeptides in isolated pea nuclei. *EMBO J.* **1**, 1493–1498.

Gilbert, D. L. (1981). "Oxygen and Living Processes." Springer-Verlag, Berlin and New York.

Giles, A. B., Grierson, D., and Smith, H. (1977). *In vitro* translation of mRNA from developing bean leaves. Evidence for the existence of stored mRNA and its light-induced mobilization onto polyribosomes. *Planta* **136**, 31–36.

Gregory, E. M., and Fridovich, I. (1973). The induction of superoxide dismutase by molecular oxygen. *J. Bacteriol.* **114**, 543–548.

Guarente, L., Lalonde, B., Gifford, P., and Alani, E. (1984). Distinctly regulated tandem upstream activation sites mediate catabolite repression of the *Cyc1* gene of *S. cerevisiae*. *Cell* **36**, 503–511.

Halliwell, B. (1974). Superoxide dismutase, catalase and glutathione peroxidase: Solutions to the problem of living with oxygen. *New Phytol.* **73**, 1075–1086.

Halliwell, B. (1982). The toxic effects of oxygen in plant tissues. *In* "Superoxide Dismutase" (L. W. Oberley, ed.), Vol. 2, pp. 90–123. CRC Press, Boca Raton, Florida.

Halliwell, B. (1984). Oxygen-derived species and herbicide action. *Physiol. Plant.* **15**, 21–24.

Halliwell, B., and Gutteridge, J. M. C. (1985). The importance of free radicals and catalytic metal ions in human diseases. *Mol. Aspects Med.* **8**, 89–193.

Hamilton, B., Hoffaner, R., and Ruis, H. (1982). Translational control of catalase synthesis by hemin in the yeast *Saccharomyces cerevisiae*. *Proc. Natl. Acad. Sci. U.S.A.* **79**, 7609–7613.

Harbour, J. R., and Bolton, J. R. (1975). Superoxide formation in spinach chloroplasts: Electron-spin resonance detection by spin trapping. *Biochem. Biophys. Res. Commun.* **64**, 803–807.

Harper, D. B., and Harvey, B. M. (1978). Mechanism of paraquat tolerance in perennial ryegrass. Role of superoxide dismutase, catalase and peroxidase. *Plant Cell Environ.* **1**, 211–215.

Hassan, H. M., and Fridovich, I. (1979). Intracellular production of superoxide radical and hydrogen peroxide by redox active compounds. *Arch. Biochem. Biophys.* **196**, 385–395.

Hatch, M. R., and Slack, C. R. (1969). NADP-specific malate dehydrogenase and glycerate kinase in leaves and evidence for their location in chloroplasts. *Bioichem. Biophys. Res. Commun.* **34**, 589–593.

Havir, A. E., and McHale, N. A. (1989). Enhanced-peroxidatic activity in specific catalase isozymes of tobacco, barley and maize. *Plant Physiol.* **91**, 812–815.

Heikkla, J. J., Papp, J. E. T., Schultz, G. A., and Bewley, J. D. (1984). Induction of heat shock messenger RNA in maize mesocotyls by water stress, abscisic acid, and wounding. *Plant Physiol.* **76**, 270–274.

Herrera-Estrella, L., Van Den Broek, G., Maenhant, R., van Montagu, M., Schell, J., Timko, M., and Cashmore, A. (1984). Light-inducible and chloroplast-associated expression of a chimaeric gene introduced into *Nicotiana tabacum* using a Ti plasmid vector. *Nature (London)* **310**, 115–120.

Huang, A. H. C., Trelease, R. N., and Moore, T. S. (1983). "Plant Peroxisomes." Academic Press, New York.

Kaiser, W. (1979). Carbon metabolism of chloroplasts in the dark. *Planta* **144**, 193–200.

Kendall, A. C., Keys, A. J., Turner, J. C., Lea, P. J., and Miflin, B. J. (1983). The isolation and characterization of a catalase-deficient mutant of barley. *Planta* **159**, 505–511.

Kimpel, J. A., and Key, J. L. (1985). Heat shock in plants. *Trends Biochem. Sci. (Pers. Ed.)* **10**, 353–356.

Kirk, M. M., and Kirk, D. L. (1985). Translational regulation of protein synthesis in response to light, at a critical stage of *Volvox* development. *Cell* **41**, 419–428.

Kitagawa, Y., Tsunasawa, S., Tanaka, N., Katsube, Y., Sakiyama, F., and Asada, K. (1986). Amino acid sequence of Cu/Zn-superoxide dismutase from spinach leaves. *J. Biochem. (Tokyo)* **99**, 1289–1298.

Kopczynski, C. C., and Scandalios, J. G. (1986). *Cat2* gene expression: Developmental control of translatable CAT-2 mRNA levels in maize scutellum. *Mol. Gen. Genet.* **203**, 185–188.

Lee, E., and Bennett, J. (1982). Superoxide dismutase: A possible protective enzyme against ozone injury in snap beans. (*Phaseolus vulgaris*). *Plant Physiol.* **69**, 1444–1449.

Lee, P. C., Bochner, B. R., and Ames, B. N. (1983). AppppA, heat-shock stress and cell oxidation. *Proc. Natl. Acad. Sci. U.S.A.* **80**, 7496–7500.

Leshem, Y. Y. (1988). Plant senescence processes and free radicals. *Free Radicals Biol. Med.* **5**, 39–49.

Lesko, S., Lorentarn, A., and Tso, R. (1980). Involvement of polycyclic aromatic hydrocarbons and reduced oxygen radicals in carcinogenesis. *Biochemistry* **19**, 3023–3028.

Levine, R., Oliver, C., Fulks, R., and Stadtman, E. (1981). Turnover of bacterial glutamine synthetase oxidative inactivation precedes proteolysis. *Proc. Natl. Acad. Sci. U.S.A.* **78**, 2120–2124.

Lorimer, G. (1981). The carboxylation and oxygenation of ribulose 1,5-bisphosphate: The primary events in photosynthesis and photorespiration. *Annu. Rev. Plant Physiol.* **32**, 349–383.

Lowry, C. V., and Zitomer, R. S. (1984). Oxygen regulation of anaerobic and aerobic genes mediated by a common factor in yeast. *Proc. Natl. Acad. Sci. U.S.A.* **81**, 6129–6133.

Matheson, I. B. C., Etheridge, R. D., Kratowich, N. R., and Lee, J. (1975). The quenching of singlet oxygen by amino acids and proteins. *Photochem. Photobiol.* **21**, 165–171.

Matters, G. L., and Scandalios, J. G. (1986a). Changes in plant gene expression during stress. *Dev. Genet.* **7**, 167–175.

Matters, G. L., and Scandalios, J. G. (1986b). Effect of the free radical-generating herbicide paraquat on the expression of the superoxide dismutase (*Sod*) genes in maize. *Biochim. Biophys. Acta* **882**, 29–38.

Matters, G. L., and Scandalios, J. G. (1986c). Effect of elevated temperature on catalase and superoxide dismutase during maize development. *Differentiation (Berlin)* **30**, 190–196.

Matters, G. L., and Scandalios, J. G. (1987). Synthesis of isozymes of superoxide dismutase in maize leaves in response to O_3, SO_2 and elevated O_2. *J. Exp. Bot.* **38**, 842–852.

McClintock, B. (1984). The significance of responses of the genome to challenge. *Science* **226**, 792–801.

McGuire, D. M., Chan, L., Smith, L. C., Towle, H. C., and Dempsey, M. E. (1985). Translational control of the circadian rhythm of liver sterol carrier protein (analysis of mRNA sequences with a specific cDNA probe). *J. Biol. Chem.* **260**, 5435–5439.

Miller, O. K., and Hughes, K. W. (1980). Selection of paraquat-resistant variants of tobacco from cell culture. *In Vitro* **16**, 1085–1091.

Minty, A., and Newark, P. (1980). Gene regulation: New, old and remote controls. *Nature (London)* **288**, 210–211.

Monk, L. S., Fagerstedt, K. V., and Crawford, R. M. (1987). Superoxide dismutase as an anaerobic polypeptide—A key factor in recovery from oxygen deprivation in *Iris pseudacorus*. *Plant Physiol.* **85**, 1016–1020.

Monk, L. S., Fagerstedt, K. V., and Crawford, R. M. (1989). Oxygen toxicity and superoxide dismutase as an antioxidant in physiological stress. *Physiol. Plant.* **76**, 456–459.

Nagy, F., Boutry, M., Hsu, M., Wong, M., and Chua, N.-H. (1987). The 5'-proximal region of the wheat *Cab-1* gene contains a 268-bp enhancer-like sequence for phytochrome response. *EMBO J.* **6**, 2537–2542.

Palatnik, C. M., Wilkins, C., and Jacobson, A. (1984). Translational control during early *Dictyostelium* development: Possible involvement of poly(A) sequences. *Cell* **36**, 1017–1025.

Perl-Treves, R., Nachmias, B., Aviv, D., Zeelon, E. P., and Galun, E. (1988). Isolation of two cDNA clones from tomato containing two different superoxide dismutase sequences. *Plant Mol. Biol.* **11**, 609–623.

Quail, P. H., and Scandalios, J. G. (1971). Turnover of genetically defined catalase isozymes in maize. *Proc. Natl. Acad. Sci. U.S.A.* **68**, 1402–1406.

Rabinowitch, H. D., and Sklan, D. (1980). Superoxide dismutase: A possible protective agent against sunscald in tomatoes. *Planta* **148**, 162–167.

Rabinowitch, H. D., Sklan, D., and Budowski, P. (1982). Photo-oxidative damage in the ripening tomato fruit: Protective role of superoxide dismutase. *Physiol. Plant.* **54**, 369–374.

Redinbaugh, M. G., Wadsworth, G. J., and Scandalios, J. G. (1988). Characterization of catalase transcripts and their differential expression in maize. *Biochim. Biophys. Acta* **951**, 104–116.

Redinbaugh, M. G., Sabre, M., and Scandalios, J. G. (1990). The distribution of catalase activity, isozyme protein, and transcript in the tissues of the developing maize seedling. *Plant Physiol.* **92**, 375–380.

Robinson, J., Smith, M., and Gibbs, M. (1980). Influence of hydrogen peroxide upon carbon dioxide photoassimilation in the spinach chloroplast. Hydrogen peroxide generated by broken chloroplasts in intact chloroplast preparation is a causal agent of the Warburg effect. *Plant Physiol.* **65**, 755–759.

Roupakias, D. G., McMillin, D. E., and Scandalios, J. G. (1980). Chromosomal location of the catalase structural genes in *Zea mays*, using B–A translocations. *Theor. Appl. Genet.* **58**, 211–218.

Sachs, M., and Ho, T. H. (1986). Alteration of gene expression during environmental stress in plants. *Annu. Rev. Plant Physiol.* **37**, 363–376.

Salin, M. L. (1988). Plant superoxide dismutase: A means of coping with oxygen radicals. *Curr. Top. Plant Biochem. Physiol.* **7**, 188–200.

Scandalios, J. G. (1965). Subunit dissociation and recombination of catalase isozymes. *Proc. Natl. Acad. Sci. U.S.A.* **53**, 1035–1040.

Scandalios, J. G. (1968). Genetic control of multiple molecular forms of catalase in maize. *Ann. N.Y. Acad. Sci.* **151**, 274–293.

Scandalios, J. G. (1974). Subcellular localization of catalase variants coded by two genetic loci during maize development. *J. Hered.* **65**, 28–32.

Scandalios, J. G. (1979). Control of gene expression and enzyme differentiation. In "Physiological Genetics" (J. G. Scandalios, ed.), pp. 63–107. Academic Press, New York.

Scandalios, J. G. (1983). Multiple varieties of isozymes and their role in studies of gene regulation and expression during eukaryote development. *Isozymes: Curr. Top. Biol. Med. Res.* **9**, 1–31.

Scandalios, J. G. (1987). The antioxidant enzyme genes *Cat* and *Sod* of maize: Regulation, functional significance, and molecular biology. *Isozymes: Curr. Top. Biol. Med. Res.* **14**, 19–44.

Scandalios, J. G., Liu, E. H., and Campeau, M. A. (1972). The effects of intragenic and intergenic complementation on catalase structure and function in maize: A molecular approach to heterosis. *Arch. Biochem. Biophys.* **153**, 695–705.

Scandalios, J. G., Tong, W. F., and Roupakias, D. G. (1980a). *Cat3*, a third gene locus coding for a tissue-specific catalase in maize: Genetics, intracellular location, and some biochemical properties. *Mol. Gen. Genet.* **179**, 33–41.

Scandalios, J. G., Chang, D. Y., McMillin, D. E., Tsaftaris, A. S., and Moll, R. (1980b). Genetic regulation of the catalase developmental program in maize scutellum: Identification of a temporal regulatory gene. *Proc. Natl. Acad. Sci. U.S.A.* **77**, 5360–5364.

Scandalios, J. G., Tsaftaris, A. S., Chandlee, J. M., and Skadsen, R. W. (1984). Expression of the developmentally regulated catalase (*Cat*) genes in maize. *Dev. Genet.* **4**, 281–293.

Scioli, J. R., and Zilinskas, B. A. (1988). Cloning and characterization of a cDNA encoding chloroplastic Cu–Zn-superoxide dismutase. *Proc. Natl. Acad. Sci. U.S.A.* **85**, 7661–7665.

Serfling, E., Jasin, M., and Schaffner, W. (1985). Enhancers and eukaryotic gene transcription. *Trends Genet.* **1**, 224–230.

Skadsen, R. W., and Scandalios, J. G. (1987). Translational control of photo-induced expression of the *Cat2* catalase gene during leaf development in maize. *Proc. Natl. Acad. Sci. U.S.A.* **84**, 2785–2789.

Skadsen, R. W., and Scandalios, J. G. (1989). Pretranslational control of the levels of glyoxysomal protein gene expression by the embryonic axis in maize. *Dev. Genet.* **10**, 1–10.

Slovin, J. P., and Tobin, E. M. (1982). Synthesis and turnover of the light-harvesting chlorophyll *a/b*-protein in *Lemna gibba* grown with intermittent red light: Possible translational control. *Planta* **154**, 465–472.

Sohal, R. S. (1987). The free radical theory of aging. A critique. *Rev. Biol. Res. Aging* **3**, 431–449.

Stuart, G. W., Searle, P. F., Chen, H. Y., Brinster, R. L., and Palmiter, R. D. (1984). A 12-base pair DNA motif that is repeated several times in metallothionein gene promoters confers metal regulation to a heterologous gene. *Proc. Natl. Acad. Sci. U.S.A.* **81**, 7318–7322.

Tanaka, K., and Sugahara, K. (1980). Role of superoxide dismutase in defense against

SO_2 toxicity and an increase in superoxide dismutase activity with SO_2 fumigation. *Plant Cell Physiol.* **21,** 601–611.

Taub, H. (1965). "Oxygen: Chemistry, Structure and Excited States." Little, Brown, Boston, Massachusetts.

Timko, M. P., Kzausch, A. P., Castresana, C., Fussler, J., Herrera-Estrella, L., Van Den Broek, G., van Montagu, M., Schell, J., and Cashmore, A. (1985). Light regulation of plant gene expression by an upstream enhancer-like element. *Nature (London)* **318,** 579–582.

Tobin, E. M., and Silverthorne, J. (1985). Light regulation of gene expression in higher plants. *Annu. Rev. Plant Physiol.* **36,** 569–593.

Tolbert, N. E. (1982). Leaf peroxisomes. *Ann. N.Y. Acad. Sci.* **386,** 254–268.

Tsaftaris, A. S., and Scandalios, J. G. (1981). Genetic and biochemical characterization of a *Cat2* catalase null mutant of *Zea mays. Mol. Gen. Genet.* **181,** 158–163.

Tsaftaris, A. S., and Scandalios, J. G. (1986). Spatial pattern of catalase (*Cat2*) gene activation in scutella during postgerminative development in maize. *Proc. Natl. Acad. Sci. U.S.A.* **83,** 5549–5553.

Tsaftaris, A. S., Bosabalidis, A. M., and Scandalios, J. G. (1983). Cell-type-specific gene expression and acatalasemic peroxisomes in a null *Cat2* catalase mutant of maize. *Proc. Natl. Acad. Sci. U.S.A.* **80,** 4455–4459.

van Loon, L. C. (1985). Pathogenesis related proteins. *Plant Mol. Biol.* **4,** 111–116.

Wadsworth, G. J., and Scandalios, J. G. (1989). Differential expression of the maize catalase genes during kernel development: The role of steady-state mRNA levels. *Dev. Genet.* **10,** 304–310.

Wadsworth, G. J., and Scandalios, J. G. (1990). Molecular characterization of a catalase null allele at the *Cat3* locus in maize. *Genetics* (in press).

White, J. A., and Scandalios, J. G. (1987). *In vitro* synthesis, importation and processing of Mn-superoxide dismutase (SOD-3) into maize mitochondria. *Biochim. Biophys. Acta* **926,** 16–25.

White, J. A., and Scandalios, J. G. (1988). Isolation and characterization of a cDNA for mitochondrial manganese superoxide dismutase (SOD-3) of maize and its relation to other manganese superoxide dismutases. *Biochim. Biophys. Acta* **951,** 61–70.

White, J. A., and Scandalios, J. G. (1989). Deletion analysis of the maize mitochondrial superoxide dismutase transit peptide. *Proc. Natl. Acad. Sci. U.S.A.* **86,** 3534–3538.

Woo, K. C., Anderson, J. M., Boardman, N. K., Downton, W., Osmond, C. B., and Thorne, S. W. (1970). Deficient photosystem II in agranal bundle sheath chloroplasts of C_4 plants. *Proc. Natl. Acad. Sci. U.S.A.* **67,** 18–25.

Youngman, R. J., and Dodge, A. D. (1980). Paraquat resistance in *Conyza. Plant Physiol.* **65S,** 59.

Zelitch, I. (1971). "Photosynthesis, Photorespiration, a d Plant Productivity." Academic Press, New York.

GENETIC ANALYSIS OF OXYGEN DEFENSE MECHANISMS IN *Drosophila melanogaster*

John P. Phillips and Arthur J. Hilliker

Department of Molecular Biology and Genetics, University of Guelph,
Guelph, Ontario, Canada N1G 2W1

I. Introduction

The univalent reduction of oxygen generates a series of unstable, active intermediates that attack cellular constituents. To counter this ever-present threat, organisms possess a diversity of defenses to intercept these reactive species of oxygen and to repair oxidative damage. But these defenses are not impenetrable and chronic toxicity of oxygen at normal atmospheric exposure has been strongly implicated in the cause and progression of many pathologies ranging from inflammatory disease to diabetes to cancer, and is almost certainly involved in the overall degenerative processes of cellular senescence and organismal aging.

Due to the extreme reactivity and short half-life of active oxygen species and to the great diversity of apparent defense mechanisms, we

43

ADVANCES IN GENETICS, Vol. 28

are still unsure of the causative relationships between oxygen toxicity and the larger biological issues of disease, aging and normal life-span. To understand these relationships more clearly will require an essentially *in vivo* experimental approach in which (1) specific components of putative defense–repair systems can be selectively manipulated, (2) oxygen stress can be measurably applied, and, most importantly, (3) biological consequences can be identified and reliably measured in the whole animal. It is precisely for these reasons that the merit of a genetic approach to the problems of oxygen defense in an experimental organism like *Drosophila melanogaster* becomes apparent.

This article is intended to serve as a prospectus for the investigation of oxygen defense mechanisms in the metazoan eukaryote, *Drosophila melanogaster*. To this end, we present a brief overview of the radical basis of oxygen toxicity and of known cellular defense components. Some of the genetic tools and approaches offered by *Drosophila* are then summarized and past and present work on the analysis of oxygen defense mechanisms in *Drosophila* is reviewed in relation to several aspects of *Drosophila* biology. Reviews of the genetic analysis of microbial oxygen defense mechanisms have been written by Touati (1988), Hassan (1988), and Christman *et al.* (1985).

II. Oxygen Toxicity

A. Radical Basis of Oxygen Toxicity

Molecular dioxygen (O_2) is a di-radical carrying two unpaired electrons in parallel spin. This spin configuration, which restricts the reactivity of O_2 with nonradicals, is important in making O_2 available to aerobic cells for respiration. However, the same electron spin restriction that renders O_2 relatively unreactive leaves it vulnerable to one-electron (univalent) reduction, through which it passes preferentially to generate the superoxide radical (O_2^-) and its protonated derivative, the hydroperoxyl radical (HO_2^-), hydrogen peroxide (H_2O_2), and the hydroxyl radical ($OH\cdot$):

$$O_2 \xrightarrow{e^-} O_2^- \xrightarrow{e^- + 2H^+} H_2O_2 \underset{H_2O}{\overset{e^- + H^+}{\longleftarrow}} OH\cdot \xrightarrow{e^- + H^+} H_2O$$

Although the spin restriction of O_2 limits its reactivity with nonradicals, it is highly reactive with atoms and molecules with only a single electron to donate. Transition metals are good examples and can change oxidation state by single-electron transfers. Hence, a close biological association has evolved between iron and copper as the redox catalysts at the active sites of oxygenases, oxidases, antioxidants, and oxygen transport and electron transport proteins. However, metal ions not complexed with such proteins can inflict damage to cells by acting as Fenton catalysts in the Haber–Weiss generation of the hydroxyl radical (OH·) from superoxide and hydrogen perioxide (for reviews, see Halliwell and Gutteridge, 1985, 1986):

$$O_2^- + Fe(III) \rightarrow O_2 + Fe(II)$$
$$Fe(II) + H_2O_2 \rightarrow Fe(III) + OH^- + OH·$$
$$\overline{O_2^- + H_2O_2 \rightarrow O_2 + OH^- + OH·} \quad \text{Net}$$

Although O_2^- is not a particularly aggressive oxidant in aqueous solution at physiological pH, its rapid reaction with Fe(III) can, in the presence of H_2O_2, lead to the generation of OH·, which is indiscriminately reactive at rates limited only by diffusion.

Many other important biological sources of active oxygen have been identified. These include autoxidations, enzyme-catalyzed reactions, the respiratory burst in phagocytic polymorphonuclear leukocytes, and "leakage" from the respiratory chain, particularly under conditions of high respiratory demand or hyperoxia (Halliwell and Gutteridge, 1985).

Most organic molecules are vulnerable to attack by reactive oxygen species, especially OH· (Sies, 1986). Two examples will serve to illustrate. One example is the oxidation of polyunsaturated fatty acids (LH) of cell and organelle membranes, an event which is widely implicated in oxygen-related cytotoxicity (for a review, see Chow, 1988a). Abstraction of a hydrogen atom (H·) from LH by OH· generates a carbon-centered radical (L·). This initiates a chain reaction in which L· rapidly reacts with O_2 to form a peroxy radical, $LO_2^·$, which in turn can abstract H· from another LH to generate another carbon-centered radical and a lipid hydroperoxide, LOOH. This free-radical chain reaction continues to propagate until two free radicals interact and destroy each other or until they are intercepted by membrane-integral scavengers such as α-tocopherol (vitamin E). Further hydroperoxidation can occur from the reaction of hydroperoxide, LOOH, with iron and copper complexes to yield alkoxy (LO·) and peroxy ($LO_2^·$) radicals.

The second example of a specific organic radical attack is the reactivity of both inorganic and organic species of reactive oxygen with DNA (Ames, 1983; Harman, 1962; Saul and Ames, 1986; Totter, 1980), which may constitute the largest single source of lethal and of heritable damage to DNA. Hydroxyl radicals (OH·) produce base damage and strand breaks in DNA (Pryor, 1976–1982; Nygaard and Simic, 1983; Scholes, 1983). Lipid hydroperoxides, such as fatty acid hydroperoxide, cholesterol hydroperoxide, endoperoxides, epoxides of cholesterol and fatty acids, enals, and other aldehydes, such as malondialdehyde, and alkoxy and hydroperoxy radicals are mutagenetic and are carcinogenic initiators and promoters (for reviews, see Ames, 1983; Saul *et al.*, 1987; Cerutti, 1985, 1986). Much of our current knowledge about the chemistry of free-radical reactions with DNA comes from studies with ionizing radiation (Scholes, 1983; Scholes and von Sonntag, 1987). The proximate cause of most of the radiation-induced changes to DNA structure comes from indirect effects, in which water molecules are stripped of electrons to produce three radical species, the hydoxyl radical, the hydrated electron, and the hydrogen atom. As will be discussed later, most radiation-induced DNA damage is attributed to the hydroxy radical (OH·).

B. Biological Consequences of Oxygen Toxicity

The central role of oxygen radicals in numerous human pathologies, such as cancer atherosclerosis, diabetes, emphysema, and arthritis, is surprising yet, in hindsight, not totally unexpected in light of the types of damage produced by oxygen radicals and their derivatives (for reviews, see Chow, 1988a; Harman, 1987; Pryor, 1987; Oberley, 1988). These pathologies, the incidence of which increases with age, are in many cases, life-span limiting. Thus, radicals may directly or indirectly be involved in processes that reduce longevity below the maximum species life-span.

In addition to life-limiting pathologies, numerous senescent changes documented at the cell and molecular levels can be accounted for by oxygen effects (see Porta, 1988, for review). The possible relationship between oxygen toxicity and aging was first formulated by Harman (1956, 1962). Over the years a vast amount of support was accumulated in favor of this relationship, although much of the evidence is indirect and inferential in nature (for reviews, see Harman, 1987; Pryor, 1987). Generally, the life-span potential of an organism is negatively correlated with specific metabolic rate (Cutler, 1984), that is, with its specific rate of O_2 utilization. Substantial correlation between antioxidant capacity and life-span further strengthens the concept that species life-span is related to oxygen utilization efficiency and to the capacity to defend against oxidative damage.

Of special interest in this regard are the now well-established correlations among free radicals, DNA damage, and aging (Ames *et al.*, 1981; Saul *et al.*, 1987). Active oxygen species, especially OH· and H_2O_2, interact with DNA in a variety of ways that have mutagenic, carcinogenic, and cytoxic effects. Among mammals, specific metabolic rates have now been shown to correlate with levels of oxidative DNA damage as assessed by rates of excretion of products of DNA repair, such as thymine glycol 8-hydroxydeoxyguanosine and 7-methylguanine (Saul *et al.*, 1987; Cathcart *et al.*, 1984; Park and Ames, 1988). Thus, DNA may be a critical molecular target in aging. Moreover, it has been proposed (Miquel and Fleming, 1986; Lints and Soliman, 1988) that aging in postmitotic cells may result primarily from radical attack on mitochondrial DNA. Mitochondrial DNA may be especially accessible to respiration-derived free-radical attack, and mitochondria may have, in addition, reduced repair capacity (Richter *et al.*, 1988). This would explain the age-related decreases in functional mitochondria, which could, in turn, contribute to decline in cellular function.

III. General Antioxidant Defense Systems

The diversity of potentially damaging radical-initiated cellular reactions necessitates a pluralistic defense in aerobic or aerotolerant cells (Chow, 1988b). The major strategies of antioxidant defense in metazoan eukaryotes appear to be as follows.

1. Preemptive scavengers, both enzymatic and nonenzymatic, which intercept reactive species before they react with other cellular substrates.
2. Reduction of hydroperoxides generated as products of radical reactions or from cellular enzymes.
3. Isolation or removal of transition metal (Fenton) catalysts.
4. Repair of oxidation damage to membranes and macromolecules (e.g., DNA).

In addition, ancillary enzymes involved in the synthesis and support of the direct agents of defense and repair can also play rate-limiting roles in the overall defense system. The primary components of these defensive systems are given in Table 1.

TABLE 1
Antioxidant Defenses[a]

Enzymatic scavengers
1. Cu–Zn superoxide dismutase: $2O_2^{\cdot-} + 2H^+ \rightarrow H_2O_2 + O_2$
2. Mn superoxide dismutase: $2O_2^{\cdot-} + 2H^+ \rightarrow H_2O_2 + O_2$
3. Catalase: $H_2O_2 \rightarrow H_2O + O_2$

Nonenzymatic scavengers
1. Glutathione (GSH): reducing agent; directly scavenges $O_2^{\cdot-}$ and $OH\cdot$; involved in regeneration of ascorbate and vitamin E
2. Vitamin E: organic/inorganic radical trap in hydrophobic domains, e.g., membranes
3. β-Carotene: radical trap; can also act as prooxidant
4. Ascorbate: radical trap; reducing agent in aqueous domain; can act as prooxidant
5. Uric acid: $OH\cdot$ scavenger; also binds Cu and Fe

Reduction of hydroperoxides
1. Catalase: $H_2O_2 \rightarrow H_2O + O_2$
2. Glutathione peroxidase: uses GSH in oxidation of H_2O_2 and organic hydroperoxides
3. Glutathione S-transferase: uses GSH in detoxification of organic hydroperoxides
4. Phospholipid hydroperoxide GSH peroxidase: directly reduces membrane fatty acid hydroperoxides

Ancillary enzymes
1. Glutathione reductase: reduces GSSH to GSH
2. Dehydroascorbate reductase: reduces dehydroascorbate to ascorbate
3. Tocopherol reductase: Regenerates oxidized vitamin E
4. Glucose-6-phosphate dehydrogenase: provides NADPH for regeneration of oxidized antioxidants
5. Phospholipase A_2: hydrolyzes fatty acyl hydroperoxides from membranes
6. Xanthine dehydrogenase: generates uric acid from xanthine/hypoxanthine
7. Aldehyde oxidase: detoxifies malonaldehyde
8. Sulfite oxidase: removes sulfite as potential source of damaging sulfite radicals arising from reaction with $O_2^{\cdot-}$

Complexing of transition metals
1. Transferrin
2. Lactoferrin
3. Ceruloplasmin
4. Albumin
5. Metallothionein

DNA repair systems

[a] Sources include Chow (1988a), Halliwell and Gutteridge (1985), and Pryor (1976–1982).

IV. Genetic Approaches in Drosophila

Drosophila melanogaster is preeminent as a metazoan eukaryote for the analysis of the genetic regulation of cellular processes. Complex biological systems, such as DNA repair, embryogenesis, behavior, or, in this instance, oxygen defense, can be genetically dissected through the isolation and analysis of mutants. Such mutants can be used to confirm the postulated *in vivo* function of a known protein or metabolite and may also reveal additional, unexpected, functions. Mutants can also serve to identify previously unknown components of a biological system and in this sense provide a tool to explore biochemically cryptic processes.

Numerous approaches can be employed to obtain mutant alleles of a gene for which the protein product is known. First, one must determine the chromosomal locus of the structural gene. This can be done by mapping nonselective variants such as electrophoretic or activity alleles. For many enzyme loci such variants can be found by screening laboratory and natural populations. This approach allowed the determination of the location of the genes for copper–zinc superoxide dismutase (Cu–ZnSOD) (Jelnes, 1971) and catalase (Bewley *et al.*, 1986) to defined positions on chromosome 3. However, in some cases such as for manganese superoxide dismutase (MnSOD) (D. Michaud and J. P. Phillips, unpublished observations) no detectable nonselective variants have been found.

To localize monomorphic loci one can attempt at least two quite different approaches. The first of these is the systematic analysis of fly segmental aneuploids. This approach utilizes Y-autosomal translocations (Lindsley *et al.*, 1972) and X–Y translocations (Stewart and Merriam, 1974). This method involves the systematic synthesis of various hyperploids and hypoploids for small portions of the euchromatic genome. These flies are then assayed for changes in enzyme activity. The assumption is that the enzyme activity level is linearly proportional to the number of copies of the structural gene. This assumption can be questioned and it is conceivable that changes in copy number of genes other than the structural gene might affect enzyme activity levels. The number of different fly stocks that must be constructed is usually very large and the assay system for the enzyme must be sufficiently sensitive and uniform to reliably detect activity level differences of 50% employing extracts obtained from a small number of flies. Nevertheless, this method has been used with great success to cytogenetically locate many specific enzyme loci, for example, catalase, an important component of oxygen defense (Lubinsky and Bewley, 1979).

A second method for localizing monomorphic loci involves the use of a DNA probe. If the gene has been cloned in another eukaryote it may be possible to identify the corresponding cloned gene in a *Drosophila* DNA library if the heterologous cloned gene has sufficient nucleotide homology to the *Drosophila* gene. Alternatively, if the amino acid sequence of the *Drosophila* protein is known, it may then be possible to identify the *Drosophila* genomic or cDNA clone by probing a library with a synthetic oligonucleotide probe corresponding to a segment of the polypeptide. Once a clone has been obtained and confirmed then the gene can be easily localized by *in situ* hybridization to polytene chromosomes. This approach was used to clone and to confirm the cytogenetic localization of the Cu–ZnSOD gene of *Drosophila* (Kirkland and Phillips, 1987; Seto *et al.,* 1989).

Once the cytogenetic location of a gene has been determined, one can then obtain mutants by use of a deletion that spans the gene of interest. If the gene is essential for viability (often not known, *a priori*), one can screen for recessive lethals uncovered by the deletion. These recessive lethals can then be sorted into complementation groups. If an enzyme is polymorphic for electrophoretic variants, then, if one starts with a defined variant on the mutagenized chromosome, lethal alleles of the enzyme gene should show loss of the electrophoretic variant; this is seen in activity-stained gels from extracts derived from heterozygotes that possess, on the nonmutagenized chromosome, an alternative electrophoretic variant. With such an approach, Campbell *et al.* (1986) recovered and identified a null allele of Cu–ZnSOD. It should be noted that in such a screen one should also look carefully for semilethals. Very often, as is the case for Cu–ZnSOD, a null allele is associated with reduced viability rather than with complete lethality. In fact, in the case of Cu–ZnSOD we were fortunate that the genetic background of the particular strains we used in the mutagenesis screen enhanced the inviability of the null allele. In other backgrounds the null allele is essentially not associated with reduced viability as assayed by eclosion ratios (Phillips *et al.,* 1989), although it could still be recovered as a male sterile. Typically, however, mutagenesis screens concentrate on lethal and visible mutations, as it is rather more laborious to screen for steriles. Thus, if one is screening for a null allele of a particular gene, one might wish to mutagenize several different and historically distinct isogenic lines (in an attempt to provide a variety of genetic backgrounds) and one might also wish to screen for lethals, semilethals, visibles, male steriles, and female steriles and possibly other phenotypes. An alternative approach is that adopted by Mackay and Bewley (1989) to obtain null and low-activity alleles of the catalase gene. Using a single fly assay system (obviously not available for all enzymes) they examined flies that were heterozygous for a mutagenized chromosome and a deletion spanning the catalase structural gene. Prior to their examination, each fly had founded an individual pedigree line. This method was quite successful in obtaining hypomorphic alleles of catalase. If catalase had an essential function, then of course this approach would not have worked, so they also screened for lethals falling within the catalase deletion, none of which proved allelic to the catalase gene.

Often one wishes to identify and tag by mutation genes specifying or regulating previously unidentified components of a complex biological system. This is done by screening for mutations associated with particular phenotypes. Such screens are most easily performed for X-linked mutations or where new mutations are associated with a dominant phenotypic effect. The latter was effective in localizing an oxygen defense-related metallothionein gene on the basis of increased resistance of a tandem duplication to dietary cadmium (Otto *et al.*, 1986; Maroni *et al.*, 1987). Usually, however, mutations affecting a particular phenotype are recessive. The screening of autosomes, which in sum constitute about 80% of the genome, is laborious. However, such systematic screens have been very successful in indentifying many gene loci important in particular biological systems, such as meiotic recombination (reviewed by Baker *et al.*, 1976), DNA repair (reviewed by Boyd *et al.*, 1987), and embryogenesis (Nusslein-Volhard *et al.*, 1984; Eberl and Hilliker, 1988). Development of an efficient screening protocol to detect genes important in oxygen defense should allow us to recover alleles of genes such as MnSOD.

The molecular product of a gene identified in a general screen is often initially obscure. However, once a mutation has been obtained the gene can ultimately be cloned (although this is no trivial pursuit) and the cDNA sequence and amino acid sequence can be determined, which may, by interphyletic comparisons, allow one to deduce the possible biochemical function. One might also find that the mutation is in a gene of known biochemical function that had previously not been thought to be related to the particular phenotype examined. Thus, mutant homozygotes can also be assayed for specific proteins that may play a role in the process under investigation.

Once mutations of particular genes have been obtained new experimental vistas arise. One can confirm the *in vivo* function hypothesized and look for additional phenotypic effects. One can use null alleles in conjunction with segmental aneuploid chromosomal aberrations, such as duplications and deletions, to examine the effects of gene dosage and activity levels on a variety of phenotypic parameters. Double, triple, etc., mutant homozygotes for different nonallelic genes affecting a system can be constructed to look for interactive effects.

Finally, the coupling of traditional genetic approaches for isolation and analysis of mutants with molecular techniques such as transposon tagging (Bingham *et al.*, 1981), transposon-mediated transformation

(Rubin and Spradling, 1981; Blackman *et al.*, 1989), and chromosome walking and jumping (Bender *et al.*, 1983) provides a formidable set of tools for investigating the genes of antioxidant defense metabolism in *Drosophila*.

V. Analysis of Oxygen Defense Mechanisms in *Drosophila*

A. Oxygen Utilization in *Drosophila*

If the *raison d'être* of oxygen defense systems is to allow access to oxygen for use in respiration, then specific defense mechanisms in Drosophila must ultimately be interpreted in terms of the general physiology of oxygen utilization. Oxygen consumption during normal Drosophila development has been investigated intermittently for many years. Of particular interest in this regard are the periods of extraordinary oxygen demand which accompany puparation and imaginal eclosion, respectively. After eclosion, oxygen demand climbs to high levels in young adults and thereafter declines slowly with advancing age (Fenn *et al.*, 1967; Fourche, 1965a, 1965b, 1967; Lints and Lints, 1968). That the periods of high oxygen demand are also periods of high oxygen stress is clearly suggested by the lethal phase of mutants with disrupted oxygen defense components (see Sections, V,B,1–3 and V,D,5).

The rate of respiration in adult *Drosophila* is proportional to ambient temperature (Lints and Lints, 1968; Miquel *et al.*, 1976) and is negatively correlated with life-span in different strains and under different environmental conditions (Miquel and Fleming, 1986), although this relationship can apparently be uncoupled in some strains selected for extended adult life-span (Arking *et al.*, 1988; Arking and Dudas, 1989). Deviation above or below normoxia (21% O_2, 1 atmosphere) shortens adult life-span, although surprisingly little effect is observed at 10% O_2 (Fenn *et al.*, 1967). In pure oxygen at 1 atmosphere, adults live about 7 days, and the reciprocal of the survival time of young adults in hyperbaric oxygen is a linear function of age (Fenn *et al.*, 1967; Williams and Beecher, 1944). Recovery of larvae and adults from anoxia induces the heat-shock response (see Section VI,B).

B. Enzymatic Defense Mechanisms

1. Copper and Zinc Superoxide Dismutases

Drosophila Cu–ZnSOD, first referred to as tetrazolium oxidase due to its property of blocking nitroblue tetrazolium (NBT) reduction on NBT-stained electrophoresis gels (Jelnes, 1971; Richmond and Powell, 1970), is a homodimer of 32 kDa (Lee *et al.*, 1981a) with extensive amino acid sequence homology to Cu–ZnSOD from other species (Lee *et al.*, 1985a; Hjalmarsson *et al.*, 1987). The developmental profile shows maximal levels of enzyme are reached in 5-day-old adults and persist through at least the midpoint of adult life-span (Graf and Ayala, 1986; Massie *et al.*, 1980). Fast (f) and slow (s) allozyme variants are found in many natural *Drosophila* populations at differing frequencies (Ayala *et al.*, 1971, 1972a,b; Richmond, 1972). Interestingly, all known allelic isozyme variants exhibit one of only two specific electrophoretic mobilities, slow (Cu–ZnSOD^s) and fast (Cu–ZnSOD^f). Although Cu–ZnSOD^s is less thermostable than Cu–ZnSOD^f, it has an intrinsic specific activity approximately $3\times$ greater than Cu–ZnSOD^f (Lee *et al.*, 1981b) and $1.5\times$ greater than Cu–ZnSOD from other eukaryote species reported in the literature (Lee *et al.*, 1981a). Amino acid analysis of Cu–ZnSOD^f and Cu–ZnSOD^s isozymes from Californian strains shows that they differ by a single amino acid: Lys$_{96}$ in Cu–ZnSOD^s is replaced by Asn$_{96}$ in Cu–ZnSOD^f (Lee *et al.*, 1981b, 1985a,b). Parallel amino acid analysis of Cu–ZnSOD isozymes from Tunisian strains shows that Cu–ZnSOD^s and Cu–ZnSOD^f differ by two amino acid substitutions in addition to the same substitution at residue 96 distinguishing the Californian strains (histidine and proline in Cu–ZnSOD^s are replaced by serine and glycine in Cu–ZnSOD^f) (Lee *et al.*, 1981b). By inference from amino acid sequences from other species (J. P. Phillips and K. Kirby, unpublished observations) the two additional substitutions are Cu–ZnSOD^f:(Glu$_{19}$; Ser$_{100}$) and Cu–ZnSOD^s:(Pro$_{19}$; His$_{100}$). By comparison with the tertiary structural models of bovine Cu–ZnSOD (Tainer *et al.*, 1982, 1983; Getzoff *et al.*, 1983), the position of these substitutions can be seen to fall outside the enzyme's critical external loop domains, which have been proposed to form the active site channel.

Recombination mapping of Cu–ZnSOD allelic isozymes placed the structural gene on the third chromosome at 32.5 (Jelnes, 1971). Seeking to isolate Cu–ZnSOD null mutants, Campbell *et al.* (1986) recovered and characterized ethyl methanesulfonate (EMS) and hybrid-dysgenesis-induced lethals uncovered by a deficiency [Df(3L) *lxd*-9] known to span both the Cu–ZnSOD and *lxd* loci (V. Finnerty, personal communication). This and subsequent mutagenesis screens (Crosby and Meyerowitz, 1986; B. Staveley, A. J. Hilliker, and J. P. Phillips, unpublished observations) have produced a detailed genetic map of the Cu–ZnSOD microregion of chromosome 3. A total of three independent Cu–ZnSOD-null alleles have been recovered from these studies.

Phenotypic characterization of Cu–ZnSOD-null mutants provides an explicit test of the role of this enzyme in oxygen defense. Homozygotes are viable as larvae and pupae but die as young adults, with a mean adult life-span at 25°C of 11 days (Phillips *et al.*, 1989). This compares to a mean adult life-span of about 60 days for the parental strain and for Cu–ZnSOD$^-$/Cu–ZnSOD$^+$ heterozygotes. Likewise, homozygotes are hypersensitive to the O_2^- radical-generating agent, paraquat, and to the transition metal, Cu(III), which indicates that the lack of Cu–ZnSOD confers an inability to scavenge O_2^-. Together, these results offer compelling evidence for the essential role of Cu–ZnSOD in O_2^- defense metabolism and demonstrate clearly the life-shortening consequences of the failure of this single component of oxygen defense.

Other biological consequences of the Cu–ZnSOD-null condition further confirm the essential role of this enzyme in *Drosophila*. The Cu–ZnSOD-null males are sterile and have markedly abnormal, immobile spermatozoa, whereas cSOD-null females are only weakly fertile (B. Staveley, A. J. Hilliker, and J. P. Phillips, unpublished observations). The Cu–ZnSOD null females appear to have elevated levels of germ line mutation as measured by X-linked recessive lethal tests (B. Duyf and D. Michaud, unpublished observations). Xu Peng *et al.* (1986) report that a hypomorphic allele of Cu–ZnSOD that produces 3.5% of the normal Cu–ZnSOD levels is hypersensitive to acute ionizing radiation exposure. This important observation confirms the role of O_2^- in radiobiology and further establishes the essential role of Cu–ZnSOD in defense against radiation-generated O_2^-.

The primary structures of the *Drosophila* Cu–ZnSOD gene and its mRNA are now known from sequence studies (Kirkland and Phillips, 1987; Seto *et al.*, 1987a,b, 1989; J. P. Phillips, and K. Kirby, unpublished observations). The *Drosophila* gene contains a single intron of 720 nucleotides (nt) separating the codons specifying amino acid residues 22 and 23, analogous to the position of the first of four introns in the human Cu–ZnSOD gene (Levanon *et al.*, 1985). The penultimate 3' codon, GTC, is not represented in the mature polypeptide and codon 97 differs by a single nucleotide in Cu–ZnSOD^f (AAC) (Seto *et al.*, 1987a,b, 1989) and Cu–ZnSOD^s (AAA) (Kirkland and Phillips, 1987). The availability of genomic and cDNA clones of Cu–ZnSOD opens the possibility of further investigation of the biological role(s) of cSOD through transposon-mediated transformation (Rubin and Spradling, 1981). Questions regarding the regulation of cSOD expression and the cell and tissue domains of its expression relative to radioprotection, aging, sensitivity to dietary radical initiators, and fertility can all be addressed in a whole-organism context through transformation in conjunction with the null mutants.

2. Manganese Superoxide Dismutase

The biochemistry of MnSOD in *Drosophila* is virtually unexplored, although its activity in extracts is detectable in electrophoretic gels (Phillips *et al.*, 1989) and by its resistance to CN. The enzyme awaits a thorough analysis including complete purification and amino acid sequence determination.

In a manner analogous to the analysis of Cu–ZnSOD, mutations for MnSOD would be exceedingly useful tools in elucidating the biological role of this enzyme. For example, questions regarding the cellular domain of MnSOD function relative to Cu–ZnSOD function could be directly addressed with such mutants, as could questions regarding the apparent oxygen-related dimunition of mitochondrial function during aging (Miquel and Fleming, 1986) and the observed high levels of oxidative damage in mitochondrial DNA relative to nuclear DNA (Richter *et al.*, 1988). The availability of both MnSOD and Cu–ZnSOD mutants would offer a formidable experimental package for directly testing many current ideas on the biological role(s) of O_2^-. Unfortunately, no MnSOD mutants have been reported. In contrast to Cu–ZnSOD, extensive searching of both natural populations and laboratory stocks (D. Michaud and J. P. Phillips, unpublished observations) has not identified any MnSOD isozyme variants. The isolation of MnSOD hypomorphs has been severely hampered by the lack of such variants and by the lack of amino acid sequence data for *Drosophila* MnSOD. However, the value of such mutants is such that we are currently following other less direct routes for their isolation.

3. Catalase

The second major class of enzymatic scavengers of active oxygen is represented by catalase (*Cat*). Work on *Drosophila* catalase (Bewley and Mackay, 1990) illustrates the role of genetic analysis in elucidating the complex relationships between a specific cellular component of oxygen defense and relevant biological aspects of the whole organism, such as longevity, fecundity, and viability. Catalase from *Drosophila* is a homotetrameric heme-containing protein of ~58 kDa (Nahmias and Bewley, 1984). Catalase activity during development shows a peak in late third larval instars and a second, larger peak during metamorphosis (Bewley *et al.*, 1983), after which it is maintained at a relatively constant level throughout adult life. The *Cat* structural gene was localized by dosage response to segmental aneuploidy to the cytogenetic interval 75D1–76A3 (Lubinsky and Bewley, 1979). Utilizing a deficiency spanning the *Cat* locus, six hypomorphic alleles of *CAT* were recovered from EMS mutagenized chromosomes (Mackay and Bewley,

1989). The six alleles, all of which (of necessity due to the screening protocol) are homozygous viable, exhibit catalase activity ranging from 0 to 5% of normal and together provide a unique opportunity to examine the biological role of catalase in oxygen defense metabolism. Using these alleles, singly and in combinations that show partial complementation, the relationship between catalase level and viability, fertility, and adult longevity were examined (Mackay and Bewley, 1989; Bewley and Mackay, 1990). As is the case for Cu–ZnSOD, catalase does not appear to be essential for cell viability per se. Viability correlates linearly with catalase levels below 3% of normal; above this level viability rapidly approaches 100%. Likewise, mean and maximum adult life-span correlate positively with catalase levels between 0 and 50% of normal. Several other age-associated indices, including locomotor activity, age, pigmentation, and onset of peak fecundity, correlate with shorter life-span and decreased catalase activity in these mutants.

The phenotypic parallels in mutants devoid of catalase or Cu–ZnSOD are striking. Most importantly, both mutants show a striking reduction of about 80% in adult life-span and neither enzyme is essential for cell viability per se. As with cSOD-null, *Cat*-null appears to significantly increase spontaneous mutation frequency (Mackay *et al.*, 1990). Whether it also exhibits increased sensitivity to oxygen stress through ionizing radiation, radical-generating xenobiotics, or transition metals remains to be determined. However, the biological role of both of these enzymes in oxygen stress metabolism is clearly delimited by these mutants. It will now be of great interest to examine the phenotypes of double-mutant combinations of various CAT and Cu–ZnSOD alleles as a way of further probing the relationships between Cu–ZnSOD and CAT in oxygen stress metabolism. Of equal interest, these double-mutant studies will allow us to determine whether screening for synthetic lethals with Cu–ZnSOD$^-$ and/or CAT$^-$ could lead to the identification of genes specifying other important and perhaps as yet biochemically unidentified components of the oxygen defense system in *Drosophila*.

4. Glutathione Peroxidase

The absence of glutathione peroxidase (GPx), which plays a prominent role in antioxidant metabolism in mammals and many other taxa, in *Drosophila* (C. Taylor and J. P. Phillips, unpublished observations), *Musca* (Sohal *et al.*, 1983), and many if not all insects (Smith and Shrift, 1979) is an enigma. The presence of glutathione in the absence of GPx in these organisms suggests that glutathione in the absence of GPx still serves as a reducing substrate by other mechanisms. The absence of

GPx (which can reduce H_2O_2 as well as organic peroxides) might mean that catalase plays a larger role in H_2O_2 reduction in *Drosophila*.

5. *Glutathione S-Transferase*

Selenium-independent glutathione transferases (GST) constitute a family of enzymes which can exhibit both glutathione peroxidase activity and glutathione transferase activity (Mannervik *et al.*, 1985). The peroxidase activity of GST may be expecially important in reduction of organic hydroperoxides (Ketterer *et al.*, 1988), whereas the transferase is largely involved in detoxification through glutathione (GSH) conjugation. The former activity may be especially important in *Drosophila* to compensate for the lack of GPx.

The major GST of *Drosophila* has been isolated and characterized (Cochrane *et al.*, 1987). The major isoform, which constitutes about 95% of total GST of *Drosophila,* is a moderately abundant protein, constituting about 0.3% of total soluble protein of the adult. A cDNA for the major GST subunit has been isolated from a cDNA expression library (Morrissey *et al.*, 1988) and will be a useful tool in the identification of the genetic locus and subsequent isolation of GST mutants.

C. Nonenzymatic Defense Mechanisms

1. *Glutathione*

Although glutathione (GSH) is present in *Drosophila* (C. Taylor and J. P. Phillips, unpublished observations) and its role in antioxidant physiology in insects has been characterized (Allen and Sohal, 1986), its specific role in oxygen defense in *Drosophila* has not been investigated to our knowledge. Note that GPx is not present in *Drosophila*.

2. *Ascorbate*

It has been reported that insects in general are unable to make ascorbate (Chatterjee, 1973), but we are not aware if this has been specifically examined in *Drosophila*. Although dietary sources could provide ascorbate, it is not known if ascorbate is essential for *Drosophila*.

3. *Vitamin E*

Vitamin E is not detectable in extracts of Drosophila adults (C. Taylor and J. P. Phillips, unpublished observations), a finding which probably relates to their relatively low content of polyunsaturated fatty acids (H. Draper and J. P. Phillips, unpublished observations). Dietary administration of vitamin E has produced slight positive enhancement of adult longevity (for a review, see Miquel and Fleming, 1986).

4. Uric Acid

Uric acid in *Drosophila* is undetectable in the mutants *rosy* (*ry*) and *maroon-like* (*ma-l*) (Glassman and Mitchell, 1958; Mitchell and Glassman, 1959) and presumably also in *cinnamon* (*cin*), all of which fail to make active xanthine dehydrogenase. The antioxidant role of urate in *Drosophila* is suggested by the negative interaction of the mutants *ry* and *ma-l* with Cu–ZnSOD null mutants (see Section V,C,4).

D. ANCILLARY ENZYMES AND PROTEINS

1. Glutathione Reductase

The presence of glutathione in *Drosophila* probably means that the corresponding reductase is present as well, although no studies of this enzyme in *Drosophila* have been reported to our knowledge. Likewise, no studies have been reported on the phospholipase-catalysed excision of peroxidized fatty acids from membranes.

2. Glucose-6-Phosphate Dehydrogenase

The identification of the gene [*Zwischenferment* (*Zw*)] specifying this enzyme and isolation of variant electromorph and activity alleles (O'Brein and MacIntyre, 1978) should permit genetic analysis of the involvement of the enzyme and of NADPH in regeneration of antioxidants, such as GSH and vitamin E.

3. Metallothionein

The only proteins identified in *Drosophila* that bind and remove potentially damaging metals are two distinct metallothioneins, Mtn (Maroni *et al.*, 1986; Lastowski-Perry *et al.*, 1985) and Mto (Mokdad *et al.*, 1987). The gene for Mtn has been cloned and sequenced and duplications have been recovered by cadmium selection (Lastowski-Perry *et al.*, 1985; Maroni and Watson, 1985; Otto *et al.*, 1986) and from natural populations (Maroni *et al.*, 1987). Unfortunately, no mutations of either Mtn or Mto have been isolated. Such mutations would be useful in examining more closely the role of transition metals in oxidative stress.

4. DNA Repair Enzymes

In view of the mounting evidence for oxidative damage to DNA as a constitutive fact of aerobic life (Saul *et al.*, 1987), systems that repair this type of damage take on a role of increasing importance. Our observations of enhanced spontaneous mutation frequency in Cu–ZnSOD-null mutants of *Drosophila* (B. Duyf and D. Michaud, unpublished observations) suggest that the DNA repair systems of *Drosophila* are inadequate to fully repair the elevated oxidative damage to DNA in these mutants. Several systems of DNA repair have been examined in *Drosophila* (reviewed by Boyd *et al.*, 1987) and mutants deficient in DNA repair as initially assayed by mutagen sensitivity and recombination defects have been identified (Baker *et al.*, 1976) and, to a limited extent, biochemically characterized (Boyd *et al.*, 1987). A productive area of investigation would be to utilize such DNA repair mutants in conjunction with mutants defective in oxygen defense to further clarify the genetic consequences of oxidative stress and the mechanisms employed to prevent and/or repair oxidative damage to DNA.

5. *Molybdoenzymes (Molybdenum Iron–Sulfur Flavohydroxylases)*

This group of apparently diverse metabolic enzymes includes xanthine dehydrogenase (XDH), aldehyde oxidase, sulfate oxidase, and pyridoxal oxidase. The genetics of these enzymes has been extensively studied in *Drosophila* (Bentley and Williamson, 1979; O'Brien and MacIntyre, 1978; Stivaletta *et al.*, 1988) through the use of mutations that reduce the activity of one or more of these enzymes. However, despite extensive genetic and biochemical investigation, the primary biological function of these enzymes is not yet understood. Nevertheless, the genetic and biochemical relatedness of these enzymes strongly suggests a common ground for their function. That the participation of these enzymes in oxygen defense may be that common ground is suggested by the following observations.

1. Uric acid, the product of XDH, is a potent scavenger of oxygen radicals (Ames *et al.*, 1981). Urate must not, however, be essential per se in this role because of the viability of the XDH-null mutants, *ry* and *ma-l*, which are devoid of detectable urate (Glassman and Mitchell, 1958; Mitchell and Glassman, 1959). The absence of urate in these mutants, however, may be the basis for their hypersensitivity to higher temperature (Hadorn and Schwink, 1956) and for their decreased adult lifespan (Shepherd *et al.*, 1989). It should be noted that urate in *Drosophila* is oxidized to allantoin by the enzyme urate oxidase (uricase). Unexpectedly, the levels of this enzyme are increased 10- to 20-fold in *ry* and *ma-l*, respectively (Friedman, 1973).

2. Malonaldehyde (MDH), a highly reactive product of lipid peroxidation, reacts via conjugation with DNA, proteins, and other cellular substrates and is both mutagenic and cytotoxic (Draper *et al.*, 1988).

MDH is a substrate for aldehyde oxidase at physiological concentrations and can be successfully detoxified by this enzyme.

3. Sulfite oxidase catalyzes the oxidation of sulfite (SO_2) to sulfate (SO_3). The univalent oxidation of sulfite initiates a free-radical chain oxidation in which the sulfite radicals, SO_3^- and SO_5^-, serve as chain propagation intermediates (Neta and Huie, 1985). Superoxide efficiently initiates SO_3 oxidation and the intermediates of this chain oxidation may account for the toxicity of sulfite (Rabinowitch et al., 1989). Thus, sulfite oxidase might mitigate against O_2^--initiated sulfite toxicity by catalyzing the nonradical oxidation of sulfite to sulfate (Rajagopalan, 1980). Sulfite is hypertoxic to sulfite oxidase-null cin mutants (V. Finnerty, personal communication). We would predict that Cu–ZnSOD-null mutants would likewise be hypersensitive to sulfite as a result of the presence of unscavenged superoxide.

Collectively these observations suggest that the molybdoenzyme system in Drosophila may serve as a multifunctional oxygen defense system. Individual components of the system are not essential for viability under conditions of normal oxidative metabolism, although the inviability of most cin alleles suggests that the loss of all three enzymatic activities generates an additive oxidative stress greater then the organism can sustain. This hypothesis predicts that loss of aldehyde oxidase, XDH, or sulfite oxidase, individually or collectively, would sensitize null mutants to oxygen stress. Such stress can be applied externally via increased temperature or by xenobiotics such as paraquat, or, preferentially, by genetically disabling primary components of the oxygen defense system. As a preliminary test of this hypothesis we have examined the phenotypic interaction of ry and ma-l with Cu–ZnSOD-null mutants. The hypothesis further predicts that ma-l, which is both XDH and aldehyde oxidase null, would have a greater negative effect on the viability of Cu–ZnSOD null mutants than would ry, which is XDH null only. We have found that both ry and ma-l are lethal in combination with Cu–ZnSOD-null, with complete mortality occurring in late larval and pupal stages (B. Duyf and J. P. Phillips, unpublished observations). To rule out other possible genetic reasons for these results, additional independently isolated alleles of ry and ma-l will have to be tested. However, we take these results as preliminary evidence for an important oxidative defense function of the molybdoenzyme family in Drosophila. It will be of interest to examine similar possible interactions with catalase null mutants. Molybdoenzyme mutations, either alone or in conjunction with catalase null or Cu–ZnSOD null mutations, may also enhance radiosensitivity, oxidative DNA damage, mutation frequency, and sensitivity to radical-generating agents.

VI. Associated Biological Phenomena

A. AGING AND LONGEVITY

Drosophila melanogaster has been a favorite tool in experimental gerontology and the copious literature in the field is covered in several excellent reviews (Arking and Dudas, 1989; Lamb, 1978; Lints and Soliman, 1988). A multitude of documented senescent changes in aging *Drosophila* (Lints and Soliman, 1988) strongly implicate oxidative stress as a major force in normal aging, and the evidence from *Drosophila* is consistent with the oxygen free-radical theory of aging (reviewed by Harman, 1987). The work by Sohal and collaborators (reviewed by Collatz and Sohal, 1986; Sohal and Allen, 1985, 1986) on antioxidant physiology relative to aging in the Dipteran relative, *Musca domestica,* establishes the fundamental role of antioxidant defense in this insect, and illuminates the current aging work on *Drosophila* despite the great phylogenetic distance between these two taxa. Certainly the current molecular and genetic work in *Drosophila* would benefit greatly from a similarly thorough and careful examination of antioxidant physiology in relation to aging and longevity.

Of all the existing work on aging and longevity in *Drosophila,* we suggest that the most important observations are the following.

1. Longevity in different strains not selected for longevity or as observed under different conditions usually varies as an inverse function of the specific metabolic rate (reviewed by Lints and Soliman, 1988).

2. Strains of *D. melanogaster* with markedly extended life-span can be constructed by genetic selection (Arking, 1987; Rose, 1984; Luckinbill *et al.,* 1984; Arking and Dudas, 1989). The specific metabolic rate in those long-lived strains that have been examined is the same as that of the founding strain. There is no corresponding decrease in the specific metabolic rate in these strains as predicted from the first observation, and the lifetime oxygen consumption is greater than the founding strain.

3. Chronic oxidative deterioration of mitochondria strongly correlates with aging in *D. melanogaster* (Miquel and Fleming, 1986; Lints and Soliman, 1988) and may relate to radical-initiated damage to mitochondrial DNA (Richter *et al.,* 1988; Saul *et al.,* 1987).

4. Strains in which specific enzymatic components of oxygen defense metabolism have been genetically disrupted have an extremely reduced adult life-span (Phillips *et al.,* 1989; Mackay and Bewley, 1989). The life-span-shortening phenotype of Cu–ZnSOD null and catalase

null mutants is correlated with an increase in spontaneous mutation frequency (see Section V,B). Mapping of the gene(s) responsible for the life-span extension in selected lines (Arking and Dudas, 1989) has placed them on the third chromosome, on which, perhaps coincidentally, the genes for Cu–ZnSOD and catalase also reside.

B. HEAT SHOCK

Environmental stress elicits a variety of responses from cells, including the well-known heat-shock response (Schlesinger *et al.*, 1982). It has been proposed that heat shock is in fact an oxidation stress and that the heat-shock response may be in part an oxygen defense mechanism (Lee *et al.*, 1983; Ropp *et al.*, 1983). The heat-shock response in *Drosophila,* as measured by puffing in larval polytene chromosomes or by the synthesis of heat-shock proteins in larvae, adults, or cultured cells, can be induced by a variety of conditions, many of which suggest an involvement of oxygen stress. In addition to induction by elevated temperature, treatment with H_2O_2, cadmium, or menadione (an oxidative quinone), agents that uncouple oxidative phosphorylation, and recovery from anoxia all induce the heat-shock response (Ropp *et al.*, 1983, and references therein; Courgeon *et al.*, 1984, 1988). Ropp *et al.* (1983) have proposed that active oxygen may serve as an agent of heat-shock induction in *Drosophila*. Temperature elevation and recovery from anoxia could possibly act in this regard through "oxygen overcharge" (Ropp *et al.*, 1983), leading to excess O_2^- production from an overburdened respiratory chain. In support of this hypothesis, it has been observed that exposure of *Drosophila* cells in culture to ecdysterone, which induces markedly increased oxygen uptake in conjunction with increased levels of catalase and superoxide dismutase, does not lead to induction of the heat-shock response (Best-Belpomme and Ropp, 1982; Ropp *et al.*, 1986; Peronnet *et al.*, 1986). It should be noted that the dinucleoside polyphosphates (e.g., Ap_4A), which have been implicated as "alarmones" initiating the heat-shock response in bacteria (Lee *et al.*, 1983), while present in *Drosophila* do not appear to be involved in the same way in the heat-shock response in *Drosophila* cells in culture (Brevet *et al.*, 1985).

C. RADIATION BIOLOGY

The classic studies by Muller (1927) on the mutagenicity of X-rays set the stage for 50 years of intensive work on the biological response of *Drosophila* to ionizing radiation (reviewed by Sankaranarayanan and

Sobels, 1976). The differential radiosensitivity of different germ-cell stages was established. Radiation-induced somatic mutation and recombination were developed as tools for investigating the nature of radiation-induced damage in somatic cells and for investigating cell lineages during development. Oxygen effects were established for both male and female gametogenesis and dose–response relations were determined for generating various types of genetic damage, including gene mutations (recessive and dominant/visible and lethal), chromosomal rearrangements (of numerous types), and chromosome loss and nondisjunction. Finally, strain differences in radiosensitivity were inidentified. These and related studies provide a wealth of information on radiation effects in *Drosophila* and also provide a basis for future studies exploring the possible relationship of the oxygen defense system to radiation mutagenesis. Indeed, chemical considerations lead us to strongly suspect that oxygen defense is highly relevant to radiation biology.

The biological effects of radiation in *Drosophila* must ultimately be interpreted in terms of energy transfer from radiation (high-energy photons) to cellular molecules (reviewed by Weiss and Kumar, 1988). The action of ionizing radiation on cells can either be direct, through excitation or ionization of cellular molecules (TH)

$$\text{Radiation} + \text{TH} \rightarrow \text{T·} + \text{H}^+ + e^-$$

or indirect, through the generation of free radicals produced initially by the interaction of radiation with intermediates

$$\text{TH} + \text{R·} \rightarrow \text{T·} + \text{RH}$$

Radiation of high linear energy transfer (LET), such as accelerated heavy ions or neutrons, produces effects arising primarily from direct action, whereas radiation of low LET, such as X rays, generates effects largely through indirect action involving water

$$\text{Radiation} + \text{H}_2\text{O} \rightarrow \text{H}_2\text{O}^+ + e^-$$
$$\text{H}_2\text{O}^+ + \text{H}_2\text{O} \rightarrow \text{H}_3\text{O}^+ + \text{OH·}$$
$$\text{H}_2\text{O} + e^- \rightarrow \text{H·} + \text{OH·}$$
$$\text{OH·} + \text{OH·} \rightarrow \text{H}_2\text{O}_2$$
$$e^- + \text{H}_2\text{O} \rightarrow e^-_{\text{aq}}$$

These radical-generating reactions occur within a period of 10^{-16}–10^{-10} seconds. Another important source of radical species is O_2 dissoved in cellular water

$$\text{O}_2 + e^-_{\text{aq}} \rightarrow \text{O}_2^{·-}$$
$$\text{O}_2 + \text{H·} \rightarrow \text{HO}_2^·$$

This dissolved oxygen may be the basis for the so-called oxygen effect and probably adds substantially to the damage inflicted on cells by ionizing radiation. The sensitivity of cells irradiated under anoxic conditions is reduced severalfold relative to cells irradiated under normoxia.

The participation of many specific components of the multifunctional oxygen defense system of aerobic cells in radioprotection has been well established (reviewed by Weiss and Kumar, 1988). In *Drosophila* the identification of radioresistant strains could offer the opportunity to further pursue mechanisms of radioresistance and in so doing could provide a novel way to further investigate oxygen defense mechanisms. Likewise radiosensitivity mutants may be a promising vehicle for investigation of oxygen defense, particularly as it relates to DNA. This approach gains strong support from the recent demonstration (Xu Peng *et al.*, 1986) that a mutant with severely reduced Cu–ZnSOD levels (3.5% of normal) is markedly hypersensitive to acute radiation exposure. In fact, it was reported that strains differing in Cu–ZnSOD alleles specifiying Cu–ZnSOD allelic isozymes with different substrate affinities for O_2^- could be differentiated on the basis of acute radiosensitivity. These important observations further establish an important role for Cu–ZnSOD in acute-exposure radioprotection in somatic as well as germ cells of *Drosophila*. Finally, the hypersensitivity of *Drosophila* mutants defective in aspects of DNA repair to ionizing radiation mutagenesis (reviewed by Boyd *et al.*, 1987) suggests that the germ-line DNA repair component of oxygen defense systems can be profitably investigated in *Drosophila*. The interaction of such DNA repair mutants with mutants defective in other specific oxygen defense components, e.g., Cu–ZnSOD null and catalase null, could illuminate further the induction versus repair components of oxidative DNA damage resulting from oxygen stress.

ACKNOWLEDGMENTS

We would like to express our gratitude to our present and former graduate students and research technicians for their important contributions to our research efforts in this field. We also thank those colleagues who generously supplied us with reprints and unpublished results that aided in this review.

REFERENCES

Allen, R. G., and Sohal, R. S. (1986). Role of glutathione in the aging and development of insects. *In* "Insect Aging: Strategies and Mechanisms" (K. G. Collatz and R. S. Sohal, eds.). Springer-Verlag, Berlin and New York.

Ames, B. N. (1983). Dietary carcinogens and anticarcinogens (oxygen radicals and degenerative diseases). *Science* **221**, 1256–1264.

Ames, B. N., Cathcart, R., Schwiers, E., and Hochstein, P. (1981). Uric acid provides an antioxidant defense in humans against oxidant- and radical-caused aging and cancer: A hypothesis. *Proc. Natl. Acad. Sci U.S.A.* **78**, 6858–6862.

Arking, R. (1987). Successful selection for increased longevity in *Drosophila*: Analysis of the survival data and presentation of a hypothesis on the genetic regulation of longevity. *Exp. Gerontol.* **22**, 199–220.

Arking, R., and Dudas, S. P. (1989). A review of genetic investigations into aging processes of *Drosophila*. *J. Am. Geriatr. Soc.* **37**, 757–773.

Arking, R., Buck, S., Wells, R. A., and Pretzlaff, R. (1988). Metabolic rates in genetically based long lived strains of *Drosophila*. *Exp. Gerontol.* **23**, 59–76.

Ayala, F. J., Powell, J., and Dobzhansky, T. (1971). Polymorphisms in continental and island populations of *Drosophila willistoni*. *Proc. Natl. Acad. Sci. U.S.A.* **68**, 2480–2483.

Ayala, F. J., Powell, J., and Tracey, M. L. (1972a). Enzyme variability in the *Drosophila willistoni* group. V. Genetic variation in natural populations of *Drosophila equinoxialis*. *Genet. Res.* **20**, 19–42.

Ayala, F. J., Powell, J., Tracey, M. L., Mourao, C., and Perez-Salas, S. (1972b). Enzyme variability in the *Drosophila willistoni* gorup. IV. Genetic variation in natural populations of *Drosophila willistoni*. *Genetics* **70**, 113–139.

Baker, B. S., Carpenter, A. T. C., Esposito, M. S., Esposito, R. E., and Sandler, I. (1976). The genetic control of meiosis. *Annu. Rev. Genet.* **10**, 53–134.

Bender, W., Spierer, P., and Hogness, D. S. (1983). Chromosome walking and jumping to isolate DNA from *Ace* and *rosy* and the bithorax complex in *Drosophila melanogaster*. *J. Mol. Biol.* **168**, 17–33.

Bentley, M. M., and Williamson, J. H. (1979). The control of aldehyde oxidase and xanthine dehydrogenase activities by the *cinnamon* gene in *Drosophila melanogaster*. *Con. J. Genet. Cytol.* **21**, 457–471.

Best-Belpomme, M., and Ropp, M. (1982). Catalase is induced by ecdysterone and ethanol in *Drosophila* cells. *Eur. J. Biochem.* **121**, 349–355.

Bewley, G. C., and Mackay, W. J. (1990). Development of a genetic model for acatalasemia: Testing the oxygen free radical theory of aging. *In* "The Genetic Effects on Aging" (D. E. Harrison, ed.). Telford, West Caldwell, New Jersey, in press.

Bewley, G. C., Nahmais, J. A., and Cook, J. L. (1983). Developmental and tissue-specific control of catalase expression in *Drosophila melanogaster*: Correlations with rates of enzyme synthesis and degradation. *Dev. Genet.* **4**, 49–60.

Bewley, G. C., Mackay, W. J., and Cook, J. L. (1986). Temporal variation for the expression of catalase in *Drosophila melanogaster*: Correlations between the rates of enzyme synthesis and levels of translatable catalase messenger RNA. *Genetics* **113**, 919–938.

Bingham, P. M., Levis, R., and Rubin, G. M. (1981). Cloning of DNA sequences from the white locus of *D. melanogaster* by a novel and general method. *Cell* **25**, 693–704.

Blackman, R. K., Macy, M., Koehler, D., Grimaila, R., and Gelbart, W. M. (1989). Identification of a fully-functional *hobo* transposable element and its use for germline transformation of *Drosophila*. *EMBO J.* **8**, 211–217.

Boyd, J. B., Mason, J. M., Yamamoto, A. H., Brodberg, R. K., Banga, S. S., and Sakaguchi, K. (1987). A genetic and molecular analysis of DNA repair in *Drosophila*. *J. Cell Sci.*, *Suppl.* **6**, 39–60.

Brevet, A., Plateau, P., Best-Belpomme, M., and Blanquet, S. (1985). Variation of Ap$_4$A and other dinucleotide polyphosphates in stressed *Drosophila* cells. *J. Biol. Chem.* **260,** 15566–15570.

Campbell, S. D., Hilliker, A. J., and Phillips, J. P. (1986). Cytogenetic analysis of the *cSOD* microregion in *Drosophila melanogaster. Genetics* **112,** 205–215.

Cathcart, R., Schwiers, E., Saul, R. L., and Ames, B. N. (1984). Thymine glycol and thymidine glycol in human and rat urine: A possible assay for oxidative DNA damage. *Proc. Natl. Acad. Sci. U.S.A.* **81,** 5633–5637.

Cerutti, P. (1985). Prooxidant states and tumor promotion. *Science* **227,** 375–381.

Cerutti, P. (1986). The role of active oxygen in tumor promotion. *In* "Biochemical and Molecular Epidemiology of Cancer" (C. C. Harris, ed.), pp. 167–176. Liss, New York.

Chatterjee, I. B. (1973). Evolution and biosynthesis of ascorbate acid. *Science* **182,** 1271–1272.

Chow, C. K., ed. (1988a). "Cellular Antioxidant Defense Mechanisms", Vols. 1–3. CRC Press, Boca Raton, Florida.

Chow, C. K. (1988b). Interrelationships of cellular antioxidant defense mechanisms. *In* "Cellular Antioxidant Defense Mechanisms" (C. K. Chow, ed.), Vol. 2, pp. 217–237. CRC Press, Boca Raton, Florida.

Christman, M. F. R., Morgan, W., Jacobson, F. S., and Ames, B. N. (1985). Positive control of a regulon for defense against oxidative stress and some heat shock proteins in *Salmonella typhimurium. Cell* **41,** 753–762.

Cochrane, B. J., Morrissey, J. J., and LeBlanc, G. A. (1987). The genetics of xenobiotic metabolism in *Drosophila.* IV. Purification and characterization of the major glutathione *S*-transferase. *Insect Biochem.* **17,** 731–738.

Collatz, K. G., and Sohal, R. S., eds. (1986). "Insect Aging: Strategies and Mechanisms." Springer-Verlag, Berlin and New York.

Courgeon, A., Maisonhaute, C., and Best-Belpomme, M. (1984). Heat shock proteins are induced by cadmium in *Drosophila* cells. *Exp. Cell Res.* **153,** 515–521.

Courgeon, A., Rollet, E., Becker, J., Maisonhaute, C., and Best-Belpomme, M. (1988). Hydrogen peroxide (H_2O_2) induces actin and some heat-shock proteins in *Drosophila* cells. *Eur. J. Biochem.* **171,** 163–170.

Crosby, M. A., and Meyerowitz, E. M. (1986). Lethal mutations flanking the *68C* glue gene cluster on chromosome 3 of *Drosophila melanogaster. Genetics* **112,** 785–802.

Cutler, R. G. (1984). Antioxidants, aging and longevity. *Free Radicals Biol.* **6,** 371–428.

Draper, H. H., Dhanakoti, S. N., Hadley, M., and Piche, L. A. (1988). Malondialdehyde in biological systems. *In* "Cellular Antioxidant Defense Mechanisms" (C. K. Chow, ed.), Vol. 2, pp. 97–109. CRC Press, Boca Raton, Florida.

Eberl, D. F., and Hilliker, A. J. (1988). Characterization of X-linked recessive lethal mutations affecting embryonic morphogenesis in *Drosophila melanogaster. Genetics* **118,** 109–120.

Fenn, W. O., Henning, M., and Philpot, M. (1967). Oxygen poisoning in *Drosophila. J. Gen. Physiol.* **50,** 1693–1707.

Fourche, J. (1965a). La respiration larvaire chez *Drosophila melanogaster.* Consommation d'oxygène au cours de la croissance. *C. R. Acad. Sci. Paris* **261,** 2965–2968.

Fourche, J. (1965b). La respiration larvaire chez *Drosophila melanogaster.* Consommation d'oxygène au cours de jeûne. *C. R. Acad. Sci. Paris* **261,** 3478–3481.

Fourche, J. (1967). La respiration chez *Drosophila melanogaster* au cours de la métamorphose. Influence de la pupaison de la mue nymphale et de l'émergence. *J. Insect Physiol.* **13,** 1269–1277.

Freidman, T. B. (1973). Observations on the regulation of uricase activity during development of *Drosophila melanogaster. Biochem. Genet.* **8,** 37–45.

Getzoff, E. A., Tainer, J. A., Weiner, P. K., Kollman, P. A., Richardson, J. S., and Richardson, D. C. (1983). Electrostatic recognition between superoxide and copper, zinc superoxide dismutase. *Nature (London)* **306**, 287–290.

Glassman, E., and Mitchell, H. K. (1958). Mutants of *Drosophila melanogaster* deficient in xanthine dehydrogenase. *Genetics* **44**, 153–162.

Graf, J., and Ayala, F. J. (1986). Genetic variation for superoxide dismutase level in *Drosophila melanogaster. Biochem. Genet.* **24**, 153–168.

Hadorn, E., and Schwinck, I. (1956). Fehlen von isoxanthopterin und nicht-autonomie in der bildung der roten augenpigmente bei einer mutante (*rosy*[2]) von *Drosophila melanogaster. Z. Vererbungsl.* **87**, 528–553.

Halliwell, B., and Gutteridge, J. M. C. (1985). "Free Radicals in Biology and Medicine." Oxford Univ. Press (Clarendon), London and New York.

Halliwell, B., and Gutteridge, J. M. C. (1986). Oxygen free radicals and iron in relation to biology and medicine: Some problems and concepts. *Arch. Biochem. Biophys.* **246**, 501–514.

Harman, D. (1956). Aging: A theory based on free radical and radiation chemistry. *J. Gerontol.* **11**, 298–300.

Harman, D. (1962). Role of free radicals in mutation, cancer, aging and the maintenance of life. *Radiat. Res.* **16**, 753–763.

Harman, D. (1987). The free-radical theory of aging. *In* "Modern Biological Theories of Aging" (H. R. Warner, ed.), pp. 81–87. Raven, New York.

Hassan, H. M. (1988). Biosynthesis and regulation of superoxide dismutases. *Free Radical Biol. Med.* **5**, 377–385.

Hjalmarsson, K., Marklund, S., Engstrom, Å., and Edlund, T. (1987). Isolation and sequence of complementary DNA encoding extracellular superoxide dismutase. *Proc. Natl. Acad. Sci. U.S.A.* **84**, 6340–6344.

Jelnes, J. E. (1971). Identification of hexokinases and localization of a fructokinase and a tetrazolium oxidase locus in *Drosophila melanogaster. Heriditas* **67**, 291–293.

Ketterer, B., Meyer, D. J., and Tan, K. H. (1988). The role of glutathione transferase in the detoxication and repair of lipid and DNA hydroperoxides. *In* "Oxygen Radicals in Biology and Medicine" (M. G. Simic, K. A. Taylor, and C. von Sonntag, eds.), pp. 669–674. Plenum, New York.

Kirkland, K. C., and Phillips, J. P. (1987). Isolation and chromosomal localization of genomic DNA sequences coding for cytoplasmic superoxide dismutase from *Drosophila melanogaster. Gene* **61**, 415–419.

Lamb, M. J. (1978). Aging. *In* "The Genetics and Biology of Drosophila" (M. Ashburner and T. R. F. Wright, eds.), Vol. 2c, pp. 43–104. Academic Press, New York.

Lastowski-Perry, D., Otto, E., and Maroni, G. (1985). Nucleotide sequence and expression of a *Drosophila* metallothionein gene. *J. Biol. Chem.* **260**, 1527–1530.

Lee, P. C., Bochner, B. R., and Ames, B. N. (1983). AppppA, heat-shock stress, and cell oxidation. *Proc. Natl. Acad. Sci. U.S.A.* **80**, 7496–7500.

Lee, Y. M., Ayala, F. J., and Misra, H. P. (1981a). Purification and properties of superoxide dismutase from *Drosophila melanogaster. J. Biol. Chem.* **256**, 8506–8509.

Lee, Y. M., Misra, H. P., and Ayala, F. J. (1981b). Superoxide dismutase in *Drosophila melanogaster*: Biochemical and structural characterization of allozyme variants. *Proc. Natl. Acad. Sci. U.S.A.* **78**, 7052–7055.

Lee, Y. M., Friedman, D. J., and Ayala, F. J. (1985). Complete amino acid sequence of copper–zinc superoxide dismutase from *Drosophila melanogaster. Arch. Biochem. Biophys.* **241**, 577–589.

Levanon, D., Leiman-Hurwitz, J., Dafni, N., Wigderson, M., Sherman, L., Bernstein, Y.,

Laver-Rudich, Z., Danciger, E., Stein, O., and Groner, Y. (1985). Architecture and anatomy of the chromosomal locus in human chromosome 21 encoding the Cu/Zn superoxide dismutase. *EMBO J.* **4**, 77–84.

Lindsley, D. L., Sandler, L., Baker, B. S., Carpenter, A. T. C., Denell, R. E., Hall, J. C., Jacobs, P. A., Miklos, G. L. G., Davis, B. K., Gethmann, R. C., Hardy, R. W., Hessler, A., Miller, S. M., Nozawa, H., Parry, D. M., and Gould-Somero, M. (1972). Segmental aneuploidy and the genetic gross structure of the Drosophila genome. *Genetics* **71**, 157–184.

Lints, F. A., and Lints, C. V. (1968). Respiration in *Drosophila*. II. Respiration in relation to age by wild, inbred and hybrid *Drosophila melanogaster* imagos. *Exp. Gerontol.* **3**, 341–349.

Lints, F. A., and Soliman, M. H., eds. (1988). "*Drosophila* as a Model Organism for Aging Studies." Blackie, Glasgow and London.

Lubinsky, S., and Bewley, G. C. (1979). Genetics of catalase in *Drosophila melanogaster*: Rates of synthesis and degradation of the enzyme in files aneuploid and euploid for the structural gene. *Genetics* **91**, 723–742.

Luckinbill, L. S., Arking, R., Clare, M. J., Cirocco, W. C., and Buck, S. A. (1984). Selection for delayed senescence in *Drosophila melanogaster*. *Evolution* **38**, 996–1004.

Mackay, W. J., and Bewley, G. C. (1989). The genetics of catalase in *Drosophila melanogaster*: Isolation and characterization of acatalasemic mutants. *Genetics* **122**, 643–652.

Mackay, W. J., Orr, W. C., and Bewley, G. C. (1990). Genetic and molecular analysis of antioxidant enzymes in *Drosophila melanogaster*. *In* "Molecular Biology of Aging," UCLA Symposium on Molecular and Cellular Biology (M. Clegg and S. O'Brien, eds.), Vol. 123. Liss, New York, in press.

Mannervik, B., Alin, P., Guthenberg, C., Jensson, H., and Warholm, M. (1985). Glutathione transferases and the detoxication of products of oxidative metabolism. *In* "Microsomes and Drug Oxidations" (A. R. Boobis, J. Caldwell, F. DeMatteis, and C. R. Elcombe, eds.). Taylor & Francis, London.

Maroni, G., and Watson, G. (1985). Uptake and binding of cadmium, copper and zinc by *Drosophila melanogaster* larvae. *Insect Biochem.* **15**, 55–63.

Maroni, G., Otto, E., and Lastowski-Perry, D. (1986). Molecular and cytogenetic characterization of a metallothionein gene of *Drosophila*. *Genetics* **112**, 493–504.

Maroni, G., Wise, J., Young, J. E., and Otto, E. (1987). Metallothionein gene duplications and metal tolerance in natural populations of *Drosophila melanogaster*. *Genetics* **117**, 739–744.

Massie, H. R., Aiello, V. R., and Williams, T. R. (1980). Changes in superoxide dismutase activity and copper during development and aging in the fruit fly *Drosophila melanogaster*. *Mech. Ageing Dev.* **12**, 279–286.

Miquel, J., and Fleming, J. (1986). Theoretical and experimental support for an oxygen radical–mitochondrial injury hypothesis of cell aging. *In* "Free Radicals, Aging and Degenerative Diseases" (J. E. Johnson, D. Harman, R. Walford, and J. Miguel, eds.), pp. 51–74. Liss, New York.

Miquel, J., Lundgren, P. R., Bensch, K. G., and Atlan, H. (1976). Effects of temperature on the life span, vitality and fine structure of *Drosophila melanogaster*. *Mech. Ageing Dev.* **5**, 347–370.

Mitchell, H. K., and Glassman, E. (1959). Hypoxanthine in *rosy* and *maroon-like* mutants of *Drosophila melanogaster*. *Science* **129**, 268.

Mokdad, R., Debec, A., and Wegnez, M. (1987). Metallothionein genes in *Drosophila* constitute a dual system. *Proc. Natl. Acad. Sci. U.S.A.* **84**, 2658–2662.

Morrissey, J. J., Tinker, R., Crocquet de Belligny, P., Gunnersen, D., and Cochrane, B. J. (1988). Molecular genetics of glutathione S-transferase of Drosophila. Genome 30 (Suppl. 1), 262 (abstr.).

Muller, H. J. (1927). Artificial transmutation of the gene. Science 66, 84–87.

Nahmias, J. A., and Bewley, G. C. (1984). Characterization of catalase purified from Drosophila melanogaster by hydrophobic interaction chromatography. Comp. Biochem. Physiol. B 77B, 355–364.

Neta, P., and Huie, R. E. (1985). Free radical chemistry of sulfite. EHP, Environ. Health. Perspect. 64, 209–217.

Nusslein-Volhard, C., Wieschaus, E., and Kluding, H. (1984). Mutations affecting the pattern of the larval cuticle in Drosophila melanogaster. I. Zygotic loci on the second chromosome. Wilhelm Roux's Arch. Dev. Biol. 193, 267–282.

Nygaard, O. F., and Simic, M., eds. (1983). "Radioprotectors and Anticarcinogens." Academic Press, New York.

Oberley, L. W. (1988). Free radicals and diabetes. Free Rad. Biol. Med. 5, 113–124.

O'Brien, S. J., and MacIntyre, R. J. (1978). Genetics and biochemistry of enymes and specific proteins of Drosophila. In "The Genetics and Biology of Drosophila" (M. Ashburner and T. R. F. Wright, eds.), Vol. 2a, pp. 396–551. Academic Press, New York.

Otto, E., Young, J. E., and Maroni, G. (1986). Structure and expression of a tandem duplication of the Drosophila metallothionein gene. Proc. Natl. Acad. Sci. U.S.A. 83, 6025–6029.

Park, J., and Ames, B. N. (1988). 7-Methylguanine adducts in DNA are normally present at high levels and increase on aging: Analysis by HPLC with electrochemical detection. Proc. Natl. Acad. Sci. U.S.A. 85, 7467–7470.

Peronnet, F., Ropp, M., Rollet, E., Becker, J. L., Becker, J., Maisonhaute, C., Pernodet, J. L., Vuillaume, M., Courgeon, A. M., Eschalier, G., and Best-Belpomme, M. (1986). Drosophila cells in culture as a model for the study of ecdysteroid action. In "Techniques in in vitro Invertebrate Hormones and Genes" (E. Kurstak and H. Oberlander, eds.), Vol. C2, pp. 1–20. Elsevier, Amsterdam.

Phillips, J. P., Campbell, S. D., Michaud, D., Charbonneau, M., and Hilliker, A. J. (1989). A null mutation of cSOD in Drosophila confers hypersensitivity to paraquat and reduced longevity. Proc. Natl. Acad. Sci. U.S.A. 86, 2761–2765.

Porta, E. A. (1988). Role of oxidative damage in the aging process. In "Cellular Antioxidant Defense Mechanisms" (C. K. Chow, ed.), Vol. 3, pp. 1–52. CRC Press, Boca Raton, Florida.

Pryor, W. A., ed. (1976–1982). Free Radicals in Biol. 1–5.

Pryor, W. A. (1987). The free-radical theory of aging revisited: A critique and a suggested disease-specific theory. In "Modern Biological Theories of Aging" (H. R. Warner, ed.), pp. 89–112. Raven, New York.

Rabinowitch, H. D., Rosen, G. M., and Fridovich, I. (1989). A mimic of superoxide dismutase activity protects Chlorella sorokiniana against the toxicity of sulfite. Free Radical Biol. Med. 6, 45–48.

Rajagopalan, K. V. (1980). Sulfite oxidases. In "Molybdenum and Molybdenum-Containing Enzymes" (M. P. Caughlan, ed.), pp. 243–272. Pergamon, New York.

Richmond, R. C. (1972). Enzyme variability in the Drosophila willistoni group. III. Amounts of variability in the superspecies Drosophila paulistorum. Genetics 70, 87–112.

Richmond, R., and Powell, J. R. (1970). Evidence of heterosis associated with an enzyme locus in a natural population. Proc. Natl. Acad. Sci. U.S.A. 67, 1264–1267.

Richter, C., Park, J., and Ames, B. N. (1988). Normal oxidative damage to mitochondrial and nuclear DNA is extensive. *Proc. Natl. Acad. Sci. U.S.A.* **85**, 7465–7467.

Ropp, M., Courgeon, A. M., Calvayrac, R., and Best-Belpomme, M. (1983). The possible role of the superoxide ion in the induction of heat-shock and specific proteins in aerobic *Drosophila* cells during return to normoxia after a period of anaerobiosis. *Can. J. Biochem. Cell Biol.* **61**, 456–461.

Ropp, M., Calvayrac, R., and Best-Belpomme, M. (1986). Ecdysterone enhances oxygen cellular consumption and superoxide dismutase activity in Drosophila aerobic cells. *In* "Superoxide and Superoxide Dismutase in Chemistry, Biology and Medicine" (G. Rotilio, ed.), pp. 313–315. Elsevier, Amsterdam.

Rose, M. R. (1984). Laboratory evolution of postponed senescence in *Drosophila melanogaster*. *Evolution* **38**, 1004–1010.

Rubin, G. M., and Spradling, A. C. (1981). Genetic transformation of *Drosophila* with transposable element vectors. *Science* **218**, 348–352.

Sankaranarayanan, K., and Sobels, F. H. (1976). Radiation genetics. *In* "The Genetics and Biology of *Drosophila*" (M. Ashburner and T. R. F. Wright, eds.), Vol. 1c, pp. 1090–1250. Academic Press, New York.

Saul, R. L., and Ames, B. N. (1986). Background levels of DNA damage in the population. *In* "Mechanisms of DNA Damage and Repair: Implications for Carcinogenesis and Risk Assessment" (M. Simic, L. Grossman, and A. D. Upton, eds.), pp. 529–536. Academic Press, Orlando, Florida.

Saul, R. L., Gee, P., and Ames, B. N. (1987). Free radicals, DNA and aging. *In* "Modern Biological Theories of Aging" (H. R. Warner, ed.), pp. 113–129. Raven, New York.

Schlesinger, M. J., Ashburner, M., and Tissières, A., eds. (1982). "Heat Shock: From Bacteria to Man." Cold Spring Harbor Lab., Cold Spring Harbor, New York.

Scholes, G. (1983). Radiation effects on DNA. *Br. J. Radiol.* **56**, 221–231.

Scholes, G., and von Sonntag, C. (1987). "The Chemical Basis of Radiation Biology." Taylor & Francis, London.

Seto, N. O. L., Hayashi, S., and Tener, G. M. (1987a). *Drosophila* Cu–Zn superoxide dismutase cDNA sequence. *Nucleic Acids Res.* **15**, 5483.

Seto, N. O. L., Hayashi, S., and Tener, G. M. (1987b). The sequence of the Cu–Zn superoxide dismutase gene of *Drosophila*. *Nucleic Acids Res.* **15**, 10601.

Seto, N. O. L., Hayashi, S., and Tener, G. M. (1989). Cloning, sequence analysis and chromosomal localization of the Cu–Zn superoxide dismutase gene of *Drosophila melanogaster*. *Gene* **75**, 85—92.

Shephard, J. C., Walldorf, U., Hug, P., and Gehring, W. J. (1989). Fruit flies with additional expression of the elongation factor EF-1α live longer. *Proc. Natl. Acad. Sci. U.S.A.* **86**, 7520–7521.

Sies, H. (1986). Biochemistry of oxidative stress. *Angew. Chem., Int. Ed. Engl.* **25**, 1085–1071.

Smith, J., and Shrift, A. (1979). Phylogenetic distribution of glutathione peroxidase. *Comp. Biochem. Physiol. B* **63B**, 39–44.

Sohal, R. S., and Allen, R. G. (1985). Relationship between metabolic rate, free radicals, differentiation and aging: A unified theory. *In* "Molecular Biology of Aging" (A. D. Woodhead, A. D. Blackett, and A. Hollaender, eds.), pp. 75–104. Plenum, New York.

Sohal, R. S., and Allen, R. G. (1986). Relationship between oxygen and metabolism, aging and development. *Adv. Free Radical Biol. Med.* **2**, 117–160.

Sohal, R. S., Farmer, K. J., Allen, R. G., and Cohen, N. R. (1983). Effect of age on oxygen consumption, superoxide dismutase, catalase, glutathione, inorganic peroxides and chloroform-soluble antioxidants in the adult male housefly, *Musca domestica. Mech. Ageing Dev.* **24,** 185–195.

Stewart, B. R., and Merriam, J. R. (1974). Segmental aneuploidy and enzyme activity as a method for cytogenetic localization in *Drosophila melanogaster. Genetics* **76,** 301–309.

Stivaletta, L., Warner, C. K., Langley, S., and Finnerty, V. (1988). Molybdoenzymes in *Drosophila*. IV. Further characterization of the *cinnamon* phenotype. *Mol. Gen. Genet.* **213,** 505–512.

Tainer, J. A., Getzoff, E. D., Beem, K. M., Richardson, J. S., and Richardson, D. C. (1982). Determination and analysis of the 2 Å structure of copper, zinc superoxide dismutase. *J. Mol. Biol.* **160,** 181–217.

Tainer, J. A., Getzoff, E. A., Richardson, J. S., and Richardson, D. C. (1983). Structure and mechanism of copper, zinc superoxide dismutase. *Nature (London)* **306,** 284–287.

Totter, J. R. (1980). Spontaneous cancer and its possible relationship to oxygen metabolism. *Proc. Natl. Acad. Sci. U.S.A.* **77,** 1763–1767.

Touati, D. (1988). Molecular genetics of superoxide dismutases. *Free Radical Biol. Med.* **5,** 393–402.

Weiss, J. F., and Kumar, K. S. (1988). Antioxidant mechanisms in radiation injury and radioprotection. *In* "Cellular Antioxidant Defense Mechanisms" (C. K. Chow, ed.), Vol. 2, pp. 163–189. CRC Press, Boca Raton, Florida.

Williams, C. M., and Beecher, H. K. (1944). Sensitivity of *Drosophila* to poisoning by oxygen. *Am. J. Physiol.* **140,** 566–573.

Xu Peng, T., Moya, A., and Ayala, F. J. (1986). Irradiation resistance conferred by superoxide dismutase: Possible adaptive role of a natural polymorphism in *Drosophila melanogaster. Proc. Natl. Acad. Sci. U.S.A.* **83,** 684–687.

DNA REARRANGEMENTS IN RESPONSE TO ENVIRONMENTAL STRESS

Christopher A. Cullis

Department of Biology, Case Western Reserve University,
Cleveland, Ohio 44106

I. Introduction

The discovery of molecular mechanisms that can rapidly alter genomic organization has led to the concept of the fluid genome. In many cases, the consequences of such mechanisms have beeen considered in light of Weismann's doctrine (Weismann, 1883) of the separation of the soma and germ line. However, in higher plants, the invalidity of this doctrine has been long known, because plant germ cells are normally produced from undifferentiated cell lineages that are not set aside early in development. The genetic and evolutionary ramifications of this fundamental difference among plants and higher animals has been frequently noted (Whitham and Slobodchikoff, 1981; Buss, 1983; Walbot and Cullis, 1983, 1985; Walbot, 1986; Cullis, 1986, 1987).

Plants can respond morphologically to changes in their environment. Examples such as sun versus shade leaves (Vogel, 1968), responses to herbivory (Hendrix, 1979; Lubchenko and Cubit, 1980), and competition (Turkington, 1983) all represent classic cases of phenotypic plasticity. However, genotypic responses to changes in the environment are

ADVANCES IN GENETICS, Vol. 28

less well documented, although there is an increasing body of evidence that the environment can influence the rate at which variation can occur. However, note must be made that the definition of phenotypic and genotypic plasticity does not imply that the responses are necessarily adaptive, but simply that there is a response to changes in the environment at both levels.

The extent of phenotypic and genotypic variability is highest under conditions of environmental stress for a variety of organisms (Parsons, 1988). This conclusion for genotypic variability can lead to the possibility that the rate at which variability may be generated, by whatever mechanism, is dependent on the level of stress in the environment. Thus, under optimal conditions the rate of variation is lowest; as the environment becomes more stressful, the rate of variability would rise in parallel. A general statement of this hypothesis would be that any shift from the environment to which an organism has become adapted should increase the rate at which underlying mechanisms generate variation.

The examples of genomic changes in plants abound and appear to occur by a variety of basic mechanisms, namely, by amplification, deletion, and transposition events (Flavell, 1980, 1982). The relative frequencies of these events have been examined in a number of instances. Whether externally defined in terms of physical parameters, or internally, in terms of the composition of the nucleus, the influence of the prevailing environment on these processes has been determined. This article will examine the structure of the plant genome and document those fractions that appear to be most frequently rearranged. These examples will be considered in regard to the concept that the plant genome may be organized into constant and fluid domains (Cullis, 1983), so that both developmental and physiological stimuli and the response to changing environmental conditions may give rise to rapid changes. The relationship between the types of changes that can occur and the physiological status of the cell in which those changes take place will be highlighted. Finally, the role of such changes in the generation of variability on which the evolutionary process can act, particularly in the relationship of the adaptive nature of these changes, will be considered.

II. Genome Size and Organization

Plants have large differences in their nuclear DNA content. Within the angiosperms there is nearly a 1000-fold range of variation and there does not appear to be any correlation between genome size and organismal complexity (Bennett and Smith, 1976). The haploid genome sizes of various flowering plants are shown in Table 1. The values have been calculated from the kinetic complexities except for *Pinus radiata,* which comes from spectrophotometric measurement.

In *Arabidopsis,* the characterization of the genome by molecular analysis has indicated that the nuclear genome consists predominantly of unique sequences and that most of the nuclear repetitive DNA is ribosomal DNA (Myerowitz, 1987). The genes of *Arabidopsis* appear to be shorter than in other higher plants and there are obviously many fewer dispersed repetitive sequences in this very small genome. Taking the value of 75% of the total genome as comprising the unique fraction of the *Arabidopsis* genome suggests that the minimum information content required for a functional higher plant is 52,500 kilobases (kb). Taking this minimum value, it is clear that the proportion of the genome necessary for coding functions in the other higher plants shown in Table 1 can be dramatically different. Thus, only 20% of the flax genome or 0.2% of the pine genome would be sufficient for the basic coding functions.

The estimates of the minimum coding information, coupled with the C-value paradox, have led to the conclusion that the vast majority of the genome in most higher plants cannot be responsible for direct coding sequences. How has this wide range of variation in DNA content arisen? Whereas a large genome size and heterogeneity seem to be indicators for evolutionary flexibility and progressivity, decline in total DNA content apparently accompanies evolutionary specialization and

TABLE 1
Haploid Genome Size in Plants[a]

Plant	Haploid genome size [in kilobase pairs (kbp)]
Arabidopsis	70,000
Flax[b]	280,000
Mung bean	470,000
Cotton	780,000
Tobacco	1,600,000
Soybean	1,800,000
Pea	4,500,000
Wheat	5,900,000
Pinus radiata[c]	10,000,000

[a] Data from Myerowitz and Pruitt (1985).

[b] Data from Cullis (1981).

[c] Data from Miksche (1985).

adaptation to certain ecological niches. As genome evolution is predominantly a molecular process governed by sequence amplification, divergence, dispersal, and loss, the final result may be determined by a balance between these events.

The relative proportions of the repetitive sequence fractions can vary widely. In flax, the vast majority of the repetitive families are arranged in tandem arrays, although these arrays may be dispersed among many sites within the genome, but few dispersed repetitive sequences. In maize, there are both tandem arrays (for example, the knob heterochromatin sequence) and many dispersed families, whereas in pea, there are few tandemly arrayed families, but the majority of the genome is composed of complex dispersed repetitive families (Cullis and Creissen, 1987). The evolutionary pressures leading to these different organizations are unknown, but the current composition could be due to differential rates of the various processes that shape the genome. Thus, in flax, tandem amplification–deletion events are much more frequent than transpositions, whereas in pea, the reverse may be true. The maize genome may be the result of a more equal frequency of these events. Whatever the mechanisms, it is clear that higher plant genomes contain a high proportion of repetitive DNA, which, on the basis of comparisons among related species, can vary rapidly when viewed in an evolutionary context. However, the species comparisons only document the present state of the genome in the examples studied, but do not give an indication of the frequency, position within the genome, and the time in the life cycle at which these events may be occurring.

III. Life History Considerations

Any change in the genome, to be recognized, has to be propagated in a sufficient number of cells, and ultimately to subsequent generations. Plants have open systems of growth and lack germ lines. Thus the persistence of altered genomes will be a function of the competitive abilities of the altered cells and the organization and growth characteristics in the meristem in which they arise. An analysis of the probabilities of this propagation and the consequences of various meristematic organizations has been made (Klekowski, 1988). In spite of the many apparent barriers to the establishment of such changes, their occurrence is widespread, as evidenced by bud sports and chimeral plants. Thus, irrespective of the potential difficulties in fixing such changes, they do occur and the examples considered here are designed to investigate the rate, and the tissues and stages of the life cycle, at which such changes can occur.

IV. Rapid Genomic Changes

There are a variety of mechanisms by which the genome can be varied. These include alterations mediated by the action of transposable elements, which not only restructure the genome by their own movement, but also cause deletions and amplifications. In addition, there are increases and decreases in the total nuclear DNA amount that appear to be confined to an unstable fraction of the genome. The way in which the genome is modified in a variety of systems is considered. The action of transposable elements and their response to the environment in which the plant is growing is one source of variability. The alteration in overall DNA content and the influence of this on the phenotype will also be considered. The process of tissue culture and regeneration (somaclonal variation) can result in alterations in nuclear and organellar DNA, although only changes in the nuclear DNA sequences will be considered here. Finally, the induction of heritable changes in flax, in which both phenotypic and genetic alterations have been observed within a single generation, will be described. The common theme within all these examples is the limitation of the alterations to a subset of the genome. This subset may be further subdivided into regions that are more or less labile depending on the physiological status of the cell or organism. Clearly, such a relationship could be important in the orchestration of any rapid genomic changes.

A. TRANSPOSABLE ELEMENTS

It has already been established that the action of transposable elements can result in the generation of substantial genetic diversity (Starlinger et al., 1985; Saedler et al., 1985; Dellaporta and Chomet, 1985). These mobile elements were first described in maize (McClintock, 1950) and have since been found in a wide variety of prokaryotes and eukaryotes. It is probable that their occurrence is ubiquitous.

The presence of these elements is generally detected by the occurrence of unstable mutants. However, the elements may not be active at all times and may be present as quiescent components of the genome. These quiescent elements can be activated following certain alterations; for example, the activation of P-elements in Drosophila requires the appropriate combination of nucleus and cytoplasm (Bingham et al., 1982). Over a number of generations, the activity of these elements can reduce until the cytotype has been converted to the P-type and the activity of the transposons ceases, only to be reactivated when an incompatible combination again occurs. In the case of maize, the elements can be inactivated, as demonstrated with the Mu element in maize (Chandler *et al.*, 1986). Here the inactivation appears to be correlated with the modification of the DNA of the element, so preventing its activity. Thus these elements, which appear to be normal components of the genome, can be activated and inactivated by the organism in which they reside.

The presence of an active transposable element system can influence the rate at which responses to selection can be achieved. Mackay (1985) carried out directional selection on the bristle number on the last abdominal tergite of *Drosophila melanogaster*. The response to selection for high and low bristle number was greater in dysgenic crosses (in which P-element transposition was expected) than in nondysgenic crosses.

The rates of transposition can be affected by the external environment and by the stage of development. The frequency with which transposable elements move in *Antirrhinum majus* can be greatly affected by the temperature at which the plants are grown (Carpenter *et al.*, 1988). The timing and frequency of maize transposable elements transposition are developmentally regulated and can exist in a number of heritable states (Banks *et al.*, 1988). Thus the activity of transposons can be controlled. However, the importance of stress, both within the cell and from without, imposed by the environment, in the mobilization of transposable elements, resulting in rapid genomic reorganization, has been proposed (McClintock, 1951, 1978, 1984; Cullis, 1983; Walbot and Cullis, 1985). The restructuring of the genome by transposable elements can occur at many levels, from small changes involving a few nucleotides, such as duplications at the insertion site, to major modifications involving large segments of chromosomes, such as duplications, deletions, inversions, and other more comlex rearrangements (McClintock, 1984). However, what is perceived by the organism as stress and how this perception is translated into the activation of these elements remain to be elucidated. In spite of these unknowns, transposable elements do provide a mechanism by which the mutation rate, and thereby the range of genetic diversity, can increase during periods of stress.

B. NUCLEAR DNA CONTENT

The variation in nuclear DNA content per genome among both plant and animal taxa has been extensively documented (Bachmann *et al.*, 1972; Sparrow *et al.*, 1972; Bennett and Smith, 1976; Price, 1976). This nuclear DNA content has not been correlated with any measure of evolutionary advancement or genetic complexity, with both increases and decreases in DNA content being observed in eukaryotic evolution (Price, 1976). The changes in nuclear DNA, apart from polyploidization and chromosome endoreduplication, affect the relative proportions of fractions of the genome. Thus, amplifications or losses of DNA usually appear to occur in the repetitive fraction of the nuclear genome, the majority of which, although it may have no specific coding capacity or function, may nevertheless be important. The DNA content by virtue of its mere bulk can affect parameters, such as nuclear volume, cell volume, and mitotic and meiotic cycle times (Bennett, 1972; Price, 1976; Cavalier-Smith, 1978). At what level of difference in total nuclear DNA do differences become of significance to the organism? Price (1988) suggests that those in the range of 5–10% may be sufficient to be subject to selection.

The evidence is accumulating that rapid changes in nuclear DNA content can accumulate within a species. What is not yet clear is the mechanism(s) by which these variations arise and their frequency. There is a latitudinal cline in the DNA content of crop plants, such that those with low DNA content are naturally adapted to equatorial latitudes, whereas those with higher DNA contents are found at higher latitudes (Bennett, 1976; Bennett and Smith, 1976). Increasing latitude and elevation may be associated with intraspecific increases in DNA in conifers (Miksche, 1967, 1968). The intraspecific variation in DNA content in the genus *Microseris* also appears to be related to environmental conditions (Price *et al.*, 1980, 1981, 1986; Price, 1988). The function and composition of the DNA fractions that alter are currently unknown, although the changes in heterochromatin are consistent with these alterations occurring in the highly repetitive sequence fractions.

In spite of the extensive variation among species and also within a species, within an inbred line, there is little evidence for quantitative variation in DNA content. One exception to this is in *Helianthus annus* (Cavallini *et al.*, 1986, 1989). Here the nuclear DNA content in the progeny of single plants from an inbred line were compared for total DNA content by microdensitometry of Feulgen-stained material. Significant differences were observed among the progeny from a single plant. However, when these individuals were inbred, the range did not continue to increase. The original distribution was recreated, so that the mean of the progeny from a high-DNA plant was lower than the parental value, whereas the mean of the progeny from a low-DNA parent was higher than the parental value. This situation has been

observed for the ribosomal RNA genes in crosses between flax genotrophs, which is described later in this section (Cullis, 1979).

The examples of DNA variations within inbred lines are few, but
there are many more examples of changes of DNA in inter- and intraspecific crosses. These changes have been observed in crosses in *Nicotiana* (Gerstel and Burns, 1966), *Microseris,* and flax. In each of the
examples, the DNA changes do not occur in every possible combination,
that is, there is some apparent genetic control over whether an interaction will occur.

Intraspecific differences in DNA content have been observed in the
diploid annuals *Microseris bigelovii* and *Microseris douglasii* (Price *et
al.,* 1980, 1981). The DNA content appeared to be related to the prevailing environment, although the phenomenon is complex. The DNA
content of both high- and low-DNA lines has remained stable over
many generations under growth chamber conditions.

In contrast to the stability of DNA in inbred lines, the inheritance of
the DNA content in both intraspecific and interspecific crosses in *Microseris* demonstrates instability over a short time scale. The inheritance of nuclear DNA content in crosses between *M. douglasii* types
that differed in their nuclear DNA content was followed (Price *et al.,*
1983). In one case, the nuclear DNA amounts did not segregate in the
F_2, whereas in a second case, they were significantly different. This was
in spite of the fact that the DNA content differences among the parents
were similar in both crosses.

The differences were even more striking in an interspecific cross
between *M. douglasii* and *M. bigelovii,* which differed in DNA content
by approximately 10% (which is the same as the difference between the
M. douglasii biotypes in the intraspecific crosses mentioned above). The
F_1 progeny did not cluster at the midparent value as expected, but
spread through the range between the parental means (Price *et al.,*
1983). The progenies from five F_1 plants were sampled, and the means
of the F_2 families corresponded to that of the selfed F_1 plant from which
the family was derived. In another interspecific cross between two
M. douglasii and *M. bigelovii* types, instability in DNA content was
seen in both the F_1 and the F_2 families (Price *et al.,* 1985). Thus,
whether instability in the total DNA content is seen, and for how many
generations this instability exists in *Microseris* spp., are dependent on
the particular parents used in the cross. However, these results demonstrate that the mechanisms for rapidly altering the nuclear DNA content exist in these species and that these mechanisms can be activated
and repressed.

A comparable set of results, but obtained from monitoring the content of a specific repetitive family, has been described in flax (Cullis,

1979). The number of ribosomal RNA genes (rDNA) was determined in three parental lines and crosses between them. In one case (L1 × S1), the F_1 showed a range exceeding the parental values (which differed by more than 2-fold). The values in F_2 progeny from selfed F_1 plants from extremes of the range again gave a range similar to that seen in the F_1. However, in a second cross using one of the same parents (L1 × L6), a narrower distribution was seen in both the F_1 and F_2 generations, with a trend toward the values of the lower parent. Thus, as in the case of *Microseris* spp., the instability in DNA, in this case a specific repetitive DNA family, was dependent on the parents used in the hybrid.

C. SOMACLONAL VARIATION

Many of the mechanisms by which the genome is reorganized have been observed in cells in tissue culture or in plants regenerated from such cultures. These genomic changes have been shown to result in aneuploidy; chromosomal rearrangements, such as translocations, inversions, deletions, gene amplification, and deamplification; activation of transposable elements; point mutations; cytoplasmic genome rearrangements; and changes in ploidy level (Larkin and Scowcroft, 1983; Orton, 1983, 1984; Evans *et al.*, 1984). There is also accumulating evidence that specific DNA sequences can be altered in copy number.

The sequence families which can be affected include both repetitive families and low and/or single copy sequences. The alterations in the sequences can be in the quantity of that family present or in the degree of modification. Thus in a crown gall callus culture of flax, the number of ribosomal RNA genes was found to have decreased (to 300 copies compared to 800 in the original plant from which the culture was initiated) (Blundy *et al.*, 1987). In contrast, in rice suspension cultures, several highly repeated sequences were amplified up to 75-fold (Zheng *et al.*, 1987). An example of a change in the methylation status of a gene family is the 5 S RNA genes in soybean, wherein there was less methylation in newly initiated callus and suspension cultures than in whole plants, and this pattern could change in some cultures with time (Quemada *et al.*, 1987). However, the methylation pattern is not always altered in culture, for example, the rDNA in the flax crown gall callus was quantitatively unchanged in methylation in spite of the massive reduction in the number of these sequences (Blundy *et al.*, 1987).

Changes in the copy number of specific genes can also be selected in tissue culture. Cells with an amplified number of genes have been selected using resistance to the herbicides phosphinothricin (Donn *et al.*, 1984) and glyphosate (Shah *et al.*, 1986; Ye *et al.*, 1987). These results show that gene amplifications can, indeed, be selected for in tissue culture systems. However, it has not been determined whether such amplifications are restricted to tissue culture systems. It has not been possible to demonstrate amplifications of many genes in cultured cells. Whether this is an absolute failure, or simply due to inappropriate conditions under which the selection was attempted, is not known.

In spite of the known instances of gene amplification, most of the DNA alterations are found in the highly repetitive sequences. Johnson *et al.* (1987) showed that the heterochromatic regions of oat root-tip chromosomes were late replicating. The occurrence of chromatid separation during anaphase, before replication was complete, would result in these heterochromatic regions being the sites of chromosome breakage. Analysis of the meiotic chromosomes of one regenerated plant did show a break, occurring in the heterochromatin. The consequence of this break was the formation of a heteromorphic chromosome pair.

The extent of chromosome abnormalities in maize cultures was dependent on the time the cells had been in culture. Meiotic analysis showed that no chromosomal abnormalities were present in plants regenerated after 3–4 months in culture, whereas about half those regenerated after 8–9 months were cytologically abnormal (Lee and Phillips, 1987). The breakpoints of these abnormalities were primarily on chromosome arms containing large blocks of heterochromatin such as knobs. Thus, in this case, the heterochromatic blocks appeared to be prime sites for the generation of chromosome abnormalities, and the number formed was dependent on the length of time the cells had been in culture.

Heterochromatin has also been implicated in the generation of chromosomal abnormalities in wheat–rye hybrids (Lapitan *et al.*, 1984). Of the 13 breakpoints involved in the deletions and interchanges characterized in this study, 12 were located in heterochromatin.

From the results such as those described above, Johnson *et al.* (1987) and Lee and Phillips (1987) have proposed that one of the mechanisms by which chromosome breaks may be induced in culture is the late replication of heterochromatin. Thus, if the replication of heterochromatin extended into anaphase, chromosome breaks could be induced. Such late replication of heterochromatin has been demonstrated in whole plants, but not in tissue cultures to date.

One of the consequences of a chromosome breakage in maize crosses is the activation of transposable elements (McClintock, 1951). Because one of the proposed possible causes of somaclonal variation is the activation of transposable elements, studies to confirm this have been undertaken. Apparent activation of transposable elements has been found in studies with maize (Peschke *et al.*, 1987) and alfalfa (Groose and Bingham, 1986). With these results in mind, Lee and Phillips (1987) have proposed a model, in maize, whereby the incomplete replication of maize knob heterochromatin can lead to the formation of chromosome bridges, leading to chromosome breakage. The end result of this breakage cycle would be the activation of transposable elements and chromosome rearrangements, duplications, and deletions, all of which could lead to somaclonal variation. It is still not clear whether the activation of transposable elements is the major cause of somaclonal variation in systems other than maize of even in maize itself.

Qualitative changes in the genome, unassociated with the action of transposable elements, can give rise to restriction fragment length polymorphisms (RFLPs). Evidence has been obtained in soybean cell cultures that these polymorphisms can arise in the process of cell culture (Roth *et al.*, 1989). One of the remarkable observations about this variation is the types of polymorphic bands that arose. In all the cases cited, the new band that was found already appeared to exist in natural populations. Thus, as in the case of the polymorphisms in the flax genotrophs, at least some of the changes are those that have already occurred. Here again, a subset of sites has shown evidence of greater lability under conditions that can be regarded as stressful. A second intriguing observation in this system was that for all the polymorphic loci, there appeared to be only two alternative forms. The cause of the alteration, in the few cases in which the alternative "alleles" have been isolated, was the insertion of a small fragment. Thus, to recreate the types of events and sizes of fragments, it appears that a very specific rearrangement must be taking place at these positions in the genome. Their contribution to the overall variation is presently unknown, but they clearly have the potential to be the cause of specific, repeatable changes within the genome.

V. Environmentally Induced Heritable Changes in Flax

In the preceding sections, a range of phenomena that result in the rapid modulation of the plant genome has been described. The prime example wherein most of these phenomena have been characterized in a single line is flax, in which heritable changes have been observed to occur within a single generation, involving changes both at the phenotypic level and in the DNA. The changes have been investigated in the flax variety Stormont Cirrus, although they are not restricted to this line.

A. Characteristics of the Induction

The induction of heritable changes in flax was first described by Durrant (1958, 1962) in response to various nutrient regimes. Seeds from an inbred line were given different combinations of nitrogen, phosphorus, potassium, and calcium for a single generation and were allowed to self. These seeds were then grown in a common environment, and there were differences in plant weight and height associated with the growth regime in the previous generation. The seeds from the treated plants were then grown under either the same conditions to which the previous generation was subjected or to a different regime. Again, the plants were self-fertilized and seeds from the various individuals were grown in a common environment. As before, there were differences among the families, but in some cases the phenotype was correlated with the treatment received two generations earlier, not with that of the immediately preceding cycle. These instances then remained stably altered for a large number of generations, irrespective of the subsequent growth conditions. However, only a minority of the conditions resulted in the stable alteration of the lines, and, in general, if any phenotypic modifications were observed they were correlated with the conditions of growth in the immediately preceding generation.

The lines in which stable changes were observed were termed genotrophs and had a number of characteristics. All the individuals of a given genotroph, generally all the progeny of a single inbred plant that had undergone an induction cycle, were identical to each other. However, each of the extremes, termed the large and small genotrophs, differed from each other and the line from which they were derived [termed plastic (Pl)] in a number of characteristics. Individuals from one induction experiment are shown in Fig. 1. These are examples of the experiments described by Cullis and Charlton (1981). The original line (Fig. 1a) is taller than either of the genotrophs, but the large type (Fig. 1c) is much more branched. The small type (Fig. 1b) is shorter than Pl and even less branched.

Another characteristic of the induction of heritable changes was the reproducibility of the phenomenon. The conditions used were not lethal, and all or most (>90%) of the seeds planted and grown survive and produce seeds. When either many seeds from a single individual plant or seeds from different plants, which have been grown under the same inducing conditions, are grown in a common environment, all are very

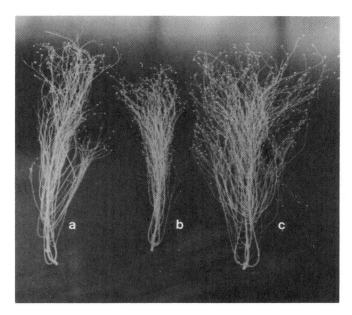

FIG. 1. Individuals grown in a common environment after a single generation of growth under inducing conditions. (a) Pl; (b) C3, a small genotroph; (c) C1, a large genotroph.

similar to each other. Examples are shown in Fig. 2, where three seeds from three different plants were grown in a common environment. These three parental plants were all grown in the same induction experiment. The plant in Fig. 2b is the same individual as in Fig. 1b. Thus the outstanding characteristic of the environmental induction of heritable changes in flax is that all the progenies from an individual inbred plant that has been grown in an inducing environment are similar to one another, and to all the progenies from any other plant that has gone through the same regime, but were different from the original line and other genotrophs, the latter having been produced by another set of environmental treatments.

The genotrophs differed from each other and the original line from which they were derived in a number of characters. These included plant weight and height (Durrant, 1962) (Fig. 1), total nuclear DNA content as determined by Feulgen staining (Evans *et al.*, 1966), hair number on the false septa of the seed capsules (Durrant and Nicholas, 1970), and the isozyme band patterns for peroxidase and acid phosphatase (Fieldes and Tyson, 1973). The alteration of the peroxidase isozyme

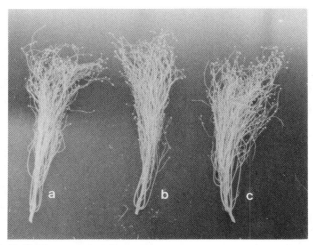

FIG. 2. Individual progenies from three different inbred individuals (a–c) that were all grown under the same inducing conditions. The individual b is the same plant as in Fig. 1b. Thus all three lines are equivalent to C3. The uniformity within each of these inbred lines is similar to that observed among the three lines. These progeny plants were grown under standard conditions in which no changes were observed on Pl.

band pattern was controlled at a single locus, with a dominant and a recessive allele. This gene showed the expected dominance in the F_1 and a 3 : 1 segregation of the dominant : recessive pattern in the F_2 in crosses between genotrophs (Cullis, 1979).

An important note in the alterations seen in both the peroxidase isozyme pattern and the seed capsule septa hair number is that the change seen in the genotrophs was from a dominant character to a recessive. However, the parental line gave no indication that it was heterozygous for either of these characters. Because the recessive phenotype shows up in the first generation after the induction treatment, the change must have taken place in both members of the homologous pair of chromosomes. Whether this change occurs by a conversion event after one of the sites has altered, or by two independent events, is not known.

B. NUCLEAR DNA VARIATION

The changes in the nuclear DNA associated with the environmentally induced heritable changes have been extensively characterized. It was shown by renaturation analysis that the majority of the variation occurred in the highly repeated fraction of the genome (Cullis, 1983). Subsequently, representatives from all the highly repeated sequence families were cloned (Cullis and Cleary, 1986a). This set of cloned probes has been used to further characterize the differences among the genomes of the genotrophs. It has been shown that all but one of the highly repetitive sequence families varied and that the variation in any one family was independent of the variation of any other family (Table 2).

The quantitative variation in the repetitive sequence families cannot be directly linked to any of the phenotypic changes observed in the genotrophs. In addition, the presence of quantitative variation does not lend itself to a determination of the mechanisms by which the variation occurs. However, the hybridization of the probes for the repetitive sequence families to various restriction enzyme digests of DNAs from the genotrophs visualized a series of low-copy-number bands that varied between the genotrophs (Cullis, 1983). One interpretation of these observations was that they represented fragments caused by the rearrangements within or at the ends of these arrays. The only way to confirm this interpretation would be by the isolation of such junction fragments from various genotrophs and their characterization. One sequence family in which this type of information is available is the 5 S RNA gene family.

5 S Ribosomal RNA Gene Family

The genes coding for the 5 S ribosomal RNA (5 S DNA) are one of the variable, highly repetitive sequences. The majority of the 5 S DNA of flax is arranged in tandem arrays with a repeating unit of 350 base pairs (bp). There can be more than 100,000 copies of this sequence in the

TABLE 2

The Quantitative Variation in Repetitive Sequence Families in Flax Genotrophs[a]

Source of DNA	Designation of repetitive family								
	pBG35 (rDNA)	pBG13 (5 S DNS)	pCL21	pCL13	pCL53	pCL2	pCL8	pBG87	pDC7
Pl	1.5	3.0	5.6	2.2	2.3	4.2	2.6	2.5	3.2
C1	1.5	2.0[b]	5.6	1.3[b]	1.5[b]	3.8	2.5	2.1	2.5
C3	0.8[b]	2.2[b]	5.6	1.2[b]	1.6[b]	2.5[b]	2.1[b]	1.9[b]	1.7[b]

[a] The values are given as a percentage of the genome comprising this family. The lines Pl, C1, and C3 are described in Cullis (1981) and are separated by only a single generation. The data are taken from Cullis and Cleary (1986a).

[b] Values which differ significantly (p<0.05) from Pl.

genome, which amounts to more than 3% of the total nuclear DNA. This sequence can vary in copy number by more than 2-fold, from 117,000 copies in Pl (the original line from which the genotrophs were generated) to 49,600 in the genotroph LH (Goldsbrough *et al.*, 1981). *In situ* hybridization has shown that the 5 S RNA genes are dispersed through the genome at many chromosomal sites (Cullis and Creissen, 1987; Schneeberger *et al.*, 1989). The dispersion of the 5 S genes at many sites should result in many junction fragments between the arrays of 5 S genes and neighboring fragments. This has been confirmed by the characterization of a dozen 5 S DNA-containing recombinants isolated from a flax library (Schneeberger *et al.*, 1989). All except one of the phage clones contained both 5 S-related sequences, and non-5 S sequences. In addition, the clones have demonstrated that there are subsets that also have repeating units of differing length, although most are based on multiples of 350 bp.

The use of a series of restriction enzymes, notably *Taq*I and *Bam*HI, led to the identification of at least two subsets of the 5 S DNAs. These two subsets appeared to be modulated differentially during the induction process, with one varying rapidly while the other remained more constant (Cullis and Cleary, 1986a). The way in which various subsets of this family are affected in the induction process can be detailed using the currently isolated variants. Even when the molecular mechanisms by which the 5 S genes are modulated are described, it will still be important to identify what features are recognized that allow the discrimination between apparently identical arrays of a repetitive sequence.

One of the subsets of 5 S genes that varies among the genotrophs has been characterized further. This family has a 700-bp repeating unit which has only two regions related to the 5 S RNA coding region within the repeat. Thus it appears to be derived from a dimer of the prevalent repeating unit. There are about 1000 copies of this particular family in the flax genome (Schneeberger and Cullis, 1990). When the organization of this family was determined in a range of genotrophs, a number of polymorphisms were observed (Fig. 3). It can be seen that the position of the new bands varied among the genotrophs, with new polymorphic bands appearing in different genotrophs. The bands that were lost from the original line were not the same in all the genotrophs. However, one consistent pattern was observed for four independently derived small genotrophs, showing that there was some common determinant in each of these sets of changes. In addition, this altered pattern remained stable within the small genotrophs through subsequent generations, and even through a cycle of tissue culture and regeneration. Prelimi-

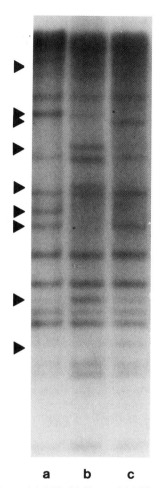

a   b   c

FIG. 3. DNA from three lines, (a) LH, (b) Sh, and (c) Pl, digested with the restriction endonuclease *Eco*RI, transferred to a nylon filter, and hybridized with a variant 5 S-related gene family. This family is present in about 1000 copies/genome. The polymorphic bands are indicated by arrowheads. The pattern seen in b is identical to that seen in all four independently produced small genotrophs.

nary analysis of the segregation of these polymorphic bands indicated that most of them were linked (less than 10 map units spanning the whole set). The next question asked was whether these same bands were present in natural populations of flax and linseed. A comparison of a number of cultivars showed that most of the polymorphic bands could be observed in a range of cultivars (Schneeberger and Cullis, 1990).

Thus it appears that during the induction of heritable changes the variation generated is part of that which can occur naturally.

The recreation of the extent of naturally occurring variation was also evident in the quantitative variation in the repetitive sequence families. The copy number changes generated in the genotrophs appeared to span the range observed in other cultivars and closely related species (Cullis and Cleary, 1986a). This cannot be taken to mean that all the differences within the range of *Linum usitatissimum* cultivars, and in *Linum bienne,* were generated in the genotrophs. The genotrophs were much more similar to each other than to any other cultivar for all the low-copy-number probes that have been tested for RFLPs (unpublished observations). It is only in that part of the genome in which changes are generated that almost the whole range of viable variability is recreated. In this regard, the polymorphisms observed in soybean tissue cultures also appeared to be the same as those observed in natural populations.

Is there any relation between the changes observed in the environmental induction in flax and somaclonal variation? The results with flax would be supportive of such a relationship. The passage of Stormont Cirrus, the original line from which all the genotrophs were derived, through a cycle of culture and regeneration resulted in both phenotypic and genomic changes (Cullis and Cleary, 1986b). The same set of repetitive sequence families was altered, as had been found in the environmentally induced changes. However, the frequency with which a given family altered was different among the two sets of results. Thus the lability of the genome, and fractions therein, may be dependent on the particular physiological status of the cell or organism at the time the stress is applied.

VI. Conclusions

The rapidity with which the plant genome can vary has been demonstrated in a number of systems. The types of changes observed include transpositions, amplifications, deletions, and rearrangements. It has also become evident that the rate at which such variation arises is also dependent on the "environment" of the DNA. The question addressed was whether these alterations, which occur at an elevated rate, were random within the genome. This question can be split into two parts. First, are all the mechanisms by which variation can occur equally susceptible to "environmental" stresses, and second, are all parts of the genome equally accessible?

The responses of transposable element families under various environmental stresses are relevant to the first part of the question. In *D. melanogaster*, transposable elements can dramatically increase mutation rates under certain conditions. In fact, environmental factors, especially temperature, can play a role in the frequency of transposition (Bregliano and Kidwell, 1983). The suppression of mutator activity generated by transposable elements, with the reactivation under certain circumstances, is consistent with the notion that the mutation rates can vary markedly at different times and under different conditions. Much less information concerning the variation in the rates of amplification, deletions, and rearrangements by mechanisms not directly attributable to the activity of transposable elements is available, with the exception of that in flax.

Not only can the rates at which variations arise be affected by environmental conditions, but also the spectrum of sequences involved. Once again, the examples of transposable elements are instructive. The activity of transposable elements in maize either when infected with brome mosaic virus (Mottinger *et al.*, 1984) or during tissue culture (Lee and Phillips, 1987) was in a subset and not in all transposable element families equally. Thus, within the genome, there is a selection for some parts that appear to be more labile. In flax, the separation is even more dramatic. Not all the repetitive sequence families were equally susceptible, but even within a repetitive sequence family different subsets were affected unequally. The relative magnitude of changes in the various repetitive sequence families varied among induction experiments involving whole plants and the changes seen in tissue culture (Cullis and Cleary, 1986b).

The relationship between environmental stress and genetic reorganizations that may underpin the evolution of new adaptations is still tenuous. In a stressful environment, under which organisms were severely restricted but able to survive for at least a limited time, a high rate of mutational events induced by the same environment could generate the genomic reorganization underlying major adaptive shifts. The limitation of these reorganization events to subsets of the genome, particularly those containing major genes controlling polygenically determined quantitative traits, would increase the number of viable variants compared with random reorganization of the genome. Once new adaptive combinations had arisen, the environment would no longer be perceived as stressful. The mechanisms by which the genomic restructuring occurred would then be quiescent, or of greatly reduced activity, promoting stability of the genome. However, the mechanisms for generating variation would still be present, to be triggered into action by further cycles of environmental stress.

A recent controversy over the origin and nature of mutants in bacteria has arisen. On the one hand, some experimental results have been interpreted as evidence for a form of directed mutation (Cairns et al., 1988; Hall, 1988). However, some objections to this interpretation have been made, and further information and experimental data will be required for the satisfactory interpretation of this phenomenon. The possibility of environmentally induced mutations in a special subset of the genome, even in bacteria, is clearly a current issue.

The variables considered as defining stress, in terms of genomic variability, have ranged from nutrients supplied to temperature to interspecific crosses. Clearly, it is vital to define as precisely as possible the nature of the stresses (their intensity, duration, and stage of the life cycle at which they act) that are associated with genomic restructuring. Are all stresses equally effective in eliciting genomic reorganization? Are any of the anthropogenic atmospheric variations likely to lead to genomic reorganization in a significant range of species? The effect of any specific perturbation would also be dependent on all the other constituent parameters of the environment. The stresses most effective at generating genomic reorganization may be those that occur at climatic and ecological margins, where a small environmental change could lead to lethality.

The exact nature and magnitude of stresses associated with genomic restructuring need to be defined as precisely as possible. The ideal environment for this is one in which fertile survivors are present yet major reorganizations occur. The best system currently available is the environmental induction of heritable changes in flax. Here reproducible changes have been documented and occur without any observable lethality. A major drawback is the lack of precise definition of the inducing conditions and the question as to which part of any stress is really responsible for the resultant reorganization. However, since the genetic material is available and because specific labile fractions of the genome can be monitored and identified, refinements in the understanding of the nature, duration, and timing of environmental stresses in the generation of genomic change are possible.

The existence of rapid reorganizations of the genome in response to environmental stress is clearly documented, but the ways in which the organism perceives the stress and responds are unknown. In flax, there is evidence that there is genetic control in the ability to respond to the specific sets of inducing conditions (Durrant and Timmis, 1983; Joarder et al., 1975; Al-Saheal and Larik, 1987). A characterization of the number of genes involved in controlling the response, and their subsequent isolation, would help in understanding how the genomic responses to environmental stresses are mediated. At another level, what is the pathway by which the environmental stress is perceived and transduced to activate the mechanisms by which genomic restructuring occurs? Are any of the physiological responses that are part of an organism's normal response to fluctuating environment involved, or is it necessary to invoke new specificities, which are activated only under conditions that produce a specific physiological state? If the latter is true, it would then suggest that not all stresses would be equally effective at inducing genomic restructuring, and some would be totally ineffective in spite of being clearly suboptimal or, in the extreme, lethal to the organism.

Genotypic and phenotypic variability are highest under conditions of environmental stress. The role of the environment in the generation of this variability is of central importance, and the evidence is accumulating that it can have a direct role. The general conclusion from these studies is that any shift in the environment to which an organism has become adapted should increase the variability. This conclusion needs to be tested further to define the genetic components involved in such responses and to assess the generality of the conclusion that environmental stresses can invoke genomic responses.

REFERENCES

Al-Saheal, Y. A., and Larik, A. S. (1987). Genetic control of environmentally induced DNA variation in flax genotrophs. *Genome* **29**, 643–646.

Bachmann, K., Goin, O. B., and Goin, C. J. (1972). Nuclear DNA amounts in vertebrates. *Brookhaven Symp. Biol.* **23**, 419–450.

Banks, J. A., Masson, P., and Federoff, N. (1988). Molecular mechanisms in the developmental regulation of the maize suppressor–mutator transposable element. *Genes Dev.* **2**, 1364–1380.

Bennett, M. D. (1972). Nuclear DNA content and minimum generation time in herbaceous plants. *Proc. R. Soc. London, B* **181**, 109–135.

Bennett, M. D. (1976). DNA amount, latitude and crop plant distribution. *Curr. Chromosome Res., Proc. Kew Chromosome Conf. 1976*, 151–158.

Bennett, M. D., and Smith, J. B. (1976). Nuclear DNA amounts in angiosperms. *Philos. Trans. R. Soc. London, B* **274**, 227–240.

Bingham, P. M., Kidwell, M. G., and Rubin, G. M. (1982). The molecular basis of P–M hybrid dysgenesis: The role of the P element, a P-strain-specific transposon family. *Cell* **29**, 995–1004.

Blundy, K. S., Cullis, C. A., and Hepburn, A. G. (1987). Ribosomal DNA methylation in a flax genotroph and a crown gall tumor. *Plant Mol. Biol.* **8**, 217–225.

Bregliano, J. C., and Kidwell, M. G. (1983). Hybrid dysgenesis determinants. *In* "Mobile Genetic Elements" (A. Shapiro, ed.), pp. 363–410. Academic Press, New York, 1983.

Buss, L. W. (1983). Evolution, development, and the units of selection. Proc. Natl. Acad. Sci. U.S.A. **80**, 1387–1391.

Cairns, J., Overbaugh, J., and Miller, S. (1988). The origin of mutants. *Nature (London)* **335**, 142–145.

Carpenter, R., Hudson, A., Robbins, T., Almeida, J., Martin, C., and Coen, E. (1988). Genetic and molecular analysis of transposable elements in *Antirrhinum majus. In* "Plant Transposable Elements" (O. Nelson, ed.), pp. 69–80. Plenum, New York.

Cavalier-Smith, T. (1978). Nuclear volume control by nucleoskeletal DNA, selection for cell volume and cell growth rate, and the solution of the DNA C-value paradox. *J. Cell Sci.* **34**, 247–278.

Cavallini, A., Zolfino, C., Cionini, G., Cremonini, R., Natali, L., Sassoli, O., and Cionini, P. G. (1986). Nuclear DNA changes within Helianthus annus L: Cytophotometric, karyological and biochemical analyses. *Theor. Appl. Genet.* **73**, 20–26.

Cavallini, A., Zolfino, C., Natali, L., Cionini, G., and Cionini, P. G. (1989). Nuclear DNA changes within Helianthus annuus L: Origin and control mechanism. *Theor. Appl. Genet.* **77**, 12–16.

Chandler, V., Rivin, C., and Walbot, W. (1986). Stable non-mutator stocks of maize have sequences homologous to the Mu1 transposable element. *Genetics* **114**, 1007–1021.

Cullis, C. A. (1979). Quantitative variation of ribosomal RNA genes in flax genotrophs. *Heredity* **42**, 237–246.

Cullis, C. A. (1981). Environmental induction of heritable changes in flax: Defined environments inducing changes in rDNA and peroxidase isozyme band pattern. *Heredity* **47**, 87–94.

Cullis, C. A. (1983). Environmentally induced DNA changes in plants. *CRC Crit. Rev. Plant Sci.* **1**, 117–129.

Cullis, C. A. (1986). Phenotypic consequences of environmentally induced changes in plant DNA. *Trends Genet.* **2**, 307–309.

Cullis, C. A. (1987). The generation of somatic and heritable variation in response to stress. *Am. Nat.* **130**, S62–S73.

Cullis, C. A., and Charlton, L. M. (1981). The induction of ribosomal DNA changes in flax. *Plant Sci. Lett.* **20**, 213–217.

Cullis, C. A., and Cleary, W. (1986a). Rapidly varying DNA sequences in flax. *Can. J. Genet. Cytol.* **28**, 252–259.

Cullis, C. A., and Cleary, W. (1986b). DNA variation in flax tissue culture. *Can. J. Genet. Cytol.* **28**, 247–251.

Cullis, C. A., and Creissen, G. P. (1987). Genomic variation in plants. *Ann. Bot.* **60** (Suppl. 4), 103–113.

Dellaporta, S. L., and Chomet, P. S. (1985). The activation of maize controlling elements. *Plant Gene Res.* **2**, 169–216.

Donn, G., Tischer, E., Smith, J. A., and Goodman, H. M. (1984). Herbicide resistant alfalfa cells: An example of gene amplification in plants. *J. Mol. Appl. Genet.* **2**, 621–635.

Durrant, A. (1958). Environmental conditioning of flax. *Nature (London)* **181**, 928–929.

Durrant, A. (1962). The environmental induction of heritable changes in *Linum. Heredity* **17**, 27–61.

Durrant, A., and Nicholas, D. B. (1970). An unstable gene in flax. *Heredity* **25**, 513–527.

Durrant, A., and Timmis, J. N. (1973). Genetic control of environmentally induced changes in *Linum. Heredity* **30**, 369–379.

Evans, D. A., Sharp, W. R., and Medina-Filho, H. P. (1984). Somaclonal and gametoclonal variation. *Am. J. Bot.* **71**, 759–774.

Evans, G. M., Durrant, A., and Rees, H. (1966). Associated nuclear changes in the induction of flax genotrophs. *Nature (London)* **212**, 697–699.

Fieldes, M.A., and Tyson, H. (1973). Activity and relative mobility of peroxidase and esterase isozymes of flax (*Linum usitatissimum*) genotrophs. I. Developing main stems. *Can. J. Genet. Cytol.* **15**, 731–744.

Flavell, R. B. (1980). The molecular characterization and organization of plant chromosomal DNA. *Annu. Rev. Plant Physiol.* **31**, 569–596.

Flavell, R. B. (1982). Amplification, deletion and rearrangement: Major sources of variation during species divergence. *In* "Genome Evolution" (G. A. Dover and R. B. Flavell, eds.), pp. 301–324. Academic Press, New York.

Gerstel, D. U., and Burns, J. A. (1966). Chromosomes of unusual length in hybrids between two species of *Nicotiana*. *Chromosomes Today* **1**, 41–56.

Goldsbrough, P. B., Ellis, T. H. N., and Cullis, C. A. (1981). Organization of the 5S RNA genes in flax. *Nucleic Acids Res.* **9**, 5895–5904.

Groose, R. W., and Bingham, E. T. (1986). An unstable anthocyanin mutation recovered from tissue culture of alfalfa (Medicago sativa). 1. High frequency of reversion upon reculture. *Plant Cell Rep.* **5**, 104–107.

Hall, B. G. (1988). Adaptive evolution that requires multiple spontaneous mutations. I. Mutations involving an insertion sequence. *Genetics* **120**, 887–897.

Hendrix, S. D. (1979). Compensatory reproduction in a biennial herb following insect defloration. *Oecologia* **42**, 107–118.

Joarder, I. O., Al-Saheal, Y., Begum, J., and Durrant, A. (1975). Environments inducing changes in the amount of DNA in flax. *Heredity* **34**, 247–253.

Johnson, S. S., Phillips, R. L., and Rines, H. W. (1987). Possible role of heterochromatin in chromosome breakage induced by tissue culture in oats. (*Avena sativa* L.) *Genome* **29**, 439–446.

Klekowski, E. J. J. (1988). "Mutation, Developmental Selection and Plant Evolution." Columbia University Press, New York.

Lapitan, N. L. V., Sears, R. G., and Gill, B. S. (1984). Translocations and other karyotypic structural changes in wheat × rye hybrids regenerated from tissue culture. *Theor. Appl. Genet.* **68**, 547–554.

Larkin, P. J., and Scowcroft, W. (1987). Somaclonal variation and crop improvement. *In* "Genetic Engineering of Plants" (T. Kosuge, C. P. Meredith, and A. Hollaender, eds.), pp. 289–314. Plenum, New York.

Lee, M., and Phillips, R. L. (1987). Genomic rearrangements in maize induced by tissue culture. *Genome* **29**, 122–128.

Lubchenko, J., and Cubit, J. (1980). Heteromorphic life histories of certain marine algae as adaptations to variations in herbivory. *Ecology* **61**, 676–687.

Mackay, T. F. C. (1985). Transposable element-induced response to artificial selection in *Drosophila melanogaster*. *Genetics* **111**, 351–374.

McClintock, B. (1950). The origin and behavior of mutable loci in maize. *Proc. Natl. Acad. Sci. U.S.A.* **36**, 344–355.

McClintock, B. (1951). Chromosome organization and genetic expression. *Cold Spring Harbor Symp Quant. Biol.* **16**, 13–47.

McClintock, B. (1978). Mechanisms that rapidly reorganize the genome. *Stadler Symp.* **10**, 25–48.

McClintock, B. (1984). The significance of responses of the genome to challenge. *Science* **26**, 792–801.

Miksche, J. P. (1967). Variation in DNA content of several gymnosperms. *Can. J. Genet. Cytol.* **9**, 717–722.

Miksche, J. P. (1968). Quantitative study of intraspecific variation of DNA per cell in

Picea glauca and *Pinus banksiana. Can. J. Genet. Cytol.* **10,** 590–600.

Miksche, J. P. (1985). Recent advances of biotechnology and forest trees. *For. Chron.* **61,** 449–453.

Mottinger, J. P., Johns, M. A., and Freeling, M. (1984). Mutations of the AdhI gene in maize following infection with barley stripe mosaic virus. *Mol. Gen. Genet.* **195,** 367–369.

Myerowitz, E. M. (1987). *Arabidopsis thaliana. Annu. Rev. Genet.* **21,** 93–111.

Myerowitz, E. M., and Pruitt, R. E. (1985). *Arabidopsis thaliana* and plant molecular genetics. *Science* **229,** 1214–1218.

Orton, T. J. (1983). Experimental approaches to the study of somaclonal variation. *Plant Mol. Biol. Rep.* **1,** 67–76.

Orton, T. J. (1984). Somaclonal variation: Theoretical and practical considerations. *In* "Genetic Manipulation and Plant Improvement" (J. P. Gustafson, ed.). Plenum, New York.

Parsons, P. A. (1988). Evolutionary rates: Effects of stress upon recombination. *Biol. J. Linn. Soc.* **35,** 49–68.

Pesche, V. M., Phillips, R. L., and Gegenbach, B. G. (1987). Discovery of transposable element activity among progeny of tissue culture-derived maize plants. *Science* **238,** 804–807.

Price, H. J. (1976). Evolution of DNA content in higher plants. *Bot. Rev.* **42,** 27–52.

Price, H. J. (1988). Nuclear DNA content variation within angiosperm species. *Evol. Trends Plants* **2,** 53–60.

Price, H. J., Bachmann, K., Chambers, K. L., and Riggs, J. (1980). Detection of intraspecific variation in nuclear DNA content in *Microseris douglasii. Bot. Gaz. (Chicago)* **141,** 195–198.

Price, H. J., Chambers, K. L., and Bachmann, K. (1981). Geographic and ecological distribution of genomic DNA content variation in *Microseris douglasii* (Asteraceae). *Bot. Gaz. (Chicago)* **142,** 415–426.

Price, H. J., Chambers, K. L., Bachmann, K., and Riggs, J. (1983). Inheritance of nuclear 2C DNA content variation in intraspecific and interspecific hybrids of *Microseris* (Asteraceae). *Am. J. Bot.* **70,** 1133–1138.

Price, H. J., Chambers, K. L., Bachmann, K., and Riggs, J. (1985). Inheritance of nuclear 2C DNA content in a cross between *Microseris douglasii* and *M. bigelovii* (Asteraceae). *Biol. Zentralbl.* **104,** 269–276.

Price, H. J., Chambers, K. L., Bachmann, K., and Riggs, J. (1986). Patterns of mean nuclear DNA content in *Microseris douglasii* (Asteraceae) populations. *Bot. Gaz. (Chicago)* **147,** 496–507.

Quemada, H., Roth, E. J., and Lark, K. G. (1987). Changes in methylation of tissue cultured soybean cells detected by digestion with the restriction enzymes HpaII and MspI. *Plant Cell Rep.* **6,** 63–66.

Roth, E. J., Frazier, B. L., Apuya, N. R., and Lark, K. G. (1989). Genetic variation in an inbred plant: Variation in tissue cultures of soybean (*Glycine max* (L.) Merrill). *Genetics* **121,** 359–368.

Saedler, H., Schwarz-Sommer, Z., and Gierl, A. (1985). The role of plant transposable elements in gene evolution. *In* "Plant Genetics" (M. Freeling, ed.), pp. 271–282. Liss, New York.

Schneeberger, R., and Cullis, C. A. (1990). Submitted.

Schneeberger, R., Creissen, G. P., and Cullis, C. A. (1989). Chromosomal and molecular analysis of 5S RNA gene organization in the flax, *Linum usitatissimum. Gene* **83,** 75–84.

Shah, D. M., Horsch, R. B., Klee, H. J., Kishore, G. M., Winter, J. A., Tumer, N. E., Hironaka, C. M., Sanders, P. R., Gasser, C. S., Aykent, S., Siegel, N. R., Rogers, S. G., and Fraley, R. T. (1986). Engineering herbicide tolerance in transgenic plants. *Science* **233**, 478–481.

Sparrow, A. H., Price, H. J., and Underbrink, A. G. (1972). A survey of DNA content per cell and per chromosome of prokaryotic and eukaryotic organisms: Some evolutionary considerations. *Brookhaven Symp. Biol.* **23**, 451–494.

Starlinger, P., Courage, U., Doring, H.-P., Frommer, W.-B., Kunze, R., Laird, A., Merckelbach, A., Muller-Neumann, M., Tillmann, E., Werr, W., and Yoder, J. (1985). Plant transposable elements—Factors in the evolution of the maize genome? *In* "Plant Genetics" (M. Freeling, ed.), pp. 251–270. Liss, New York.

Turkington, R. (1983). Plasticity in growth and patterns of dry matter distribution of two genotypes of Trifolium repens grown in different environments of neighbours. *Can. J. Bot.* **61**, 2186–2194.

Vogel, S. (1968). "Sun leaves" and "shade leaves": Differences in convective heat dissipation. *Ecology* **49**, 1203–1204.

Walbot, V. (1986). On the life strategies of plants and animals. *Trends Genet.* **1**, 165–169.

Walbot, V., and Cullis, C. A. (1983). The plasticity of the plant genome—Is it a requirement for success? *Plant Mol. Biol. Rep.* **1**, 3–11.

Walbot, V., and Cullis, C. A. (1985). Rapid genomic change in higher plants. *Annu. Rev. Plant Physiol.* **36**, 367–396.

Weismann, A. (1883). "On Heredity." Clarendon, Oxford.

Whitham, T. G., and Slobodchikoff, C. N. (1981). Evolution by individuals, plant–herbivore interactions, and mosaics of genetic variability: The adaptive significance of somatic mutations in plants. *Oecologia* **49**, 287–292.

Ye, J., Hauptman, R. M., Smith, A. G., and Widholm, J. M. (1987). Selection of a *Nicotiana plumbaginifolia* universal hybridizer and its use in intergenic somatic hybrid formation. *Mol. Gen. Genet.* **208**, 474–480.

Zheng, K. L., Castiglione, S., Biasini, M. G., Biroli, A., Morandi, C., and Sala, F. (1987). Nuclear DNA amplification in cultured cells of *Oryza sativa* L. *Theor. Appl. Genet.* **74**, 65–70.

MOLECULAR GENETICS OF COLD ACCLIMATION IN HIGHER PLANTS

Michael F. Thomashow

Department of Crop and Soil Sciences and
Department of Microbiology and Public Health,
Michigan State University, East Lansing, Michigan 48824

I. Introduction

Plants have evolved mechanisms that allow them to acclimate to a variety of environmental "stresses" (Key and Kosuge, 1984; Lange *et al.*, 1981; Levitt, 1980). Cold temperature is one such condition. Perhaps the most dramatic manifestation of cold acclimation, or cold hardening, is the increased freezing tolerance that occurs in many plant species (see Levitt, 1980; Sakai and Larcher, 1987). The woody perennials birch and dogwood offer an extreme example. Whereas nonacclimated trees are severely injured or killed by temperatures of about −10°C, trees that are fully cold acclimated can survive experimental freezing temperatures of −196°C; in their natural environments, these trees are often exposed to, and survive, temperatures of −40 to −50°C. Other plants, including a number of important crop species, can attain intermediate levels of frost hardiness. For example, nonacclimated wheat and rye are killed at temperatures between −5 and −10°C, but after hardening, wheat can survive temperatures of about −15 to −20°C and hardened rye can survive from −25 to −30°C.

A number of biochemical changes have been shown to occur in plants during cold acclimation. Common examples include alterations in lipid

ADVANCES IN GENETICS, Vol. 28

composition, the appearance of new isozymes, increased sugar and soluble protein content, and increased levels of proline and other organic acids (see Levitt, 1980; Sakai and Larcher, 1987; Steponkus, 1984). Some of these changes, such as the alterations in lipid composition, appear to have roles in bringing about the increased frost tolerance of acclimated plants (see Steponkus and Lynch, 1989). These and the other changes also potentially comprise or mediate modifications that increase the overall fitness of the plant for low-temperature survival. In most cases, however, the precise role that each change has in the cold acclimation process is not yet certain.

The biochemical, biophysical, and physiological changes that occur in plant cells during cold acclimation could be brought about by preexisting macromolecules and structures—enzymes, structural proteins, lipids, membranes—that undergo changes in their physical properties at low temperatures. It is also possible, as first proposed by Weiser (1970), that cold acclimation involves changes in gene expression. The finding that cycloheximide, an inhibitor of protein synthesis, can prevent the increased frost tolerance that normally occurs in cold-treated *Brassica napus* (Kacperska-Palacz *et al.*, 1977), wheat (Trunova, 1982), and *Solanum commersonii* (Chen *et al.*, 1983) is consistent with this notion. Moreover, *in vitro* translation studies and the isolation of cDNA clones containing inserts from *cold-regulated* (*cor*) genes have provided direct evidence that changes in gene expression occur during cold acclimation (see Sections III,B,3 and III,B,4). Thus, low temperatures, like the other environmental conditions covered in this volume, can induce "genomic responses" in plants. Determining how these responses are triggered by low temperatures and establishing the roles that they have in cold acclimation are two major questions that are now beginning to be addressed.

The identification and characterization of genes that have roles in cold acclimation are fundamental to our knowledge and understanding of the cold-hardening process. In this article, I will review what is known about the genetics of cold acclimation, focusing primarily on the recent molecular genetic studies regarding cold-acclimation-associated changes in gene expression. These studies, although only in their infancy, have yielded significant results and suggest future areas of research. I will begin, however, with an overview of the biochemistry of cold acclimation, as this information is required for discussion of potential *cor* gene function. For comprehensive treatments of the biochemistry and physiology of cold acclimation the reader is referred to a number of excellent recent review articles (Burke *et al.*, 1976; Graham and Patterson, 1982; Steponkus, 1984; Steponkus and Lynch, 1989) and book volumes (Levitt, 1980; Li and Sakai, 1978, 1982; Lyons *et al.*, 1979; Sakai and Larcher, 1987).

II. Biochemistry of Cold Acclimation

Much of the research on cold acclimation has been directed toward understanding the mechanism(s) responsible for the increased frost tolerance of acclimated plants. Toward this end, many of the initial studies focused on determining the critical forms of injury that occur in plants during a freeze–thaw cycle. In 1912, Maximov suggested that disruption of cellular membranes, particularly the plasma membrane, was the primary cause of freezing injury in plants. Subsequent studies have supported this view (see Levitt, 1980; Steponkus, 1984). Further, they have indicated that membrane damage results primarily from the severe dehydration that occurs during a freeze–thaw cycle. Thus, tolerance to freezing must include a tolerance to dehydration stress. What then are the forms of membrane damage that occur during freeze-induced dehydration and what mechanisms are responsible for tolerance?

"Expansion-induced lysis," one form of cell and membrane injury, occurs in response to relatively high freezing temperatures, about -3 to $-7°C$ (see Steponkus, 1984). As temperatures drop below $0°C$, the extracellular water of the plant begins to freeze, resulting in a lowered water activity and an increased solute concentration in the extracellular spaces. In response to these changes in chemical and osmotic potentials, water moves out of the cells, causing severe dehydration and cell shrinkage. When the extracellular ice melts, the cells rehydrate and expand. If the cells are to survive, their plasma membranes must be able to withstand the efflux and influx of water. This occurs in cells that can frost harden, but nonacclimated cells lyse.

A critical difference between the membranes from cold-acclimated and nonacclimated plant cells appears to be that membrane material is "conserved" during a freeze–thaw cycle in hardened cells. In an elegant series of experiments, Steponkus and colleagues (Gordon-Kamm and Steponkus, 1984a,b; Dowgert and Steponkus, 1984; Dowgert *et al.*, 1987) showed that protoplasts prepared from acclimated rye seedlings had a much greater potential to reexpand after a freeze–thaw cycle than did protoplasts isolated from nonacclimated plants. This difference resulted from a change in the cryobehavior of the membranes in the acclimated cells. Upon freezing to about $-5°C$, the nonacclimated protoplasts dehydrated and the plasma membranes formed endocytotic vesicles. When the protoplasts were thawed, the vesicular material was not reincorporated into the plasma membranes and, consequently, rehydration resulted in an intolerable osmotic pressure and the protoplasts burst. When the cold-acclimated protoplasts were frozen, they, too, dehydrated, but instead of forming endocytotic vesicles, the plasma membranes formed exocytotic extrusions. The important difference was that these membrane extrusions remained in association with plasma membrane and were reincorporated during rehydration. Thus, the protoplasts could swell to their original size.

Whether the plasma membranes from whole plant cells behave in precisely the same way as those from protoplasts is open to question (see, e.g., Pearce and Willison, 1985; Johnson-Flanagan and Singh, 1986; Singh et al., 1987). For example, instead of extracellular extrusions, hardened whole cells of B. napus, alfalfa, and rye produced many more osmotically active plasmalemma strands during plasmolysis than did nonhardened cells (see, e.g., Johnson-Flanagan and Singh, 1986; Singh et al., 1987). These strands appeared to tether the protoplast to the cell wall in the plasmolyzed cells. The critical point, however, was that the plasmalemma strands, like extracellular extrusions, were reincorporated back into the expanding protoplasts during deplasmolysis. Thus, whereas the nonacclimated cells suffered expansion-induced lysis, the acclimated cells did not.

Other forms of damage have also been shown to occur in the plasma membranes of nonacclimated cells as they are frozen to subzero temperatures of $-10°C$ and below. At these temperatures, the dehydration experienced by the cells is, of course, more severe than freezing to $-5°C$, and injury is generally manifested as a loss of osmotic responsiveness (see Steponkus, 1984; Steponkus and Lynch, 1989). Examination of the membranes from such cells has revealed the formation of lateral phase separations, aparticulate lamellae, lamellar-to-hexagonal-II phase transitions, and multilamellar vesicles (see, e.g., Gordon-Kamm and Steponkus, 1984a,b; Pearce and Willison, 1985; Johnson-Flanagan and Singh, 1986; Singh et al., 1987). Similar membrane damage is suffered by nonacclimated cells subjected to equivalent degrees of dehydration induced by suspension in hypertonic solutions at 0°C, indicating that the injury is due to dehydration rather than low temperature per se. In cold-acclimated cells, these forms of injury are greatly reduced. For example, whereas osmotic nonresponsiveness occurs when protoplasts from nonacclimated rye leaves are frozen to $-10°C$, protoplasts isolated from acclimated leaves can be frozen to $-10°C$ without injury (Gordon-Kamm and Steponkus 1984b). Protoplasts from acclimated rye leaves do become osmotically nonresponsive when frozen to about -25 to $-30°C$, but even when frozen to $-35°C$, the plasma membranes do not undergo lamellar-to-hexagonal-II phase transitions (Gordon-Kamm and Steponkus, 1984b).

The differences in the biophysical properties of plasma membranes from cold-acclimated and nonacclimated plants suggests that the biochemical composition of the plasma membrane is altered during cold hardening. This is indeed the case (see Steponkus, 1984; Steponkus and Lynch, 1989). For example, Uemura and Yoshida (1984) and Yoshida and Uemura (1984), working with rye seedlings and orchard grass, respectively, found that plasma membranes from acclimated and nonacclimated plants had different protein profiles and that those from acclimated plants had significant increases in phospholipid-to-protein ratios. Lynch and Steponkus (1987) made a detailed analysis of the lipids from acclimated and nonacclimated rye seedlings and found that there were increases in both free sterols (from 33 to 44 mol%) and phospholipids (from 32 to 41 mol%) and decreases in steryl glucosides (from 15 to 6 mol%), acylated steryl glucosides (from 4 to 1 mol%), and glucocerebrosides (from 16 to 7 mol%). In addition, the levels of diunsaturated molecular species of phosphatidylcholine and phosphatidylethanolamine doubled in cold-acclimated cells.

Efforts are now being directed toward establishing which biochemical changes are responsible for the differences in the freezing tolerance of plasma membranes isolated from cold-acclimated and nonacclimated plant cells. Toward this end, Steponkus et al. (1988) have made exciting progress. They have shown that the freezing tolerance of protoplasts isolated from nonacclimated rye leaves can be increased by modifying the lipid composition of the plasma membrane. In particular, they found that they could eliminate expansion-induced lysis in nonacclimated protoplasts by elevating the levels of monounsaturated and diunsaturated molecular species of phosphatidylcholine. Significantly, the modified protoplasts formed extracellular extrusions instead of endocytotic vesicles following moderate dehydration. Whether this is the only plasma membrane alteration that can abolish expansion-induced lysis, and whether this alteration has a role in cold-acclimation-induced frost tolerance in other plant species, remain to be determined. It should also be noted that the enrichment of mono- and diunsaturated molecular species of phosphatidylcholine did not impart freezing tolerance at $-10°C$, the point where dehydration is severe enough to cause lateral phase separations, lamellar-to-hexagonal-II phase transitions, and osmotic unresponsiveness. Thus, the freezing tolerance of plasma membranes would appear to involve multiple alterations, each having different effects. Nevertheless, the results of Steponkus *et al.* (1988) provide strong evidence that changes in plasma membrane lipid composition are causally related to at least one facet of the cold acclimation process.

One additional point regarding freezing-induced cell and membrane damage is that other forms of stress may also contribute to injury besides severe dehydration. Specifically, Olien (1973, 1974) has presented a thermodynamic analysis of equilibrium freezing in complex interfaces between ice and hydrophilic substances and concluded that competition for interfacial liquid water causes an "energy of adhesion" to develop. It was proposed that this energy would be of sufficient magnitude to cause significant injury to cells when temperatures fell to $-10°C$ and below. Evidence in support of this hypothesis has been presented (Olien and Smith, 1977). Further, it has been proposed that freeze-inhibitor polysaccharides produced by various cereals afford protection against this form of freezing injury (Olien, 1965; Shearman *et al.*, 1973).

Finally, studies on the biochemistry of cold acclimation have indicated that a number of changes other than membrane alterations also occur in acclimated plants. The appearance of new isozymes, increased sugar and soluble protein content, and increased levels of proline and other organic acids are common examples (see Levitt, 1980; Li and Sakai, 1982; Sakai and Larcher, 1987). The precise role that each of these changes has in the cold acclimation process, however, is not certain. Some changes may contribute directly to the frost tolerance of acclimated cells. Indeed, Volger and Heber (1975) have reported that certain soluble polypeptides that accumulate in cold-acclimated spinach have cryoprotective effects, as do proline and a number of simple sugars (Crowe and Crowe, 1986; Carpenter and Crowe, 1988; Caffrey *et al.*, 1988). Whether these molecules have important cryoprotective roles *in vivo* and contribute significantly to freezing tolerance remains to be determined. Other biochemical changes may contribute to the overall fitness of the plant for low-temperature environments, which in turn could indirectly affect frost hardiness: new isozymes might have properties that better balance the activities of metabolic pathways at low temperature, increased concentrations of sugars could provide critical energy sources, and changes in membrane composition could improve photosynthesis, respiration, and transport. Again, however, the direct cause-and-effect relationships of these changes and their significance to low-temperature survival and fitness remain to be established.

III. Genetics of Cold Acclimation

A. INHERITANCE OF FROST HARDINESS

The inheritance of the ability to frost harden has been examined in a wide range of plants, with the most detailed body of knowledge existing for wheat. The first studies were those of Nilsson-Ehle (1912), who reported the results of a cross between two varieties of winter wheat that were intermediate in their abilities to frost harden. The results indicated transgressive segregation for the frost hardiness character and it was concluded that frost hardiness was a quantitative trait controlled by several genes. Subsequent studies with wheat conducted by Hayes and Aamodt (1927), Worzella (1935), and Quisenberry (1931) have confirmed this notion.

Additional, more detailed genetic analyses have indicated that the gene interactions responsible for frost hardiness in wheat are quite complex. Gullord (1975) and Gullord *et al.* (1975) examined the material from two complete diallels of winter wheat (*Triticum aestivum* L.), one having six parental genotypes and the other four genotypes. They assessed the freezing tolerance of cold-acclimated plants using two different protocols, a "low-intensity" freeze (the crowns of the plants were kept dry and frozen to -15 to $-22°C$) and a "high-intensity" freeze (the crowns of the plants were submerged in water and frozen to -11 to $-15°C$). From the analysis of the data, the investigators concluded that frost tolerance involved the action of several partially dominant genes that were mostly additive in their effects. In addition, the data suggested that the genes controlling frost tolerance under high-intensity freezing were different from the genes controlling freezing hardiness under low-intensity freezing. This latter conclusion is intriguing in light of the work of Steponkus *et al.* (1988) discussed previously, indicating that frost tolerance in acclimated rye protoplasts involves at least two different mechanisms.

In studies analogous to those by Gullord and colleagues, Sutka (1981, 1984) analyzed two complete diallels of winter wheat, one having 6 parental genotypes and the other 10 genotypes. Again, the analyses indicated that the inheritance of frost hardiness involved multiple genes that were mostly additive in their effect. However, the variance–covariance graphical analysis indicated that frost sensitivity was partially dominant. Similar conclusions regarding the direction of dominance were reached by Puchkov and Zhirov (1978) and Schafer (1923). Other studies (for a review, see Orlyuk, 1985), for example, those by

Gullord (1975), suggested that frost tolerance was in the direction of dominance. The reason for these differing results is not known. However, the results of Gullord and colleagues (Gullord, 1975; Gullord *et al.*, 1975) suggest that it may be due, in part, to differences in freezing protocols used in the various investigations. Indeed, Sutka and Veisz (1988) have found that a gene(s) on chromosome 5A switches direction of dominance depending on the temperature at which freezing tolerance is measured. At high freezing temperatures ($-10°C$) frost resistance was dominant, whereas at low freezing temperatures ($-14°C$) frost sensitivity was dominant.

A number of studies have been conducted to determine which wheat chromosomes contain genes that affect frost hardiness. The results are consistent with the notion that frost tolerance is a complex character; 11 of the 21 chromosomes of hexaploid wheat have been reported to have effects. Table 1 summarizes the literature on this subject [this list is an updated version of the summary previously presented by Roberts (1986)]. It can be seen from Table 1 that chromosomes 5A and 5D have been implicated most frequently. Further, in many of these studies, 5A and 5D were found to have major effects on frost hardiness. Why there has been so much variation in the results, however, is not known. It might have to do with the freezing tests used, or the particular cultivars used in the studies. In addition, it is clear that hardening conditions can influence the results.

Roberts (1986) found, in examining 42 reciprocal chromosome substitution lines of two spring wheats, Cadet and Rescue, that chromosomes 2A, 5A, and 5B carried loci affecting frost hardiness measured after an 8-week day (6°C) and night (4°C) cycle. In contrast, chromosomes 6A, 3B, 5B, and 5D were found to affect frost hardiness when plants were hardened for 15 weeks in the dark—7 weeks at 0.8°C followed by 8 weeks at $-5°C$. Similarly, Veisz and Sutka (1989) found that substitution of chromosomes 5A, 5B, 5D, 4B, and 7A from the highly frost-hardy winter wheat Cheyenne into the relatively frost-sensitive spring wheat Chinese Spring resulted in varying levels of increased frost resistance. However, the effects observed depended on the duration of the hardening period. Thus, the investigators concluded that the frost-tolerance genes had their effects at different times in the cold-acclimation process.

Inheritance of frost hardiness has also been studied in oats and barley. The results with oats (*Avena* spp.) indicate that genes determining frost resistance are mainly additive in nature, although significant nonadditive effects were observed in some crosses (Amirshashi and Patterson, 1956; Jenkins, 1969; Pfahler, 1966). Jenkins (1969) found

TABLE 1
Wheat Chromosomes That Can Affect Frost Hardiness: Summary of Literature

Chromosome	Reference(s)
1A	Fletcher (1976), Musich and Bondar' (1981)
1B	Surkova (1978), Sutka and Rajki (1978, 1979), Goujon *et al.* (1968), Puchkov and Zhirov (1978)
1D	Goujon *et al.* (1968)
2A	Fletcher (1976), Roberts (1986)
2B	Musich and Bondar' (1981), Sutka and Rajki (1978), Goujon *et al.* (1968)
2D	Surkova (1978), Goujon *et al.* (1968)
3B	Sutka and Rajki (1978), Roberts (1986)
4B	Sutka (1981), Sutka and Rajki (1978, 1979), Puchkov and Zhirov (1978), Veisz and Sutka (1989)
4D	Fletcher (1976), Law and Jenkins (1970), Sutka (1981), Sutka and Rajki (1978)
5A	Musich and Bondar' (1981), Surkova (1978), Bondar' (1981), Cahalan and Law (1979), Sutka and Rajki (1979), Sutka (1981), Goujon *et al.* (1968), Poysa (1984), Puchkov and Zhirov (1978), Sutka and Kovács (1985), Roberts (1986), Veisz and Sutka (1989), Sutka and Veisz (1988)
5B	Sutka (1981), Poysa (1984), Roberts (1986), Veisz and Sutka (1989)
5D	Fletcher (1976), Sutka and Rajki (1978), Law and Jenkins (1970), Cahalan and Law (1979), Sutka (1981), Poysa (1984), Puchkov and Zhirov (1978), Roberts (1986), Veisz and Sutka (1989)
6A	Puchkov and Zhirov (1978), Roberts (1986)
6B	Sutka and Rajki (1978, 1979)
6D	Fletcher (1976)
7A	Goujon *et al.* (1968), Sutka and Rajki (1978), Law and Jenkins (1970), Surkova (1978), Sutka (1981), Veisz and Sutka (1989)
7D	Fletcher (1976)

that frost tolerance was largely determined by recessive genes and, like others (Coffman, 1962; Finkner, 1966), observed transgressive segregation of the frost-hardiness character (F_3 progenies that were more frost tolerant than the original parents were obtained in many crosses). As for barley, winter survival (freezing tolerance being the major factor) was found to be controlled primarily by genes with additive effects (Eunus *et al.*, 1962; Rhode and Pulham, 1960). Both dominant and recessive genes contributed to winter hardiness, and the direction of dominance (toward frost tolerance or frost sensitivity) appeared to depend on the severity of the freezing conditions.

Finally, the inheritance of frost hardiness has also been examined in a variety of woody plants, including fruit trees, tea plants, pine trees, citrus trees, grapes, and roses (see Sakai and Larcher, 1987). Although the studies are not as extensive as those conducted with wheat and the other cereals, the results generally indicate, as with the cereals, that frost hardiness is a complex genetic trait. Some exceptions may exist, however, such as with the frost tolerance of tea stems (Toyao, 1982) and winterkill of roses (Svejda, 1979), wherein only a few genes may be involved.

B. Gene Expression during Cold Acclimation

1. Isozyme Composition

Temperature can have a variety of effects on the structures and functions of enzymes (Brandts, 1967; Somero, 1975; Tappel, 1966). Of course, it affects the rate at which covalent bonds are made or broken during enzyme catalysis; generally, the rate of catalysis under conditions of saturating substrate concentration (V_{max}) approximately doubles for each 10°C increase in temperature. In addition, temperature affects the "weak" chemical bonds—hydrogen bonds, electrostatic interactions, and hydrophobic interactions—that stabilize the higher orders of protein structure (secondary, tertiary, and quaternary structure) and enzyme–ligand interactions. For example, hydrogen bonds and electrostatic interactions, because they form exothermically, are stabilized by reductions in temperature. Hydrophobic interactions, which form endothermically, are favored by increases in temperature. Thus, small changes in temperature commonly affect the K_m values and activation energies for enzymes, whereas temperature extremes, both high and low, can result in enzyme denaturation.

Given the effects of temperature on protein structure and function, one might expect that certain enzymes would be modified, or that different forms might be synthesized, in cold-acclimated plants. That is, different isozymes, better suited to low-temperature environments, might be produced during cold acclimation. There is evidence to indicate that this is the case. Huner and Macdowall (1976a,b, 1979a,b) isolated and characterized the ribulose bisphosphate carboxylase/oxygenase (Rubisco) from acclimated and nonacclimated rye plants and found that there were different isozymes in the two types of tissues. Moreover, they presented evidence indicating that the "acclimated enzyme" had twice the specific activity of the "nonacclimated enzyme" (measured at 25°C), and that the apparent K_m for CO_2 of the two enzymes differed: at temperatures below 5°C, the enzyme from the acclimated plants had a lower K_m, but at 25°C, the nonacclimated enzyme had the lower K_m (the crossover point was 10°C). From these and additional data, Huner and Macdowall (1979b) suggested that the changes that occurred in the physical and enzymatic properties of the Rubisco enzyme during cold acclimation better suited the enzyme, and thus the plant, to low-temperature environments.

A number of other examples of cold-acclimation-associated changes in various isozyme profiles have been documented (see, e.g., DeJong, 1973; Guy and Carter, 1984; Hall *et al.*, 1970; Krasnuk *et al.*, 1976a,b; Makinen and Stegemann, 1981; McCown *et al.*, 1969; Roberts, 1967, 1975; Shomer-Ilan and Waisel, 1975; and see references cited by Sakai and Larcher, 1987), and in many of these cases, it was suggested that the "hardened" isozymes might better enable the plant to survive in low-temperature environments. Roberts (1979, 1982), in fact, found a high rank-order correlation among the abilities of 12 varieties of bread wheat (*T. aestivum*) to frost harden and the changes that occurred in the ratios of two invertase isozymes during acclimation. Whether the differences in isozyme profiles that were observed in most of these studies resulted from posttranslational modifications of the "nonhardened" enzyme or from changes in gene expression is not clear. An apparent exception is that of the cold-acclimation-associated Rubisco isozyme of cabbage described by Shomer-Ilan and Waisel (1975). In this case, the amino acid compositions of both the nonacclimated and acclimated Rubisco isozymes were determined and proved to be significantly different. Thus, differential gene expression was probably involved. Interestingly, the hydrophobicity of the enzyme from the acclimated plants was less than that from the nonacclimated plants: the "acclimated" enzyme contained more hydrophilic amino acids (e.g., aspartic acid, arginine, and glycine) and less hydrophobic residues (e.g., leucine, isoleucine, and phenylalanine). As discussed previously, hydrophobic interactions are less favorable at low temperatures. Thus, as suggested by the authors, this change might have contributed to the stability of the acclimated enzyme at low temperature, which in turn might have contributed to the increased CO_2 fixation capacity of the cold-hardened plants.

2. Protein Synthesis

A common, though not universal, observation has been that the soluble protein content of plants increases during cold acclimation (see Levitt, 1980; Sakai and Larcher, 1987). Examples include studies on black locust (Siminovitch *et al.*, 1968) and wheat (Trunova, 1982), in which it was found that soluble protein levels in acclimated plants were about 50 and 300% greater, respectively, than in nonacclimated plants. Moreover, it has been shown that changes in polypeptide composition occur during the cold acclimation process. The polypeptide profiles of tissues from a variety of cold-acclimated and nonacclimated plants have been examined by SDS–PAGE and *in vivo* radiolabeling (Table 2). The results indicate that cold acclimation is associated with both the appearance of new polypeptides and increases and/or decreases in others. The *in vivo* radiolabeling experiments suggest that many, if not most, of these changes involve modifications in polypeptide synthesis. However, because the radiolabeling periods used in most of the studies were relatively long (usually 6–24 hours), it is possible that posttranslational processes such as polypeptide turnover also have roles.

Although it is clear that a number of changes in polypeptide composition occur during cold acclimation, it is significant to note that, unlike the heat-shock and anaerobic stress responses, wherein extensive changes in protein synthesis occur (see Kimpel and Key, 1985; Sachs and Ho, 1986), the changes in protein synthesis during cold acclimation appear to be relatively modest. That is, although a number of changes occur, the overall patterns of protein synthesis in cold-acclimated and nonacclimated plants appear to be quite similar. For example, Guy and Haskell (1987) examined the synthesis of polypeptides during cold acclimation in spinach (*Spinacia oleracea* L.) by radiolabeling plant leaves *in vivo* with [^{35}S]methionine and analyzing the polypeptides by two-dimensional SDS–PAGE. Using this procedure, the investigators could resolve about 500 different polypeptides. They found that most of the polypeptides synthesized by the control and cold-acclimated leaves were the same. There were, however, certain reproducible differences. Exposure of the plants to 5°C resulted in the appearance of three new high-molecular-weight cold-regulated (COR) polypeptides (the polypeptides were referred to by the investigators as "CAPs," cold-acclimation proteins) having molecular weights of 160K, 117K, and 85K. A dramatic increase in a fourth high-molecular-weight COR polypeptide, molecular weight 79K, was also consistently observed. An additional 18 polypeptides, ranging in molecular weight from 19K to 48K, were detected in acclimated leaves in some experiments, but not in others. Nine polypeptides, in the range of 22–55 kDa, were present in nonacclimated plants but were absent in acclimated plants.

Results similar to those of Guy and Haskell (1987) have been obtained by other investigators working with other plants (Table 2). Kurkela *et al.* (1988) used *in vivo* radiolabeling and one-dimensional SDS–PAGE to compare the polypeptide profiles of nonacclimated and acclimated *Arabidopsis thaliana* and found that seven polypeptides dramatically increased in the acclimated plants; a few others were severely reduced. As with spinach, however, the overall polypeptide profiles of the acclimated and nonacclimated plants were very similar. Meza-Basso *et al.* (1986) used *in vivo* labeling and two-dimensional SDS–PAGE analysis to examine the polypeptides in acclimated and nonacclimated *B. napus* and found that 14 polypeptides reproducibly increased in acclimated plants while 6 polypeptides reproducibly decreased. Mohapatra *et al.* (1987a), using *in vivo* radiolabeling and one-dimensional SDS–PAGE, compared cold-acclimated and nonacclimated alfalfa (*Medicago falcata* cv. Anik) and found that 11 polypeptides were unique to the acclimated plants. Other studies documenting cold-acclimation-associated changes in protein composition include reports on *A. thaliana* (Gilmour *et al.*, 1988), *Bromus inermis* cell suspension cultures (Robertson *et al.*, 1987, 1988), *Citrus sinenis* (Guy *et al.*, 1988), *Dactylis glomerata* (Yoshida and Uemura, 1984), *Medicago sativa* (Mohapatra *et al.*, 1987b), *S. oleracea* (Guy *et al.*, 1985, 1988), *Secale cereale* (Uemura and Yoshida, 1984), and *T. aestivum* (Perras and Sarhan, 1989).

Where it has been examined closely, the general finding has been that the synthesis of most COR polypeptides closely parallels the freezing tolerance of plants. That is, the appearance of the COR polypeptides generally coincides with the onset of freezing tolerance; their synthesis continues for as long as the plants are kept at cold-acclimating temperatures, and their synthesis declines quickly during deacclimation. The results of Guy and Haskell (1987) clearly demonstrate these points. First, it was established that the freezing tolerance of spinach leaves increased after only 24 hours at low temperature (5°C), and that maximum freezing tolerance occurred in about 7 days (nonacclimated leaves had an LT_{50} of about -5°C and acclimated leaves had an LT_{50} of -10°C). It was also found that returning the acclimated plants to warm temperatures (25°C) resulted in quick deacclimation; a loss in freezing tolerance was detected after 24 hours, and by 3 days, freezing tolerance was the same as for nonacclimated plants. The *in vivo* radiolabeling studies revealed that the 160-, 117-, and 85-kDa COR polypeptides were synthesized within 24 hours of exposing the plants to the cold temperatures, and that they continued to be synthesized for as long as the plants were kept in the cold. Further, it was found that return of the plants to 25°C resulted in a quick loss in synthesis of these polypeptides; within 24 hours, their level of production fell dramatically, and after 3 days, their synthesis was barely detectable. Similar correlations of COR polypeptide synthesis with acclimation and deacclimation have been obtained with alfalfa (Mohapatra *et al.*, 1987a,b).

TABLE 2

Changes in Gene Expression Associated with Cold Acclimation

Plant species	Increased (or decreased) expression of polypeptides observed using different techniques			
	Coomassie- or silver-stained SDS–PAGE[a]	In vivo radiolabeling of polypeptides[b]	In vitro translation of mRNA[c]	Reference(s)
Arabidopsis thaliana	—	150[d], 85, 69, 60, 45, 30, 24 (a few noted)[e]	150, 45, 24	Kurkela et al. (1988)
	—	160[f], 47[f]	160, 47, 24, 15	Gilmour et al. (1988), Thomashow et al. (1990)
Citrus sinensis	160 and six other low-molecular-weight polypeptides (eight polypeptides)	—	—	Guy et al. (1988)
Bromus inermis (cell suspensions)	200, 190, 165, 25	48, 47, 24.5 (23, 22)	—	Robertson et al. (1987)
Brassica napus	—	Fourteen polypeptides ranging from 14 to 50 kDa (six polypeptides ranging from 18 to 50 kDa)	Nine polypeptides ranging from 14 to 80 kDa (eight polypeptides ranging from 20 to 80 kDa)	Robertson et al. (1988), Meza-Basso et al. (1986)
	—	—	20, 17	Johnson-Flanagan and Singh (1987)

Species				Reference
Medicago falcata	90, 43, 38, 27, 23, 22, 18, 16, 13, 11	95, 90, 68, 63, 38, 27, 23, 22, 16, 13, 11	90, 86, 72, 48, 38, 36, 33	Mohapatra *et al.* (1987a)
Medicago sativa	180, 145, 135, 90, 80, 76, 70, 53	90, 80, 70, 60, 38, 27, 11, (30)	90, 60, 33, 27, 16	Mohapatra *et al.* (1987b)
Spinacia oleracea	—	110, 82, 66, 55, 13	180, 82, 72, 43, 31, 19 (68, 23, 13)	Guy *et al.* (1985)
	160, 85	160, 117, 85, 79 (nine low-molecular-weight polypeptides)	—	Guy and Haskell (1987)
	160, 85, and 10 others that showed variation from experiment to experiment (eight polypeptides)	—	—	Guy *et al.* (1988)
Triticum aestivum	200, 180, 77, 74, 68, 52, 43, 38, 31 (157, 42, 34)	200,150, 64, 52, 45, 44, 36 (60, 53, 46, 41, 39, 38, 33)	—	Perras and Sarhan (1989)

[a] Polypeptides were fractionated by one- or two-dimensional SDS–PAGE and detected by either Coomassie blue or silver staining.

[b] Plant tissues were radiolabeled with [^{35}S]methionine or other radioactive amino acid, and the polypeptides were fractionated by one- or two-dimensional SDS–PAGE and were detected by autoradiography or fluorography.

[c] RNA [usually poly(A)$^{+}$] was translated *in vitro* using [^{35}S]methionine or other radioactive amino acid and either a rabbit reticulocyte lysate or a wheat germ system. The polypeptides were fractionated by either one- or two-dimensional SDS–PAGE, and the polypeptides were detected by autoradiography or fluorography.

[d] The numerical values are the polypeptides in kilodaltons.

[e] Values in parentheses refer to decreased expression of polypeptides.

[f] These experiments were designed to specifically look for changes in the synthesis of the 160- and 47-kDa polypeptides. Other changes may have also occurred.

3. Messenger RNA Populations

It is now clear that at least some of the alterations in polypeptide synthesis that occur during cold acclimation involve changes in mRNA populations. The first evidence along these lines came from the work by Guy *et al.* (1985). In their studies of spinach, they used *in vitro* translation assays to show that poly(A)$^+$ RNA from 2-day acclimated plants produced polypeptides of 180 and 82 kDa that were not synthesized by poly(A)$^+$ RNA isolated from nonacclimated plants. Additional differences were observed in the translation products of poly(A)$^+$ RNA from 8-day acclimated plants: new polypeptides of 72, 43, 31, and 19 kDa were detected as were decreases in polypeptides of 68, 23, and 13 kDa. It was also clear from the data, however, that the overall *in vitro* translation patterns of poly(A)$^+$ RNA isolated from the acclimated and nonacclimated plants were very similar, a result that is in agreement with the *in vivo* radiolabeling experiments discussed previously.

Cold-acclimation-associated changes in mRNA populations have been examined in a number of other plant species using *in vitro* translation assays; results were similar to those obtained with spinach (Table 2). We (Gilmour *et al.*, 1988; Thomashow *et al.*, 1990) have obtained evidence for increases in translatable mRNAs encoding polypeptides of 160, 47, 24, and 15 kDa in cold-acclimated *Arabidopsis*. These changes occur rapidly, increases being detectable for all of the mRNAs within 24 hours of exposing plants to cold-acclimating conditions. Kurkela *et al.* (1988) have also studied the mRNA populations of *Arabidopsis* and have obtained similar results: they observed increases in translatable mRNAs for polypeptides of 150, 45, and 24 kDa in 24-hour cold-acclimated plants (these polypeptides presumably correspond to the 160-, 47-, and 24-kDa polypeptides that we have detected). Mohapatra *et al.* (1987a,b) found increases in translatable mRNAs for seven polypeptides in cold-acclimated *M. falcata* and five polypeptides in cold acclimated *M. sativa*. In cold-acclimated *B. napus*, Meza-Basso *et al.* (1986) reported increases in mRNAs encoding nine polypeptides ranging in molecular weight from about 25 to 80 kDa and Johnson-Flanagan and Singh (1987) found increases in mRNAs encoding polypeptides of 20 and 17 kDa.

4. Complementary DNA and Genomic Clones of Cold-Regulated Genes

The *in vitro* translation data discussed in Section III,B,3 strongly suggest that changes in the concentrations of certain mRNAs occur during cold acclimation. This was not certain, however, because it is theoretically possible, and in at least one case Skadsen and Scandalios (1987) have presented evidence that indicates that mRNAs subject to translational control *in vivo* are not always translated *in vitro.* Direct evidence for increased levels of specific RNAs in cold-acclimated alfalfa and *Arabidopsis,* however, has now been obtained by Northern blot analysis using isolated cDNA and genomic clones of *cor* genes.

Mohapatra *et al.* (1988b, 1989) have reported the isolation of cDNA clones for four *cor* genes in alfalfa (they designated the genes "*cas,*" for cold-acclimation-specific) (Table 3). Northern blot analysis indicated that the levels of the transcripts hybridizing with clone pSM1409 increased greater than 10-fold in cold-acclimated plants, and the transcripts for the other clones, pSM784, pSM2358, and pSM2201, increased at least 20-fold. Interestingly, two of the clones, pSM784 and pSM2358, hybridized with more than one transcript: pSM784 hybridized with three transcripts (1.4, 1.0, and 0.4 kb) and pSM2358 hybridized with two transcripts (1.2 and 0.9 kb). The investigators hypothesized that the genes represented by these cDNA clones might be members of multigene families. The fact that a large number of genomic restriction fragments were found to be hybridized with pSM784 and pSM2358 (six and seven *Eco*RI fragments, respectively) was consistent with this notion. It is also interesting to note that pSM784, pSM2358, and pSM2201 hybridized to high percentages of the cDNA library—6, 3.3, and 1.5% of the clones, respectively.

TABLE 3
cDNA and Genomic Clones of *Cold-Regulated* Genes

Plant species	Plasmid designation	Type of clone	Insert size (bp)	Transcripts (kb)	Reference
Medicago falcata	pSM1409	cDNA	—	1.1	Mohapatra *et al.* (1988b)
	pSM784	cDNA	740	1.4, 1.0, 0.4	Mohapatra *et al.* (1989)
	pSM2358	cDNA	860	1.2, 0.9	Mohapatra *et al.* (1989)
	pSM2201	cDNA	720	0.9	Mohapatra *et al.* (1989)
Arabidopsis thaliana	pHH7.2	cDNA	1100	1.4	Hajela *et al.* (1990)
	pHH28	cDNA	1300	2.5	Hajela *et al.* (1990)
	pHH29	cDNA	600	0.6	Hajela *et al.* (1990)
	pHH67	cDNA	700	0.7	Hajela *et al.* (1990)
	pMEH34	Genomic	—	0.45	M. Franck and S. Kurkela (personal communicationn)

cDNA and genomic copies of *cor* genes have also been isolated from *Arabidopsis*. We (Hajela *et al.*, 1990) isolated cDNA clones by constructing a cDNA library of poly(A)$^+$ RNA isolated from cold-acclimated plants and screening it by differential hybridization to cDNA probes prepared against RNA isolated from acclimated and nonacclimated plants. Initially 25 "cold-regulated clones" were identified that, upon further analysis, were found to fall into one of four groups. Each group, represented by either pHH7.2, pHH28, pHH29, or pHH67, contained cDNA inserts of different *cor* genes (Table 3). Northern blot analysis of total RNA isolated from acclimated and nonacclimated plants indicated that pHH7.2, pHH28, pHH29, and pHH67 hybridized to transcripts of 1.4, 2.5, 0.6, and 0.7 kb, respectively. The levels of these transcripts increased dramatically (10-fold or greater) in cold-acclimated plants and returned to low levels in deacclimated plants. Time-course studies indicated that the levels of all four *cor* transcripts increased markedly within 4 hours, continued to increase up to about 12 hours, and then remained at elevated levels for as long as the plants were kept at low temperatures (the longest period tested was 14 days). Returning the plants to warm temperatures resulted in rapid decreases in the levels of all four transcripts; within 4–8 hours, they fell to very low or undetectable levels. Nuclear run-on transcription assays indicated that the temperature-regulated accumulation of the *cor* transcript represented by pHH7.2, pHH28, and pHH29 was controlled primarily at the posttranscriptional level, while accumulation of the *cor* transcript represented by pHH67 was regulated largely at the transcriptional level.

M. Franck and S. Kurkela (personal communication) have isolated a genomic clone of an *Arabidopsis cor* gene. The gene codes for a 0.45-kb transcript that increases greater than 10-fold in acclimated plants. The level of the transcript remains high for as long as the plants are kept in the cold and falls to low or undetectable levels within 14 hours of returning the plants to control temperatures.

C. CHARACTERIZATION OF COLD-REGULATED GENES AND POLYPEPTIDES

Experiments designed to determine the mechanism(s) responsible for *cor* gene regulation and the functions encoded by these genes are now getting under way in a number of laboratories. Results to date are primarily of an introductory nature. Nevertheless, many of the initial studies have yielded interesting results.

1. Correlation between Cold-Regulated Gene Expression and Freezing Tolerance

Mohapatra *et al.* (1989) have addressed the question of whether *cor* gene expression correlates with the degree of freezing tolerance attained by cold-acclimated alfalfa cultivars. The conclusion is that, at least in some cases, it does. The experiment was done as follows. Total RNA was isolated from acclimated and nonacclimated alfalfa plants, and the level of *cor* gene expression was monitored using the cDNA clones pSM784, pSM2358, and pSM2201 (Table 3). A cDNA clone, pSM355, representing a gene that was expressed equally in acclimated and nonacclimated plants was included as a control. The alfalfa cultivars tested were Anik (*M. falcata*), Iroquois (*Medicago media*), Algonquin (*M. media*), and Trek (*M. sativa*), which, when cold acclimated, had LT_{50} values of -14.6, -11.8, -11.5, and $-9.7°C$, respectively. The results indicated that the relative levels of expression of the pSM2201 *cor* gene in acclimated Anik, Iroquois, Algonquin, and Trek were 100, 53, 43, and 23%, respectively. The correlation coefficient between the levels of expression and the degree of freezing tolerance attained by the cultivars was 0.993. Similar high positive correlation coefficients were obtained for the two other *cor* gene probes, pSM784 and pSM2358. In contrast, there was no correlation between the level of freezing tolerance of the cultivars and either the expression of the control gene in the acclimated plants, or expression of the *cor* genes in nonacclimated plants.

The results of Mohapatra *et al.* (1989) indicate that a link exists between the expression of at least some alfalfa *cor* genes and the level of freezing tolerance obtained by certain alfalfa cultivars. Whether this correlation will also be observed with other alfalfa *cor* genes and other alfalfa cultivars remains to be determined. More importantly, the nature of the observed link between *cor* gene expression and freezing tolerance remains to be established. Does it mean that the *cor* genes have direct roles in freezing tolerance or is it reflective of a less direct relationship? The isolation of the *cor* genes should facilitate studies to address these issues. It will also be interesting to determine whether the cloned *cor* genes can be used as tools to aid breeding efforts designed to increase the freezing tolerance of alfalfa.

2. Expression of Cold-Regulated Genes in Response to Abscisic Acid

A number of studies have been directed at determining whether the expression of *cor* genes is regulated by the phytohormone abscisic acid (ABA). The reason why this is of interest is that exogenous application of ABA to whole plants of box elder (*Acer negundo*) (Irving, 1969), alfalfa (Mohapatra *et al.*, 1988b; Rikin *et al.*, 1975; Waldman *et al.*, 1975), and *Arabidopsis* (Lång et al., 1989), as well as stem cultures of *S. commersonii* (Chen *et al.*, 1979), has been shown to induce increased frost tolerance in the absence of cold treatment. In addition, ABA-induced frost tolerance has been observed with callus and suspension cultures of a wide variety of plant species (Chen *et al.*, 1979; Chen and Gusta, 1982, 1983; Orr *et al.*, 1986; Reaney and Gusta, 1987). One of the most thorough studies was that by Chen and Gusta (1983). These investigators established that ABA could induce even greater levels of frost tolerance in suspension cultures of wheat, rye, and bromegrass than could low-temperature treatment: cultures grown at control temperatures (20°C) were killed at about −9°C, cold-acclimated cultures were killed at about −17°C, and cultures grown at control temperatures in the presence of 7.5×10^{-5} M ABA were killed at about −30°C. The results also indicated a positive correlation between the abilities of low temperature and ABA to induce increased frost tolerance in cultures of 10 different plant species. All of the species in which low temperatures induced increased freezing tolerance (*M. sativa, B. inermis, Daucus carota, S. cereale,* and *T. aestivum*) also became more frost tolerant in response to ABA. Those species that did not "frost harden" in response to low temperature (*Datura innoxia, Catharanthus roseus, Glycine max, Vicia hajastana,* and a cell line of *Triticale*) did not harden in response to ABA.

A fundamental question raised by the ABA experiments is whether any of the *cor* genes are also regulated in response to ABA. If such genes existed, they might be considered to be prime candidates for encoding frost-tolerance polypeptides. Indeed, certain *cor* genes are regulated by ABA. Mohapatra *et al.* (1988b, 1989) have shown this in alfalfa. In their initial experiments, they radiolabeled nonacclimated, cold-treated, and ABA-treated plants *in vivo* with [^{35}S]methionine and analyzed the polypeptides by SDS–PAGE (in control experiments, they showed that both the low-temperature and ABA treatments resulted in increased frost tolerance). The data indicated that the ABA treatment, like the cold treatment, resulted in changes in protein synthesis. Approximately 19 new polypeptides were observed in the ABA-treated plants, as compared to the nonacclimated control, 11 of which also appeared to be synthesized in response to low-temperature treatment. Direct evidence that ABA could affect the expression of *cor* genes was then obtained by Northern blot analysis using the four alfalfa *cor* cDNA clones that the investigators had isolated (Table 3). Of these genes, transcripts corresponding to pSM1409 were found to increase dramatically in response to ABA treatment in a cultivar that was capable of attaining relatively high levels of frost tolerance. Interestingly, RNA levels for this gene were not significantly increased by ABA treatment in a relatively frost-sensitive cultivar.

Certain *cor* genes of *Arabidopsis* are also subject to ABA regulation. Lång *et al.* (1989) analyzed the soluble protein profiles of control, cold-treated, and ABA-treated *Arabidopsis* by *in vivo* protein labeling and found that several polypeptides produced in cold-treated and ABA-treated plants were not produced in control plants and that a subset of these "induced" polypeptides was present in both cold-treated and ABA-treated plants. We (Hajela *et al.*, 1990) have analyzed the expression of the four *Arabidopsis cor* genes for which we have obtained cDNA clones (Table 3) and have found that the transcript levels for each increase in response to ABA application. Similarly, M. Franck and S. Kurkela (personal communication) have found that the transcript levels for the *cor* gene that they have isolated (Table 3) increase in response to ABA. The results of Robertson *et al.* (1987, 1988) and Johnson-Flanagan and Singh (1987) suggest that that ABA can affect the expression of *cor* genes in bromegrass and *B. napus,* respectively.

While it is clear that the expression of certain *cor* genes is responsive to ABA treatment, the link between low-temperature- and ABA-regulated gene expression remains to be established. It is possible, as proposed by Chen *et al.* (1983), that low temperature triggers an elevation of endogenous ABA levels which in turn induces the expression of some or all of the *cor* genes. The finding that endogenous ABA levels increase in certain plants in response to low temperature is consistent with this notion (Chen *et al.*, 1983; Daie and Campbell, 1981; Lalk and Dörffling, 1985). However, with *Solanum commersonii*, the increase in ABA is only transient (Chen *et al.*, 1983), and in alfalfa, ABA levels do not appear to increase in response to low temperature (Waldman *et al.*, 1975). Clearly, additional experimentation is required to determine whether low-temperature regulation of *cor* genes is mediated through the action of ABA or whether *cor* gene expression is affected by distinct low-temperature and ABA regulatory mechanisms.

3. Expression of Cold-Regulated Genes in Response to Drought

Determining whether *cor* gene expression is affected by drought is of interest for at least two reasons. First, it has been generally observed that ABA concentration increases in water-stressed plants. Thus, *cor* genes that are responsive to ABA might also be responsive to drought stress. Second, there are studies indicating that severe drought stress can induce increased frost tolerance in at least some plants, including cabbage (Cox and Levitt, 1976), wheat (Cloutier and Siminovitch, 1982; Siminovitch and Cloutier, 1983), and rye (Cloutier and Siminovitch, 1982; Siminovitch and Cloutier, 1982, 1983). The potential link between drought and freezing tolerance, as previously mentioned, is that tolerance to both stresses requires tolerance to dehydration. Certainly, the degree of dehydration suffered as a result of freezing is generally much more severe than that caused by drought (at $-10°C$, the water potential of the unfrozen water in the extracellular spaces of plant tissue would be less than -100 bars). Nevertheless, it is reasonable to hypothesize that genes that have roles in drought tolerance might also have roles in freezing tolerance.

Expression of *cor* genes in response to water stress has only been examined to a limited degree in alfalfa and *Arabidopsis*. Indeed, the results indicate that at least some of the *cor* genes are responsive to this form of stress. Mohapatra *et al.* (1988b, 1989) found that the mRNA levels for the alfalfa *cor* gene represented by cDNA pSM1409 (Table 3) increased about 10-fold in water-stressed plants; the other three *cor* genes tested were not responsive. Not surprisingly, the pSM1409 cDNA was also "induced" by ABA application (see Section III,C,2). We (Hajela *et al.*, 1990) have examined the effect of water stress on the expression of the four *Arabidopsis cor* genes for which we have isolated cDNA clones (Table 3) and found that that the RNA levels for all four genes increased in drought-stressed plants. Each of these four *cor* genes was also found to be responsive to ABA.

4. Cold-Regulated Polypeptides of Arabidopsis Are "Boiling Stable"

The *cor* genes are members of a related gene family in that they are all subject to low-temperature regulation. Further, we have recently found (Lin *et al.*, 1989; 1990) that at least four of the *cor* gene products, the 160-, 47-, 24-, and 15-kDa COR polypeptides, share an unusual biochemical property: they are "boiling stable"—that is, whereas most polypeptides are denatured and form a precipitate upon boiling, these major COR polypeptides of *Arabidopsis* remain soluble after such treatment. The full significance of the polypeptides sharing this unusual biochemical property is not known. Perhaps it is reflective of an underlying amino acid composition that favors stable structure and function at low temperatures. As mentioned earlier, hydrogen bonds and electrostatic interactions are generally more stable at lower temperatures, whereas hydrophobic interactions are favored at higher temperatures. Perhaps the COR polypeptides are composed of a high proportion of hydrophilic amino acids. This could result in high water solubility and could potentially prevent precipitation by boiling.

Another intriguing possibility is that the boiling-stable property might relate to a common function shared by the polypeptides. In this context it is interesting to note that three families of polypeptides with properties that could potentially have roles in cold acclimation have also been shown to be boiling stable. First, there are the antifreeze proteins of fish (see Feeney and Yeh, 1978; DeVries, 1983). These proteins are very hydrophilic and at least some have been shown to be boiling stable. Second, there are the dehydrins described by Jacobsen, Chandler, and colleagues (Jacobsen and Shaw, 1989; Close *et al.*, 1989). These investigators have shown that barley and other grasses, in response to both water stress and ABA treatment, accumulate certain proteins that are very hydrophilic and do not precipitate upon boiling. These proteins are hypothesized to protect plant seedlings from the dehydration stress that occurs during drought. Similarly, the late-embryogenesis-abundant (LEA) proteins described by Dure and colleagues (Baker *et al.*, 1988; Galau *et al.*, 1986) are also very hydrophilic and, as suggested by the investigators, may have desiccation-protective effects. The production of desiccation-protective proteins during cold acclimation could potentially help plant cells tolerate the severe dehydration that occurs during a freeze–thaw cycle.

Last, there are the cryoprotective proteins described by Heber and colleagues (Volger and Heber, 1975; Heber *et al.*, 1979; D. Hincha, personal communication). These investigators have found that cold-acclimated, but not nonacclimated, spinach plants produce polypeptides (in the range of 10–20 kDa) that are, on a molar basis, more than 1000 times more effective in protecting thylakoid membranes against freezing damage than are low-molecular-weight cryoprotectants, such as sucrose and glycerol. Biochemical characterization of these polypeptides has indicated that they have a high content of polar amino acids, that they are very hydrophilic, and that they do not precipitate on boiling. Clearly, it will be important to determine whether the *Arabidopsis* COR polypeptides have antifreeze, desiccation-protective, or cryoprotective effects, and if they do, to determine whether these properties contribute significantly to the increased freezing tolerance of cold-acclimated plants.

5. Relationship between Cold Acclimation and Heat Shock

The results of the *in vivo* radiolabeling and *in vitro* translation experiments described in the preceding sections indicate that gene expression during cold acclimation is not dramatically different from gene expression at normal control temperatures. A number of changes certainly do occur, but, by and large, it would appear that the genes that are expressed at normal temperatures continue to be expressed at cold acclimation temperatures. This overall picture is in sharp contrast to that of the heat-shock response of plants (see Kimpel and Key, 1985; Sachs and Ho, 1986). With this environmental stress, as with anaerobic stress (see Sachs and Ho, 1986), extensive changes in gene expression occur: there is generally a suppression of translation of preexisting mRNAs and an induction of transcription and translation of mRNAs for the heat-shock genes. Thus, heat shock and cold acclimation would appear to be distinct responses. Indeed, where it has been examined, the results indicate that *cor* gene expression is not induced by heat shock (Guy and Haskell, 1987; Hajela *et al.*, 1990; Kurkela *et al.*, 1988; Mohapatra *et al.*, 1988b, 1989). Further, whereas the expression of heat-shock proteins is generally transient, the expression of *cor* genes remains at high levels for as long (days to weeks) as the plants are kept in the cold (Gilmour *et al.*, 1988; Guy and Haskell, 1987; Hajela *et al.*, 1990; Kurkela *et al.*, 1988; Mohapatra *et al.*, 1989).

6. Physical Location of Cold-Regulated Polypeptides

There is a paucity of information available regarding the distribution of the COR polypeptides within the various tissues of cold-acclimated plants and their physical location within plant cells. Perras and Sarhan (1989) found that cold acclimation was associated with changes in gene expression in three different tissues of wheat: leaves, crowns, and roots. Some of the identified COR polypeptides were synthesized in all three types of tissues and others appeared to be tissue specific. Guy *et al.* (1988) reported that the 160- and 80-kDa COR polypeptides of spinach were present in the leaves, but not in the roots, of cold-acclimated plants. Interestingly, spinach leaves became more frost tolerant as a result of acclimation but the roots did not. A strict correlation between the presence of the 160- and 80-kDa COR polypeptides and frost tolerance was not observed, however, because both polypeptides appeared to be constitutively expressed in hypocotyls (although the levels of the two polypeptides increased about 3-fold in acclimated hypocotyls).

As for the cellular location of COR polypeptides, the only conclusion that can be drawn is that some appear to be located in the membranes of cells. The most detailed information is that of Yoshida and Uemura (1984; Uemura and Yoshida, 1984), who showed that the plasma membrane protein profiles of orchard grass and winter rye seedlings, respectively, were different in cold acclimated and nonacclimated plants. This result is quite interesting given the prominent role of the plasma membrane in freezing tolerance. Other reports (Mohapatra *et al.*, 1988a; Robertson *et al.*, 1988) also indicate that COR polypeptides are located in membrane fractions, but precisely which membranes was not determined.

7. Conservation of Cold-Regulated Genes

Are the *cor* genes highly conserved among plants? Are they expressed in response to low temperatures in all plant species? Are *cor*-related genes found in organisms other than plants? The isolation of *cor* genes and preparation of antibodies to specific COR polypeptides should help investigators address these basic questions. As for now, however, there is little evidence one way or the other regarding these issues. Although it is true that a number of the COR polypeptides that have been identified have similar molecular weights and p*I* values, it is also true that most of these molecular weights and p*I* values are quite common. Thus, with one possible exception, it is premature to suggest potential relationships. The one exception is the 160-kDa COR polypeptide of spinach (Guy and Haskell, 1987). This polypeptide has both an unusually high molecular weight and an unusually low isoelectric point (p*I* 4.6). These attributes combine to place this polypeptide in an otherwise unoccupied region of two-dimensional SDS–PAGE gels. As discussed above, we have reported (Gilmour *et al.*, 1988) that *Arabidopsis* produces a 160-kDa COR polypeptide and have found that it, too, has a p*I* of 4.6. Further, Guy *et al.* (1988) reported the presence of COR polypeptides having similar molecular weights and p*I* values in citrus (*C. sinensis*) and petunia. The uncommon sizes, p*I* values, and cold-responsive expression of these polypeptides suggest that they may be related. Proof of this, however, must await further experimentation.

IV. Concluding Remarks

Our knowledge of the molecular genetics of cold acclimation is primitive in comparison to what is known about this aspect of a number of other plant responses triggered by environmental signals, e.g., heat shock, anaerobic stress, and light. Nevertheless, significant points have been established. In terms of gene regulation, it is now clear that reproducible changes in gene expression occur in cold-acclimated plants, that both increases and decreases in *cor* gene expression take place, and that at least some of the changes in gene expression that occur are at the level of mRNA accumulation. Further, initial studies with *Arabidopsis* indicate that the cold-regulated accumulation of *cor* transcripts involves both transcriptional and posttranscriptional control mechanisms. It is also clear that the cold-acclimation response is distinct from the heat-shock response. Where studied, *cor* genes have been found not to be responsive to heat stress and, unlike heat shock, the changes in gene expression that accompany cold acclimation are not transient and are relatively modest (of course, a change in expression of only 1% of the genes in a higher plant would still be more than 100 genes). Major issues that now need to be addressed include detailing the transcriptional and posttranscriptional regulatory mechanisms involved in controlling *cor* gene expression, determining how plants "sense" cold temperatures and relay this information into altered gene expression, and establishing the link between low-temperature- and ABA-regulated expression of *cor* genes. The isolation of cDNA and genomic clones of *cor* genes is an important step toward these goals.

As for the functions of *cor* genes and the roles that they have in cold acclimation, virtually nothing is known. There is, in fact, no direct evidence for or against the hypothesis that COR polypeptides have significant roles in cold hardening. This hypothesis, as first proposed by Weiser (1970), however, is reasonable, especially in light of what is known about the changes in gene expression that occur in response to other environmental stresses such as heat shock and anaerobic stress. Further, as discussed above, a number of observations have been made that are consistent with the hypothesis: there is the finding in alfalfa that a high positive correlation exists between the levels of expression of three *cor* genes and the frost hardiness of four different cultivars; where it has been examined in detail, *cor* gene expression has been found to closely parallel plant acclimation and deacclimation; it has been concluded in studies with wheat, *B. napus*, and *S. commersonii* that inhibitors of protein synthesis hinder cold acclimation; and it has been shown that certain *cor* genes are responsive to both ABA application and drought stress, two treatments that can increase the freezing tolerance of plants in the absence of low temperatures. There is also the intriguing observation that certain COR polypeptides of *Arabidopsis* share the unusual biochemical property of not being precipitated by boiling. It would seem unlikely that the sharing of this property, along with cold regulation, would be mere happenstance. More likely, it is reflective of some common underlying biochemical property, perhaps even shared function. Determining the functions of these and other *cor* genes and determining whether the COR polypeptides have important roles in cold acclimation are major issues to be addressed. Again, the isolation of cDNA and genomic clones of *cor* genes should help in such endeavors. I would anticipate that significant progress will be made regarding these issues over the coming few years.

Acknowledgments

I wish to thank Andrew Hanson, Peter Steponkus, Charlie Guy, and Jas Singh for insightful, informative discussions regarding cold acclimation and plant stress; Marianne Franck, Peter Chandler, Tim Close, and O. Veisz for sharing unpublished results and preprints of articles; and Peter Steponkus, Suzanne Hugly, Sarah Gilmour, Ravindra Hajela, and David Horvath for critical comments on the manuscript. Research conducted in the author's laboratory was supported by grants from the USDA Competitive Grants Program, the Michigan Research Excellence Fund, and the Michigan State Agricultural Experiment Station.

References

Amirshashi, M. C., and Patterson, F. L. (1956). Cold resistance of parent varieties, F_2 populations and F_3 lines of 20 oat crosses. *Agron. J.* **48**, 184–188.

Baker, J., Steele, C., and Dure, L., III (1988). Sequence and characterization of 6 *Lea* proteins and their genes from cotton. *Plant Mol. Biol.* **11**, 277–291.

Bondar', G. P. (1981). Results of a study of winter hardiness in the wheat variety Odessa 16 by means of monosomic analysis. *Ekol. Genet. Rast. Zhivotn. Tezisy Dokl. Vses. Konf., 2, Kishinev, Moldavian, SSR*, 70–71.

Brandts, J. F. (1967). Heat effects on proteins and enzymes. *In* "Thermobiology" (A. Rose, ed.), pp. 25–72. Academic Press, New York.

Burke, M. J., Gusta, L. V., Quamme, H. A., Weiser, C. J., and Li, P. H. (1976). Freezing and injury in plants. *Annu. Rev. Plant Physiol.* **27**, 507–528.

Caffrey, M., Fonseca, V., and Leopold, A. C. (1988). Lipid–sugar interactions. Relevance to anhydrous biology. *Plant Physiol.* **86**, 754–758.

Cahalan, C., and Law, C. N. (1979). The genetical control of cold resistance and vernalization requirement in wheat. *Heredity* **42**, 125–132.

Carpenter, J. F., and Crowe, J. H. (1988). Modes of stabilization of protein by organic solutes during desiccation. *Cryobiology* **25**, 459–470.

Chen, P. M., and Gusta, L. V. (1982). Cold acclimation of wheat and smooth brome-grass cell suspensions. *Can. J. Bot.* **60**, 1207–1211.

Chen, T. H. H., and Gusta, L. V. (1983). Abscisic acid-induced freezing resistance in cultured plant cells. *Plant Physiol.* **73**, 71–75.

Chen, H. H., Gavinlertvatana, P., and Li, P. H. (1979). Cold acclimation of stem-cultured plants and leaf callus of *Solanum* species. *Bot. Gaz. (Chicago)* **140**, 142–147.

Chen, H. H., Li, P. H., and Brenner, M. L. (1983). Involvement of abscisic acid in potato cold acclimation. *Plant Physiol.* **71**, 362–365.

Close, T., Kortt, A. A., and Chandler, P. (1989). A cDNA based comparison of dehydration induced proteins (dehydrins) in barley and corn. *Plant Mol. Biol.* **13**, 95–108.

Cloutier, Y., and Siminovitch, D. (1982). Correlation between cold- and drought-induced frost hardiness in winter wheat and rye varieties. *Plant Physiol.* **69**, 256–258.

Coffman, F. A. (1962). Increased winter hardiness in oats now available. *Agron. J.* **54**, 489–491.

Cox, W., and Levitt, J. (1976). Interrelations between environmental factors and freezing resistance of cabbage leaves. *Plant Physiol.* **57**, 553–555.

Crowe, J. H., and Crowe, L. M. (1986). Stabilization of membranes in anhydrobiotic organisms. *In* "Membranes, Metabolism, and Dry Organisms" (A. C. Leopold, ed.), pp. 188–209. Cornell Univ. Press (Comstock), Ithaca, New York.

Daie, J., and Campbell, W. F. (1981). Response of tomato plants to stressful temperature. Increase in abscisic acid concentrations. *Plant Physiol.* **67,** 26–29.

DeJong, D. W. (1973). Effect of temperature and daylength on peroxidase and malate (NAD) dehydrogenase isozyme composition in tobacco leaf extracts. *Am. J. Bot.* **60,** 846–852.

DeVries, A. L. (1983). Antifreeze peptides and glycopeptides in cold water fishes. *Annu. Rev. Physiol.* **45,** 245–260.

Dowgert, M. F., and Steponkus, P. L. (1984). Behavior of the plasma membrane of isolated protoplasts during a freeze–thaw cycle. *Plant Physiol.* **75,** 1139–1151.

Dowgert, M. F., Wolfe, J., and Steponkus, P. L. (1987). The mechanics of injury to isolated protoplasts following osmotic contraction and expansion. *Plant Physiol.* **83,** 1001–1007.

Eunus, A. M., Johnson, L. P. V., and Askel, R. (1962). Inheritance of winter hardiness in an eighteen-parent diallel cross of barley. *Can. J. Genet. Cytol.* **4,** 356–376.

Feeney, R. E., and Yeh, Y. (1978). Antifreeze proteins from fish bloods. *Adv. Protein Chem.* **32,** 191–283.

Finkner, V. C. (1966). Transgressive segregation for increased winter survival in oats (*Avena sativa* L.). *Crop Sci.* **6,** 297–298.

Fletcher, R. J. (1976). Physiological and genetic studies of wheat characters associated with frost resistance. Ph.D. thesis, Univ. of Sydney, Sydney, New South Wales, Australia.

Galau, G. A., Hughes, D. W., and Dure, L., III (1986). Abscisic acid induction of cloned late embryogenesis abundant (*Lea*) mRNAs. *Plant Mol. Biol.* **7,** 155–170.

Gilmour, S. J., Hajela, R. K., and Thomashow, M. F. (1988). Cold acclimation in *Arabidopsis thaliana. Plant Physiol.* **87,** 745–750.

Gordon-Kamm, W. J., and Steponkus, P. L. (1984a). The behavior of the plasma membrane following osmotic contraction of isolated protoplasts: Implications in freezing injury. *Protoplasma* **123,** 83–94.

Gordon-Kamm, W. J., and Steponkus, P. L. (1984b). Lamellar-to-hexagonal$_{II}$ phase transitions in the plasma membrane of isolated protoplasts after freeze-induced dehydration. *Proc. Natl. Acad. Sci. U.S.A.* **81,** 6373–6377.

Goujon, C., Maia, N., and Doussinault, G. (1968). Réistance au froid chez le blé. II. Réactions au stade coléoptile étudiées en conditions artificielles. *Ann. Amelior. Plant.* **18,** 49–57.

Graham, D., and Patterson, B. D. (1982). Responses of plants to low nonfreezing temperatures: Proteins, metabolism and acclimation. *Annu. Rev. Plant Physiol.* **33,** 347–372.

Gullord, M. (1975). Genetics of freezing hardiness in winter wheat (*Triticum aestivum* L.). Ph.D. dissertation, Michigan State Univ., East Lansing, Michigan.

Gullord, M., Olien, C. R., and Everson, E. H. (1975). Evaluation of freezing hardiness in winter wheat. *Crop Sci.* **15,** 153–157.

Guy, C. L., and Carter, J. V. (1984). Characterization of partially purified glutathione reductase from cold-hardened and unhardened spinach leaf tissue. *Cryobiology* **21,** 454–464.

Guy, C. L., and Haskell, D. (1987). Induction of freezing tolerance in spinach is associated with the synthesis of cold acclimation induced proteins. *Plant Physiol.* **84,** 872–878.

Guy, C. L., Niemi, K. J., and Brambl, R. (1985). Altered gene expression during cold acclimation of spinach. *Proc. Natl. Acad. Sci. U.S.A.* **82,** 3673–3677.

Guy, C. L., Haskell, D., and Yelenosky, G. (1988). Changes in freezing tolerance and polypeptide content of spinach and citrus at 5°C. *Cryobiology* **25,** 264–271.

Hajela, R. K., Horvath, D. P., Gilmour, S. J., and Thomashow, M. F. (1990). Molecular cloning and expression of *cor* (cold-regulated) genes in *Arabidopsis thaliana*. *Plant Physiol.* (in press).

Hall, T. C., McLeester, R. C., McCown, B. H., and Beck, G. E. (1970). Enzyme changes during deacclimation of willow stem. *Cryobiology* **7**, 130–135.

Hayes, H. K., and Aamodt, O. S. (1927). Inheritance of winter hardiness and growth habit in crosses of "Marquis" with "Minhardi" wheats. *J. Agric. Res.* **35**, 223–236.

Heber, U., Volger, H., Overbeck, V., and Santarius, K. A. (1979). Membrane damage and protection during freezing. *Adv. Chem.* **180**, 159–189.

Huner, N. P. A., and Macdowall, F. D. H. (1976a). Chloroplast proteins of wheat and rye grown at warm and cold-hardening temperatures. *Can. J. Biochem.* **54**, 848–853.

Huner, N. P. A., and Macdowall, F. D. H. (1976b). Effect of cold adaptation of Puma rye on properties of RUDP carboxylase. *Biochem. Biophys. Res. Commun.* **73**, 411–420.

Huner, N. P. A., and Macdowall, F. D. H. (1979a). Change in the net charge and subunit properties of ribulose bisphosphate carboxylase–oxygenase during cold hardening of Puma rye. *Can. J. Biochem.* **57**, 155–164.

Huner, N. P. A., and Macdowall, F. D. H. (1979b). The effects of low temperature acclimation of winter rye on catalytic properties of its ribulose bisphosphate carboxylase. *Can. J. Biochem.* **57**, 1036–1041.

Irving, R. M. (1969). Characterization and role of an endogenous inhibitor in the induction of cold hardiness of *Acer negundo*. *Plant Physiol.* **44**, 801–805.

Jacobsen, J. V., and Shaw, D. C. (1989). Heat-stable proteins and abscisic acid action in barley aleurone cells. *Plant Physiol.* **91**, 1520–1526.

Jenkins, G. (1969). Transgressive segregation for frost resistance in hexaploid oats (*Avena* spp.). *J. Agric. Sci.* **73**, 477–482.

Johnson-Flanagan, A. M., and Singh, J. (1986). Membrane deletion during plasmolysis in hardened and non-hardened plant cells. *Plant, Cell Environ.* **9**, 299–305.

Johnson-Flanagan, A. M., and Singh, J. (1987). Alteration of gene expression during the induction of freezing tolerance in *Brassua napus* suspension cultures. *Plant Physiol.* **85**, 699–705.

Kacperska-Palacz, A., Dlugokecka, E., Breitenwald, B. W, and Wciślińska, B. (1977). Physiological mechanisms of frost tolerance: Possible role of protein in plant adaptation to cold. *Biol. Plant.* **19**, 10–17.

Key, J. L., and Kosuge, T. (1984). "Cellular and Molecular Biology of Plant Stress." Liss, New York.

Kimpel, J. A., and Key, J. L. (1985). Heat shock in plants. *Trends Biochem. Sci. (Pers. Ed.)* **10**, 353–357.

Krasnuk, M., Jung, G. A., and Witham, F. H. (1976a). Electrophoretic studies of several dehydrogenases in relation to the cold tolerance of alfalfa. *Cryobiology* **13**, 375–393.

Krasnuk, M., Witham, F. H., and Jung, G. A. (1976b). Electrophoretic studies of several hydrolytic enzymes in relation to the cold tolerance of alfalfa. *Cryobiology* **13**, 225–242.

Kurkela, S., Franck, M., Heino, P., Lång, V., and Palva, E. T. (1988). Cold induced gene expression in *Arabidopsis thaliana* L. *Plant Cell Rep.* **7**, 495–498.

Lalk, I., and Dörffling, K. (1985). Hardening, abscisic acid, proline and freezing resistance in two winter wheat varieties. *Physiol. Plant.* **63**, 287–292.

Lång, V., Heino, P., and Palva, E. T. (1989). Low temperature acclimation and treatment with exogenous abscisic acid induce common polypeptides in *Arabidopsis thaliana* (L.) Heynh. *Theor. Appl. Genet.* **77**, 729–734.

Lange, O. L., Nobel, P. S., Osmond, C. B., and Ziegler, H., eds. (1981). "Physiological Plant

Ecology. I. Responses to the Physical Environment," Vol. 12A. Springer-Verlag, Berlin and New York.

Law, D. N., and Jenkins, G. (1970). A genetic study of cold resistance in wheat. *Genet. Res.* **15**, 197–208.

Levitt, J. (1980). "Responses of Plants to Environmental Stress: Chilling, Freezing, and High Temperature Stresses," 2nd ed. Academic Press, New York.

Li, P. H., and Sakai, A., eds. (1978). "Plant Cold Hardiness and Freezing Stress: Mechanisms and Crop Implications." Academic Press, New York.

Li, P. H., and Sakai, A., eds. (1982). "Plant Cold Hardiness and Freezing Stress: Mechanisms and Crop Implications," Vol. 2. Academic Press, New York.

Lin, C., Hajela, R., Horvath, D., Gilmour, S., and Thomashow, M. (1989). Identification of cold-inducible genes coding for heat-stable polypeptides in *Arabidopsis thaliana. Plant Physiol., Suppl.* **89**, 134.

Lin, C., Guo, W. W., Everson, E., and Thomashow, M. F. (1990). Cold acclimation in *Arabidopsis* and wheat: A response associated with expression of related genes encoding "boiling-stable" polypeptides. Submitted.

Lynch, D. V., and Steponkus, P. L. (1987). Plasma membrane lipid alterations associated with cold acclimation of winter rye seedlings (*Secale cereale* L. cv Puma). *Plant Physiol.* **83**, 761–767.

Lyons, J. M., Graham, D., and Raison, J. K., eds. (1979). "Low Temperature Stress in Crop Plants: The Role of the Membrane." Academic Press, New York.

Makinen, A., and Stegemann, H. (1981). Effects of low temperature on wheat leaf proteins. *Phytochemistry* **20**, 379–382.

Maximov, N. A. (1912). Chemische Schulzmittel der Pflanzen gegan Erfrieren. *Ber. Dtsch. Bot. Ges.* **30**, 52–65.

McCown, B. H., McLeester, R. C., Beck, G. E., and Hall, T. C. (1969). Environment-induced changes in peroxidase zymograms in the stems of deciduous and evergreen plants. *Cryobiology* **5**, 410–412.

Meza-Basso, L., Alberdi, M., Raynal, M., Ferrero-Cadinanos, M.-L., and Delseny, M. (1986). Changes in protein synthesis in rapeseed (*Brassica napus*) seedlings during a low temperature treatment. *Plant Physiol.* **82**, 733–738.

Mohapatra, S. S., Poole, R. J., and Dhindsa, R. S. (1987a). Changes in protein patterns and translatable messenger RNA populations during cold acclimation of alfalfa. *Plant Physiol.* **84**, 1172–1176.

Mohapatra, S. S., Poole, R. J., and Dhindsa, R. S. (1987b). Cold acclimation, freezing resistance, and protein synthesis in alfalfa (*Medicago sativa* L. cv. Saranac). *J. Exp. Bot.* **38**, 1697–1703.

Mohapatra, S. S., Poole, R. J., and Dhindsa, R. S. (1988a). Alterations in membrane protein-profile during cold treatment of alfalfa. *Plant Physiol.* **86**, 1005–1007.

Mohapatra, S. S., Poole, R. J., and Dhindsa, R. S. (1988b). Abscisic acid-regulated gene expression in relation to freezing tolerance in alfalfa. *Plant Physiol.* **87**, 468–473.

Mohapatra, S. S., Wolfraim, L., Poole, R. J., and Dhindsa, R. S. (1989). Molecular cloning and relation to freezing tolerance of cold-acclimation-specific genes of alfalfa. *Plant Physiol.* **89**, 375–380.

Musich, V. N., and Bondar', G. P. (1981). Study of frost resistance in the winter wheat Odessa 16 by means of monosomic analysis. *Biol. Aspekty Izuch. Ratsional'n. Ispol'z. Zhivotn. Rastitel'n Mira. Tez. Dokl. Konf. Molod. Uchenbiologov. Riga, Latvian SSR*, 70–71.

Nilsson-Ehle, H. (1912). Zur Kenntnis der Erblichkeitsverhältnisse der Eigenschaft Winterfestigkeit beim Weizen. *Z. Pflanzenzuecht.* **1**, 3–12.

Olien, C. R. (1965). Interference of cereal polymers and related compounds with freezing. *Cryobiology* **2**, 47–54.

Olien, C. R. (1973). Thermodynamic components of freezing stress. *J. Theor. Biol.* **39**, 201–210.

Olien, C. R. (1974). Energies of freezing and frost desiccation. *Plant Physiol.* **53**, 764–767.

Olien, C. R., and Smith, M. V. (1977). Ice adhesions in relation to freeze stress. *Plant Physiol.* **60**, 499–503.

Orlyuk, A. P. (1985). The problems of frost- and winter-hardiness in genetic studies of winter wheat. *Genetika (Moscow)* **21**, 15–22.

Orr, W., Keller, W. A., and Singh, J. (1986). Induction of freezing tolerance in an embryogenic cell suspension culture of *Brassica napus* by abscisic acid at room temperature. *J. Plant Physiol.* **126**, 23–32.

Pearce, R. S., and Willison, J. H. M. (1985). Wheat tissues freeze-etched during exposure to extracellular freezing: Distribution of ice. *Planta* **163**, 295–303.

Perras, M., and Sarhan, F. (1989). Synthesis of freezing tolerance proteins in leaves, crown, and roots during cold acclimation of wheat. *Plant Physiol.* **89**, 577–585.

Pfahler, P. L. (1966). Small grain improvement by breeding and selection. *Rep. Fla. Agric. Exp. Stn.* p. 52.

Poysa, V. W. (1984). The genetic control of low temperature, ice-encasement, and flooding tolerances by chromosomes 5A, 5B, and 5D in wheat. *Cereal Res. Commun.* **12**, 135–141.

Puchkov, Y. M., and Zhirov, E. G. (1978). Breeding of common wheat varieties with high frost resistance and genetic aspects of it. *World Sci. News* **15**, 17–22.

Quisenberry, K. S. (1931). Inheritance of winter hardiness, growth habit, and stem-rust reaction in crosses between Minhardi and H-44 spring wheats. *U.S., Dep. Agric., Tech. Bull.* **218**, 45 pp.

Reaney, M. J. T., and Gusta, L. V. (1987). Factors influencing the induction of freezing tolerance by abscisic acid in cell suspension cultures of *Bromus inermis* Leyss and *Medicago sativa* L. *Plant Physiol.* **83**, 423–427.

Rhode, D. R., and Pulham, C. F. (1960). Heritability estimates of winter hardiness in winter barley determined by the standard unit method of regression analysis. *Agron. J.* **52**, 584–586.

Rikin, A., Waldman, M., Richmond, A. E., and Dovart, A. (1975). Hormonal regulation of morphogenesis and cold-resistance. I. Modifications by abscisic acid and gibberellic acid in alfalfa (*Medicago sativa* L.) seedlings. *J. Exp. Bot.* **26**, 175–183.

Roberts, D. W. A. (1967). The temperature coefficient of invertase from the leaves of cold-hardened and cold-susceptible wheat plants. *Can. J. Bot.* **45**, 1347–1357.

Roberts, D. W. A. (1975). The invertase complement of cold-hardening and cold-sensitive wheat leaves. *Can. J. Bot.* **53**, 1333–1337.

Roberts, D. W. A. (1979). Changes in the proportions of two forms of invertase associated with the cold acclimation of wheat. *Can. J. Bot.* **57**, 413–419.

Roberts, D. W. A. (1982). Changes in the forms of invertase during the development of wheat leaves growing under cold-hardening and non-hardening conditions. *Can. J. Bot.* **60**, 1–6.

Roberts, D. W. A. (1986). Chromosomes in 'Cadet' and 'Rescue' wheats carrying loci for cold hardiness and vernalization response. *Can. J. Genet. Cytol.* **28**, 991–997.

Robertson, A. J., Gusta, L. V., Reaney, M. J. T., and Ishikawa, M. (1987). Protein synthesis in bromegrass (*Bromus inermis* Leyss) cultured cells during the induction of frost tolerance by abscisic acid or low temperature. *Plant Physiol.* **84**, 1331–1336.

Robertson, A. J., Gusta, L. V., Reaney, M. J. T., and Ishikawa, M. (1988). Identification of proteins correlated with increased freezing tolerance in bromegrass (*Bromus inermis* Leyss. cv Manchar) cell cultures. *Plant Physiol.* **86**, 344–347.

Sachs, M. M., and Ho, T.-H. D. (1986). Alterations of gene expression during environmental stress in plants. *Annu. Rev. Plant Physiol.* **37**, 363–376.

Sakai, A., and Larcher, W. (1987). "Frost Survival of Plants: Responses and Adaptations to Freezing Stress." Springer-Verlag, Berlin and New York.

Schafer, E. G. (1923). Inheritance studies. *Bull.—Wash. Agric. Exp. Stn.* **180** (Annu. Rep. 33), 31.

Shearman, L. L., Olien, C. R., Marchetti, B. L., and Everson, E. H. (1973). Characterization of freezing inhibitors from winter wheat cultivars. *Crop Sci.* **13**, 514–519.

Shomer-Ilan, A., and Waisel, Y. (1975). Cold hardiness of plants: Correlation with changes in electrophoretic mobility, composition of amino acids and average hydrophobicity of fraction-1-protein. *Physiol. Plant.* **34**, 90–96.

Siminovitch, D., and Cloutier, Y. (1982). Twenty-four hour induction of freezing and drought tolerance in plumules of winter rye seedlings by desiccation stress at room temperature in the dark. *Plant Physiol.* **69**, 250–255.

Siminovitch, D., and Cloutier, Y. (1983). Drought and freezing tolerance and adaptation in plants: Some evidence for near equivalences. *Cryobiology* **20**, 487–503.

Siminovitch, D., Rheaume, B., Pomeroy, K., and Lepage, M. (1968). Phospholipid, protein and nucleic acid increases in protoplasm and membrane structures associated with development of extreme freezing resistance in black locust tree cells. *Cryobiology* **5**, 202–225.

Singh, J., Iu, B., and Johnson-Flanagan, A. M. (1987). Membrane alterations in winter rye and *Brassica napus* cells during lethal freezing and plasmolysis. *Plant, Cell Environ.* **10**, 163–168.

Skadsen, R. W., and Scandalios, J. G. (1987). Translational control of photo-induced expression of the *Cat 2* catalase gene during leaf development in maize. *Proc. Natl. Acad. Sci. U.S.A.* **84**, 2785–2789.

Somero, G. N. (1975). Temperature as a selective factor in protein evolution: The adaptational strategy of compromise. *J. Exp. Zool.* **194**, 175–188.

Steponkus, P. L. (1984). Role of the plasma membrane in freezing injury and cold acclimation. *Annu. Rev. Plant Physiol.* **35**, 543–581.

Steponkus, P. L., and Lynch, D. V. (1989). Freeze/thaw-induced destabilization of the plasma membrane and the effects of cold acclimation. *J. Bioenerg. Biomembr.* **21**, 21–41.

Steponkus, P. L., Uemura, M., Balsamo, R. A., Arvinte, T., and Lynch, D. V. (1988). Transformation of the cryobehavior of rye protoplasts by modification of the plasma membrane lipid composition. *Proc. Natl. Acad. Sci. U.S.A.* **85**, 9026–9030.

Surkova, L. I. (1978). The genetical control of frost resistance and winter hardiness in winter cereal crops. A review. *Skh. Rubezhom* **7**, 18–21.

Sutka, J. (1981). Genetic studies of frost resistance in wheat. *Theor. Appl. Genet.* **59**, 145–152.

Sutka, J. (1984). A ten-parental diallel analysis of frost resistance in winter wheat. *Z. Pflanzenzuecht.* **93**, 147–157.

Sutka, J., and Kovács, G. (1985). Reciprocal monosomic analysis of frost resistance on chromosome 5A in wheat. *Euphytica* **34**, 367–370.

Sutka, J., and Rajki, E. (1978). A Mironovszkaja 808 búzafjta fagytüröképességenek monoszómás F$_2$ analizise. *Novenytermeles* **27**, 185–192.

Sutka, J., and Rajki, E. (1979). A cytogenetic study of frost resistance in the winter wheat variety, Rannyaya 12 by F$_2$ monosomic analysis. *Cereal Res. Commun.* **7**, 281–288.

Sutka, J., and Veisz, O. (1988). Reversal of dominance in a gene on chromosome 5A controlling frost resistance in wheat. *Genome* **30**, 313–317.

Svejda, F. (1979). Inheritance of winter hardiness in roses. *Euphytica* **28**, 309–314.

Tappel, A. L. (1966). Effects of low temperatures and freezing on enzymes and enzyme systems. *In* "Cryobiology" (H. T. Meryman, ed.), pp. 163–172. Academic Press, New York.

Thomashow, M., Gilmour, S., Hajela, R., Horvath, D., Lin, C., and Guo, W. (1990). Studies on cold acclimation in *Arabidopsis thaliana*. *In* "Horticultural Biotechnology" (A. B. Bennett, ed.), in press. Liss, New York.

Toyao, T. (1982). Inheritance of cold hardiness of tea plants in crosses between var. *Sinensis* and var. *Assamica*. *In* "Plant Cold Hardiness and Freezing Stress: Mechanisms and Crop Implications" (P. H. Li and A. Sakai, eds.), Vol. 2, pp. 591–603. Academic Press, New York.

Trunova, T. I. (1982). Mechanism of winter wheat hardening at low temperature. *In* "Plant Cold Hardiness and Freezing Stress: Mechanisms and Crop Implications" (P. H. Li and A. Sakai, eds.), Vol. 2, pp. 41–54. Academic Press, New York.

Uemura, M., and Yoshida, S. (1984). Involvement of plasma membrane alterations in cold acclimation of winter rye seedlings (*Secale cereale* L. cv Puma). *Plant Physiol.* **75**, 818–826.

Veisz, O., and Sutka, J. (1989). The relationships of hardening period and expression of frost resistance in chromosome substitution lines of wheat. *Euphytica* **43**, 41–45.

Volger, H. G., and Heber, U. (1975). Cryoprotective leaf proteins. *Biochim. Biophys. Acta* **412**, 335–349.

Waldman, M., Rikin, A., Dovrat, A., and Richmond, A. E. (1975). Hormonal regulation of morphogenesis and cold resistance. II. Effect of cold acclimation and exogenous abscisic acid on gibberellic acid and abscisic acid activities in alfalfa (*Medicago sativa* L.) seedlings. *J. Exp. Bot.* **26**, 853–859.

Weiser, C. J. (1970). Cold resistance and injury in woody plants. *Science* **169**, 1269–1278.

Worzella, W. W. (1935). Inheritance of cold resistance in winter wheat, with preliminary studies on the technique of artificial freezing tests. *J. Agric. Res.* **50**, 625–635.

Yoshida, S., and Uemura, M. (1984). Protein and lipid compositions of isolated plasma membranes from orchard grass (*Dactylis glomerata* L.) and changes during cold acclimation. *Plant Physiol.* **75**, 31–37.

INFLUENCE OF ENVIRONMENTAL FACTORS ON PHOTOSYNTHETIC GENES

Luis Herrera-Estrella and June Simpson

Centro de Investigación y Estudios, Avanzados del I. P. N.,
36500 Irapuato, Guanajuato, Mexico

I. Introduction

Photosynthesis is the most important mechanism of energy input in the living world and the most important biochemical process in plants. The photosynthetic capacity of a plant is determined by its genetic nature and its interaction with the environment. The maximum photosynthetic capacity of a given plant is determined by its efficiency in fixing CO_2, its foliar area, and many other genetically determined characteristics; however, the percentage of this maximum capacity that it is reached on a daily basis is influenced by many environmental factors, such as light intensity, water and nutrient availability, and temperature. The biochemistry and physiology of photosynthesis and its interaction with environmental factors have been studied for many decades. More recently the molecular biology of photosynthesis and the mechanisms by which the expression of genes encoding photosynthetic proteins is regulated have been investigated. Among the effects caused by environmental factors that influence photosynthesis, by far the best analyzed at the molecular level are those triggered by light; the effects of other important factors, such as nutrient deficiencies, water

ADVANCES IN GENETICS, Vol. 28

availability, or temperature, are just starting to be investigated or have been somewhat overlooked. In this article, the effect of environmental factors on the expression of photosynthetic genes and on the accumulation of the corresponding polypeptides will be reviewed.

II. Effect of Light on Photosynthetic Genes

Probably the most important environmental stimulus to which plants react is light. Light is not only the driving force behind photosynthesis but is also involved in many developmental and regulatory processes. From germination through growth, flowering and fruiting, light is essential for the initiation and regulation of all these functions. These light responses are mediated by at least four photoreceptors: protochlorophylide, phytochrome, one or more blue light photoreceptors, and an ultraviolet B (UVB) photoreceptor.

Light affects plants from the macroscopic level, resulting in visible movements, to the level of metabolic pathways that are essential for the life of the plant. Despite the great interest of many scientists in unraveling the molecular mysteries of photosynthesis, for many decades it remained unknown how light affects cells at the molecular level. The technical limitations that hampered these types of investigations have been superseded by the development of recombinant DNA technology and the availability of methods for the genetic transformation of plants.

The effect of light on gene expression was first observed in experiments showing that different polypeptides were produced by *in vitro:* translation of mRNA extracted from light- or dark-grown plants (Thompson and Cleland, 1972; Tobin and Klein, 1975; Apel and Kloppstech, 1978). Photoregulation of gene expression was confirmed by run-off transcription (Smith and Ellis, 1978; Gollmer and Apel, 1983; Nelson *et al.*, 1984) and Northern blot analysis (Tobin, 1981; Gallagher and Ellis, 1982; Gallagher *et al.*, 1985), indicating that indeed the steady-state mRNA and transcription levels of many genes are directly affected by light.

Table 1 gives a list of genes whose expression is known to be regulated by light and the corresponding photoreceptor involved in this process. As shown, both chloroplast and nuclear gene expression are regulated by light. It has been observed that the effect of light on the molecular mechanisms that control the production of nuclear and plastid-encoded proteins is different, thus they will be described separately.

TABLE 1
Examples of Light-Regulated Genes

Gene	Regulation[a]	Receptor involved		
		Phytochrome	Blue light	UVB
Plastid genes				
r16	+	+		
pS1A1	+	+		
psbA	+	+		
cfo III	+	+		
rbcL	+	+		
cf1A	+	+		
Nuclear genes				
rbcs	+	+	+	
cab	+	+		
pep	+			
gdph	+			
phytochrome	−	+/−		
chs	+	+/−		+

[a] Positive or negative.

A. EFFECT OF LIGHT ON PLASTID GENES

The most significant morphogenetic change during leaf formation is the development of plastids from proplastids to etioplasts and finally to photosynthetically active chloroplasts. The principal effector in this crucial morphogenetic change is light (for a review, see Tobin and Silverthorne, 1985). The first easily detectable event during this process is the rapid synthesis and assembly of the thylakoid membranes. The components of the thylakoid membranes, in addition to other photosynthesis-related proteins, which are undetectable in proplastids, appear during the greening process. How the synthesis of photosynthetic proteins is regulated during this process is not completely understood, but several elegant pieces of work have shown that it is a very complex process. Initially, during development from proplastid to etioplast in dark-grown cotyledons or leaves, the mesophyll cells are already prepared to respond to a light stimulus and form the functional photosynthetic apparatus. During this light-independent process of plastid development, the mRNAs of some plastid genes that encode photosynthetic proteins are already present, such as those encoding the large subunit of ribulose bisphosphate carboxylase (rbcL), the 32-kDa protein of photosystem II (psbA), and the β subunit of the ATP synthase complex (atpB), albeit at a lower level than that at which they are found in active chloroplasts. Other mRNAs are present at similar levels in light- or dark-grown tissue, such as that encoding the 65- to 68-kDa chlorophyll apoproteins of photosystem I. However, despite the presence of these mRNAs, no detectable levels of the corresponding proteins are found in etioplasts. It has been proposed that in the absence of light, a posttranscriptional regulatory mechanism is active that prevents protein accumulation, probably at the level of mRNA translation of protein turnover, and that this negative regulation of protein accumulation is overcome by light (Klein and Mullet, 1987a,b).

Following illumination, the mRNAs of several plastid photogenes rapidly accumulate, although the increase in each individual mRNA is variable (Bedbrook et al., 1978; Rodermel and Bogorad, 1985). The clearest example of this light-inducible increase in mRNA is that of the 32-kDa quinone-binding protein of photosystem II, which increases up to 20-fold in several plant species. As an example, an estimated 19% of the maize plastid DNA is photoregulated. These genes are not clustered but are found in scattered groups throughout the maize chloroplast genome. Some of these groups are transcribed as polycistronic messengers, whereas others are transcribed individually. At least some of these genes have been found to be negatively photoregulated (Rodermel and Bogorad, 1985).

The observation that illumination produces changes in the steady-state level of mRNA of many chloroplast genes led to the conclusion that transcriptional regulation is a major factor controlling plastid light-inducible gene expression (Bedbrook et al., 1978; Herrman et al., 1985). However, the lack of an in vitro transcription system in which induction by light can still be studied has hampered the possibilities of directly testing this hypothesis.

Nevertheless, transcription experiments with isolated maize plastids from dark- and light-grown plants have shown that ribonucleotide triphosphates can be incorporated into newly synthesized RNA equally well in both etioplasts and chloroplasts (Klein and Mullet, 1987b). More recently, a plastid runon transcription system has been developed that allows the quantification of transcripts for individual plastid genes (Deng et al., 1987). Using this transcription runon system, Deng and Gruissem (1987) have demonstrated in spinach cotyledons that the relative transcriptional activity of 10 plastid genes (among them psbA, rbcL, rsaA, psbB, petB, petD, and atpB) is maintained during plastid development and is independent of light. Because light has an effect on determining the mRNA steady-state level of photosynthetic genes, this experiment suggests that light has an important role at the posttranscriptional level, probably increasing the stability of mRNAs in light-grown versus dark-grown tissue, or inactivating a mechanism for degradation of specific mRNAs present in dark-grown cells. As previously mentioned, the mRNA of photosynthetic genes is expressed at a considerable level in etioplasts, suggesting that in addition to light the morphogenetic processes that regulate cell differentiation in plants play a crucial role in determining the expression of photosynthetic genes.

The influence of light at the level of translation has been investigated for the plastid *rbcL* gene in peas and amaranth and for the *psbA* gene in barley (Berry *et al.*, 1988; Klein and Mullet, 1987b; Inamine *et al.*, 1985). In both cases it was found that the corresponding transcripts accumulate in dark-grown seedlings but that no detectable translation products of these transcripts are present. When these seedlings are transferred to normal light conditions or are given a pulse of red light, the rbcL subunit in pea and the translation product of *psbA* increase in concentration 20- to 50-fold, whereas the level of their respective transcripts increase only 3-fold. The fact that neither of these two proteins is present in dark-grown seedlings indicates that the rbcL transcripts in pea and amaranth and the psbA transcript in barley are not translated in the absence of light or that their translation products are very rapidly degraded under these conditions. Moreover, the large increase in the accumulation of these proteins cannot be explained by the modest accumulation of their transcripts. These two observations strongly suggest that in these cases the effect of light on the accumulation of these two photosynthetic proteins is mainly at the translational or protein stability level. Furthermore, light does not seem to have an influence on the transcriptional level of these genes, because the increase in the accumulation of their corresponding transcripts can be explained by an increase in the number of plastid genomes per cell; therefore, it is a gene dosage effect.

An important conclusion from this work is that to be able to separate the influence of light on plastid gene expression from the influence of light on plastid and leaf development, with its associated increase in plastid and plastid genome numbers per leaf, the results obtained from experiments determining the transcript level of plastid genes should be reported on a per-plastid volume or plastid number basis.

There are limitations to extending these studies on the molecular mechanisms by which light regulates the production of polypeptides in plastids; one limitation is the lack of a chloroplast transformation system, which would allow a clear understanding of the role of transcriptional regulation. Recently a major breakthrough has been achieved by the development of the biolistic technique of transformation. This system, although still to be perfected, does not seem to have any limitations for the transformation of any eukaryotic cell and can even be used to transform organelles (Sanford, 1988). The system has been shown to be effective in generating *Chlamydomonas* cells with transformed chloroplasts. Both homologous and foreign genes (flanked by plastid sequences) have been successfully introduced into the chloroplast genome of *Chlamydomonas* (Boynton *et al.*, 1988; Blowers *et al.*, 1989). In both cases the introduced DNA was stably maintained and integrated into the plastid genome by homologous recombination. It is still unknown whether plant cells have an intrinsic plastid genome limitation that would make plastid transformation impossible, but, in principle, plant chloroplast transformation and other basic studies that will uncover the molecular mechanisms that modulate gene expression in chloroplasts are already being carried out.

In summary, it can be concluded that the regulation by light of plastid gene expression is rather complex and that light can influence not only transcription initiation, but probably RNA processing, RNA stability, and translation of existing mRNAs. It appears that in this case the most significant effects of light are at the posttranscriptional level. Observed effects at the level of transcription could be explained in some cases by the increase in the number of plastid genomes per cell. This to some extent contrasts, as will be seen later, with the effect of light on nuclear gene expression, in which the principal effect, at least in most plant systems studied to date, seems to be at the transcriptional level. Certain exceptions have been found, as in the case of the ribulose 1,5-bisphosphate carboxylase genes *(rbcS)* in several plant species, which are regulated in a post transcriptional manner. Studies of chloroplast gene expression have also given conflicting results and it will be important to study chloroplast gene expression in several plant systems before a general model for regulation can be determined.

B. EFFECT OF LIGHT ON NUCLEAR GENES

Using an *in vitro* translation system and two-dimensional gel electrophoresis techniques, De Vries *et al.*, (1982) showed that there are differences in mRNA patterns in etiolated and light-grown pea seedlings. They determined that total mRNA isolated from light-grown pea seedlings produced approximately 25 translation products not detected by *in vitro* translation of mRNA from etiolated seedlings. In addition, 9 translation products of etiolated seedlings were not found in light-grown plants. Of the 25 light-induced mRNAs, 10 were shown to be under phytochrome control. A later report by De Vries *et al.*, (1983) reaches essentially the same conclusions by means of RNA–DNA hybridization kinetics.

Increasingly, the corresponding photoregulated genes are being characterized at the molecular level. However, by far the best-characterized photoregulated genes also encode the most abundant mRNA transcripts in green leaves; these are the genes encoding the small subunit of ribulose bisphosphate carboxylase *(rbcS)* and the chlorophyll *a/b*-binding proteins *(cab)*. We will concentrate on these two sets of genes and the detailed molecular information now available on their mode of regulation as an example of how light-regulated gene expression is being studied.

Both cab and rbcS proteins are encoded by small multigene families in the plant cell nucleus (Timko and Cashmore, 1983; Broglie *et al.*, 1983; Cashmore, 1983: Dean *et al.*, 1987; Dunsmuir, 1985), and both proteins are active in the chloroplasts. The rbcS subunit combines with the rbcL subunit and is involved in the initial reaction of the Calvin cycle (for reviews, see Ellis, 1979; Lorimer, 1981), whereas the cab proteins are involved in the light-harvesting reactions within the thylakoid membranes (for reviews, see Boardman *et al.*, 1978; Arntzen, 1978). Gallagher and Ellis (1982) showed by runoff transcription experiments that *rbcS* and *cab* gene transcriptions are increased 18- and 9-fold, respectively, in light-grown pea nuclei. This light inducibility has now been shown for several other plants, such as lemna, tobacco, petunia, soybean, and wheat for the *cab* genes and lemna, tobacco, soybean, and petunia for the *rbcS* genes. Interestingly, in several plants, such as barley and cucumber (Apel and Kloppstech, 1978; Greeland *et al.*, 1987), very little difference is observed between light- and dark-grown levels of rbcS mRNA.

1. Ribulose Bisphosphate Carboxylase Small Subunit Gene Expression

a. Transcriptional Regulation   As mentioned before, the abundance of transcripts corresponding to the *rbcS* gene family increases markedly following exposure to light of cells that contain chloroplasts (Corruzi *et al.*, 1984; Fluhr *et al.*, 1986a,b; Kaufman *et al.*, 1984). Studies using nuclear runon experiments strongly suggest that light regulation of the *rbcS* gene family is at the transcriptional level in pea and lemna. Several studies have shown that red light induces the expression of the *rbcS* gene family and this induction is reversed when the cells are immediately exposed to far-red light, demonstrating the involvement of phytochrome in the regulatory mechanism of the *rbcS* family. This has been demonstrated in pea, lemna, barley, maize, and soybean (Sasaki *et al.*, 1983; Bennett *et al.*, 1984; Tobin, 1981; Silverthorne and Tobin, 1984; Berry-Lowe and Meagher, 1985). However, two reports using 6-day-old seedlings found that red light is not sufficient to fully induce expression of *rbcS* genes, indicating that perhaps another photoreceptor is involved in this photoresponse (Tobin, 1981; Kaufman *et al.*, 1984). Kaufman *et al.* (1985) have shown that maize seedlings, when exposed to enough red light to saturate the phytochrome system, only express low levels of *rbcS* transcripts, and that when these seedlings are exposed to blue light in addition to red light, *rbcS* transcripts are obtained at levels comparable to those present in seedlings illuminated with white light. This suggests that not only phytochrome but also a blue light photoreceptor is involved in the regulation of *rbcS* genes. The role of a blue light photoreceptor mediating the increase of *rbcS* transcripts in photosynthetic cells has been confirmed in pea seedlings (Fluhr and Chua, 1986). In most plant species the rbcS polypeptide is encoded by a gene family, thus it is possible that individual genes respond differently to phytochrome and blue light. Moreover, experiments using chimeric gene constructs expressed in transgenic plants have shown that phytochrome is not involved in the light-regulated expression of at least one of the members of the pea *rbcS* gene family *(ss3.6)* (Simpson *et al.*, 1986a). These results indicate that the clear-cut picture of the expression of gene families, as derived from experiments measuring transcription or steady-state levels of transcripts of whole gene families, may turn out to be much more complex when the expression of each member of a gene family is analyzed.

Although at the beginning of the present decade it had already been clearly shown that light regulates gene expression in plants, many fundamental questions remained to be answered. Are the mechanisms controlling light-inducible gene expression conserved among different plant species? Where do the DNA sequences responsible for light-inducible expression reside? With the development of gene transfer techniques, many of these questions are now being answered. One of the first fundamental questions was resolved using gene transfer techniques, when a pea *rbcS* gene was transferred to tobacco and functionally expressed and light regulated, thus demonstrating that the light regulatory mechanism of gene expression is conserved among different plant species (Broglie *et al.*, 1984).

The first experiments demonstrating that the 5' flanking sequences of the *rbcS* genes are responsible for the light-inducible regulation of these genes involved the construction of a chimeric gene composed of the 5' flanking sequence of the pea *ss3.6* gene, the coding sequence of the bacterial chloramphenicol acetyltransferase gene *(cat)*, and the 3' flanking sequences of the constitutive nopaline synthase *(nos)* gene. When this chimeric gene was transferred to tobacco cells it was observed that in chloroplast-containing cells the production of the chloramphenicol acetyltransferase protein was light inducible, thus demonstrating that the sequences responsible for light induction reside in the 5′flanking sequences of the *rbcS* genes and that the effect of light is at the level of transcription initiation (Herrera-Estrella *et al.*, 1984).

To understand how a plant modifies the transcriptional rate of photosynthetic genes after a light stimulus, it is necessary to study the different steps involved in the mechanism of transduction of this environmental signal. Therefore, it is essential to identify the precise DNA sequences that mediate light-inducible gene expression, the proteins that interact with these sequences, and the molecular mechanisms that activate transcription. Because the discovery that the DNA sequences that determine light regulation reside in the 5′ flanking region of the *rbcS* gene, the precise cis-acting sequences responsible for this property have been intensively studied.

Nuclease-generated deletion analysis of the 5′ flanking sequences of pea *rbcS* genes showed that in order to have full levels of expression more than 1000 base pairs (bp) of upstream sequence are required (Morelli *et al.*, 1985; Timko *et al.*, 1985a,b). The same studies produced controversial results, leading to a confusion as to which nucleotide sequence was really responsible for the light-inducible properties of the *rbcS* genes, (Morelli *et al.*, 1985). This point was clarified in experiments in which the 5′ flanking sequence from −92 to −973 of the pea *rbcS* gene, *ss3.6* (containing no TATA and CAAT boxes), was placed upstream of the constitutively expressed *nos* promoter. The fused *rbcS/nos* promoter then controlled transcription of the *cat* reporter gene in a light-inducible fashion. This experiment demonstrated that the sequences responsible for light inducibility reside within this 861-bp DNA fragment, and because this sequence functions independently of its orientation, in effect it acts as an enhancer element, the first to be described in plant systems (Timko *et al.*, 1985a). One difference between this plant enhancer and most enhancers described in animal cell systems is that it has no effect when placed 3' to the heterologous gene construct, resembling more the yeast upstream regulatory elements (Struhl, 1987).

The light-inducible enhancer sequence has been narrowed down to 280 bp in the *rbcS3A* gene. Based on these experimental results and the analysis of homology between the 5′flanking sequences of all pea *rbcS* genes, three highly conserved sequences have been identified and named boxes I, II, and III (Fluhr *et al.*, 1986a). Green *et al.* (1987) have used gel retardation assays and DNase I footprinting experiments to identify a trans-acting protein factor that binds specifically to these conserved regions.This DNA-binding protein (GT-I) binds to boxes II and III and was shown to be present in both light-grown and dark-grown plants. Further analysis of the *in vivo* function of boxes II and III and their interaction with GT-I has revealed that there is a core sequence, GGTTAA, within box II which is critical for both light-inducible gene expression and GT-I binding. Within this GGTTAA sequence, a mutation where the GG was substituted by CC in box II both abolishes *in vivo* expression regulated by a 170-bp fragment of the 5' flanking sequence of the *rbcS3A* gene and binding of GT-I to this fragment (Green *et al.*, 1988; Kuhlemeir *et al.*, 1988). Although the simplest interpretation of these results would suggest a light-inducible activator function for GT-I, other results showing that GT-I binds to the 5' flanking sequences of genes that are not responsive to light, and even to the 5' flanking sequences of a phytochrome gene known to be negatively light regulated, suggest that this protein is a general transcription factor (Kuhlemeir *et al.*, 1988). Taking this into account, one could propose that boxes II and III in the presence of GT-I work as enhancers and that the DNA sequence responsible for light inducibility resides elsewhere or that a second trans-acting factor binds to the same sequences and its presence was masked by GT-I in the *in vitro* binding assays. In the first case, the effect of this light-inducible sequence would then only be seen when an active enhancer is sitting nearby. Other possible explanations are that GT-I functions as a platform for the interaction of other specific trans-acting factors, or that several alternative modified states of GT-I can bind to different cis-acting elements and determine differential gene expression.

Another conserved nucleotide sequence, ACGTGGCA(G box), found in the 5' flanking sequences of the *rbcS* genes of tomato, tobacco, soybean, petunia, pea, and *Arabidopsis,* has been proposed to be involved in light regulation. The DNA-binding proteins that interact with the G box have been also identified (Giuliano *et al.*, 1988). This so-called GBF DNA-binding protein is present in the nuclear extracts of tomato and *Arabidopsis* and binds sequences from tomato, pea, and *Arabidopsis* genes. Although in the case of the G box its conservation during evolution suggests that it may play an important role in regulation, no information on the *in vivo* function of this sequence is available. The importance of these DNA-binding proteins (incorrectly called trans-acting factors, until their function is proved *in vivo* or *in vitro*) in regulating gene expression remains to be determined. At the time of this writing there is not really sufficient information to propose a molecular mechanism by which light regulates the expression of *rbcS* genes, but from the data available it looks like there are several cis-acting sequences that are involved in the light-inducible properties of photosynthetic genes. It might be that several more protein-binding sequences are necessary to confer this property, some of them involved in the transduction mechanism and others having the properties of tissue specificity or general enhancer. Alternative sequences with similar light-inducible characteristics but different nucleotide sequences might also be present in the regulatory region of these genes. If this turns out to be correct, the activation of transcription by light could be the result of the interaction of several trans-acting factors that directly or indirectly allow the RNA polymerase to be more active on transcribing the photosynthetic genes.

    b. Postranscriptional Regulation   It has been conclusively shown that one of the major mechanisms by which light regulates the accumulation of nuclear-encoded photosynthetic polypeptides is at the level of transcription initiation. However, although not much information is available so far, posttranscriptional mechanisms have been suggested to play an important role in the overall process by which light regulates photosynthetic genes. One of the few examples of the effect of light at a posttranslational level is its effect on the accumulation of the rbcS polypeptide in amaranth cotyledons. Berry *et al.* (1985, 1986, 1988) have shown that the rbcS mRNA accumulates to significant levels in dark-grown seedlings, but that no accumulation of the corresponding polypeptide is observed. It was also shown that when light-grown seedlings are transferred to darkness, the level of rbcS polypeptide decreases rapidly without a corresponding change in the level of mRNA. These experiments demonstrate that an important component in the regulation of rbcS accumulation by light is at a posttranscriptional level. Using pulse-labeling experiments, it was shown that the decrease in the level of rbcS is not due to a reduced stability of this polypeptide in dark-grown plants as compared to those grown in the light. *In vitro* translation of rbcS transcripts obtained from light- and dark-grown plants showed that both are equally functional as templates for translation in this system. Isolation of polysome-bound rbcS mRNA showed that in dark-grown amaranth seedlings this mRNA is associated with ribosomes and that translation is blocked by a still unknown mechanism. Furthermore, rbcS polysomes isolated from dark-grown seedlings are able to synthesize the rbcS precursor in cell-free translation systems supplemented with inhibiters of translation initiation, suggesting that light regulates the accumulation of rbcS at the level of translation elongation in amaranth seedlings (Berry *et al.*, 1988).

2. Chlorophyll a/b-Binding Protein Genes

Two distinct photosystems, photosystem I (PSI) and photosystem II (PSII) are found in the thylakoid membranes of chloroplasts. Each photosytem contains chlorophylls *a* and *b* and chlorophyll *a/b*-binding (cab) proteins, which bind the chlorophylls. Two distinct types of cab proteins bind to PSI and PSII. Among the cab proteins that bind to PSII, two subfamilies have been found that are encoded by small gene families (Dunsmuir, 1985; Castresana *et al.*, 1987; Leutwiler *et al.*, 1986; Karlin-Neumann *et al.*, 1985). Only two PSI *cab* genes (type 1) have been characterized and mapped in the tomato genome (Hoffman *et al.*, 1987) and a full cDNA clone from petunia (Stayton *et al.*, 1987). The *cab* genes, similar to the *rbcS* genes, are regulated by light. Light-inducible increases in *cab* mRNA transcription have been demonstrated in lemna, barley, and pea (Tobin, 1981; Apel and Kloppstech, 1978; Gallagher *et al.*, 1985). This light inducibility is mediated by phytochrome (Bennet *et al.*, 1984; Silverthorne and Tobin, 1984; An, 1989; Tobin *et al.*, 1984; Karlin-Neumann *et al.*, 1988) and, in contrast to RbcS genes, gene activation often occurs at very low fluences of red light. Experiments similar to those carried out to investigate the process by which light activates *rbcS* gene expression (experiments involving gene transfer techniques and RNA studies) have also been carried out the the *cab* genes.

Chimeric constructs fusing 5' flanking sequences of the pea *ab80* gene to the *nptII* reporter gene showed that sequences 400 bp upstream of the start point of transcription are sufficient to confer light-inducible regulation and that sequences further upstream are necessary to give higher levels of expression (Simpson *et al.*, 1985). A 2.5-kbp fragment of 5' flanking sequences of these gene was shown to contain all the sequences involved in phytochrome-mediated regulation (Simpson *et al.*, 1986a). In a similar experiment 1.8-kbp fragment of the wheat *cab* gene *(cab-1)* was shown to be sufficient for light-inducible/phytochrome-mediated regulation of a chimeric *cab–cat* construct (Nagy *et al.*, 1986a). In *Arabidopsis,* sequences 800–950 bp upstream of the start point of transcription of the *cab* genes are sufficient to control light-induced expression of chimeric constructs (An, 1989). In the case of the *cab3 (ab180* in Tobin's nomenclature), it was found that 111 bp of 5' flanking sequences are sufficient to direct light-inducible and tissue-specific gene expression to low-level expression and that 156 bp are necessary for high-level expression (Mitra *et al.*, 1989). However, these latter constructs have not yet been tested for phytochrome responses. Whether the fact that the *cab* genes of *Arabidopsis* seem to require less upstream sequences to have full levels of expression reflects the fact that the genome of these plants is much smaller than is found in most angiosperms remains to be determined. In the case of all the *cab* genes tested to date, no significant effect of blue light has been detected (Kaufman *et al.*, 1984, 1985).

Simpson *et al.* (1986b) have defined a 247-bp fragment from -100 to -347 of the pea gene *ab80;* this fragment contains an enhancer/silencer sequence that acts both in a negative and a positive regulatory manner, stimulating expression of a heterologous, constitutively expressed *nos–nptII* gene in light-grown plants and repressing expression in the roots of these plants. It is interesting to note that the sequence from -336 to -112 of the *ab80* gene has close homology with the pea RbcS sequences from -141 to -166. This includes box I and part of box II as defined by Fluhr *et al.* (1986a). These boxes have been shown to contain overlapping positively acting sequences. Furthermore comparing the *rbcS* genes of pea, soybean, tobacco, and petunia with the *ab80* sequences indicates reasonably good homology with the remainder of box II (Kuhlemeir *et al.*, 1987). The -336 to -300 sequence therefore shows very strong homology with defined *rbcS* regulatory elements. The functional significance of these sequences of course has to be determined. An enhancer has also been delineated for the *cab-1* gene of wheat extending from -90 to -354. This sequence confers light (phytochrome) regulation and tissue specificity on heterologous chimeric gene functioning in both orientations and at the 3' end of the heterologous system (Nagy *et al.*, 1986b, 1987). Surprisingly, no obvious homology can be found between the monocotyledonous and dicotyledonous enhancer sequences, although the monocotyledon sequence functions efficiently in dicotyledons.

A report (Castresana *et al.*, 1988) on studies of regulatory sequences of the *cabE* gene of *Nicotiana plumbaginifolia* has described two positively regulating elements (PREs). PRE1 is contained in the sequence from -1554 to -1182 and PRE2 is found between -747 to -516. These elements were identified using chimeric fusions to the *cat* reporter gene and studying expression in transgenic tobacco plants. In addition, a negatively regulating element (NRE), or silencer, has been determined between nucleotides -973 and -747. Finally, a light-regulating element (LRE) has been determined between -396 and -186. This element confers light inducibility on a truncated nopaline synthase promoter.

Again, no obvious homology can be found between the 5' flanking sequences of the *cabE* gene from *N. plumbaginifolia,* the *ab80* gene from pea, or the *cab-1* gene from wheat. However, the sequence

ACGTGGCA or G box, which is conserved in most *rbcS* genes sequenced to date (Giuliano *et al.*, 1988), is present in the *cabE* regulatory region at position −241 from the cap site, and an inverted sequence of the same box is present in the *Arabidopsis cab3* gene (Mitra *et al.*, 1989). Deletion of the inverted G box in the *cab3* genes reduces the level of expression, but it does not eliminate light induction and tissue specificity (Mitra *et al.*, 1989). In summary, despite all the work that has been done attempting to pinpoint the cis-acting elements responsible for light inducibility, no consensus has been obtained and it might turn out that there are different nucleotide sequences that can confer light inducibility or that this property is the result of the interaction of several trans-acting proteins that bind to different sites in the regulatory region of photosynthetic genes.

It is also interesting to point out that plant development could have an important role in determining the degree of response of different members of a gene family to environmental changes. In this context, it has been shown that in *Arabidopsis* seedlings the *cab3* genes are induced by light to much higher levels than are the two other members of this gene family, whereas in plants containing primary leaves all three genes seem to be equally induced by light (Karlin-Neumann *et al.*, 1988). Developmental factors seem to be involved in determining the level of light induction of *cab* genes in corn; those genes that are preferentially expressed in mesophyll cells have a higher degree of light inducibility than do those preferentially expressed in bundle sheath cells (Sheen and Bogorad, 1986, 1987).

III. Effect of Plastid Factors on Photosynthetic Genes

Without doubt it has been demonstrated that environmental factors have an important role at different molecular levels in determining the amounts of photosynthetic proteins that are produced in a cell; however, the receptors of these stimuli (except for phytochrome for light) and the primary processes they influence still remain unknown. Because many of these factors in one way or another alter plastid development, we shall review the role of plastid development in the regulation of photosynthetic genes.

Several reports using either chlorophyll-deficient mutant plants or herbicide-bleached plants, including pea, maize, and mustard (Mayfield and Taylor, 1984; Muller *et al.*, 1980; Oelmuller and Mohr, 1986; Oelmuller *et al.*, 1986; Mayfield *et al.*, 1986; Burgess and Taylor, 1987, 1988), have shown that in bleached plants mRNAs for the *cab* and *rbcS* genes are not produced. However, in chlorophyll-deficient plants, which retain plastid integrity, the genes are expressed. In mutants or herbicide-bleached plants, in which the chloroplasts are developmentally arrested, no or very low levels of these mRNAs are detected. Simpson *et al.*, (1986a) have reported similar experiments using transgenic tobacco seedlings containing *rbcS* and *cab* chimeric constructs. They have shown that these chimeric constructs also fail to be expressed in herbicide-bleached plants, therefore suggesting that control is at the level of transcription, because only 5′ flanking sequences of *rbcS* and *cab* genes are involved in these constructs. The lack of chlorophyll and carotenoids, and the effect of the herbicide per se, have been taken into account and these factors have no direct effect on the expression of the genes (Mayfield and Taylor, 1984; Oelmuller and Mohr, 1986; Mayfield *et al.*, 1986; Burgess and Taylor, 1987). The other factor that could be involved in this effect is phytochrome. However, experiments have indicated that all other phytochrome effects are normal in these albino plants (Mayfield and Taylor, 1984; Mayfield *et al.*, 1986). This has led to the suggestion that the developmental stage of the chloroplast is crucial for the expression of these genes and that a chloroplast factor could be involved in nuclear gene regulation.

Results have also indicated that this factor is perhaps not a protein, because in the absence of a functional chloroplast translation system, stimulation of certain genes still occurs (Bradbeer *et al.*, 1979; Borner, 1986). RNA species have previously been proposed as the most likely candidates, but the nature of these factors remains a mystery. Because regulatory sequences for these genes contain negatively acting elements, it is possible that in roots or other chloroplast-deficient tissue a repressor is acting and that a factor produced during chloroplast development eliminates the effect of the repressor, allowing expression to occur. Therefore, light-regulated nuclear gene expression could be dependent on the light-regulated development of chloroplasts. Plastid development is one of the candidate processes through which environmental factors could regulate the expression of both nuclear and plastid-encoded photosynthetic genes.

IV. Diurnal and Circadian Effects
on Photosynthetic Genes

In higher plants, it has been observed that several biochemical and physiological oscillations are the result of endogenous rhythms (for a review, see Vince-Prue, 1983). This type of periodical process occurs even when all environmental cues associated with the solar day are excluded. The most common rhythms observed in nature are related to 24-hour periods and are called circadian rhythms. Among many other effects, the rate of photosynthesis has been shown to be influenced by circadian rhythms. The molecular mechanisms by which these cycles affect gene expression and their interaction with environmental factors have just begun to be studied. The first experiments showing that circadian rhythms regulate the expression of photosynthetic genes were carried out in pea (Kloppstech, 1985). It was observed that the transcript level of three genes involved in photosynthesis, namely, *cab*, RbcS, and one encoding a 24-kDa protein, varies considerably during the day and that these variations are maintained even when the plants are grown in constant darkness for several days (Kloppstech, 1985). These experiments demonstrated the existence of diurnal and circadian rhythms regulating the expression of photosynthetic genes and that light is not the only factor that determines the level of their expression. It was subsequently demonstrated in a more careful analysis that similar types of circadian rhythms regulate the expression of *cab* genes in pea (Spiller *et al.*, 1987), petunia (Stayton *et al.*, 1989), maize (Taylor, 1989), wheat (Nagy *et al.*, 1988), and tomato (Piechulla, 1988, 1989; Piechulla and Gruissem, 1987). A study of 10 different monocot and dicot plants supports the notions that the circadian regulation of *cab* expression is highly conserved (Meyer *et al.*, 1989). In most cases the transcript level of *cab* genes reaches a maximum 2 to 4 hours after illumination and starts to decrease before the onset of darkness, reaching a minimum after 4 hours of darkness; cab mRNA starts to accumulate again to high levels 2 hours before illumination. In petunia, two distinct patterns of expression have been observed for two classes of *cab* genes. One of them follows the pattern of expression described above and the other reaches a maximum 2 hours before darkness, drops to the lowest level of transcription during the dark period, and shows no increase in transcription until the onset of the next light period (Stayton *et al.*, 1989). Recent results have shown that the expression of *rbcS* compared to the *cab* genes is subjected to much less diurnal variation but to some extent is also subjected to circadian rhythms (Stayton *et al.*, 1989; Paulsen and Bogorad, 1988). Genes encoding proteins that are not involved in photosynthesis, such as phosphoenolpyruvate carboxylase and the cytosolic form of aldolase, show no diurnal changes in transcription in maize plants (Taylor, 1989).

Two observations suggest that the expression of the *cab* genes is controlled primarily by an internal circadian clock with control via phytochrome as a secondary mechanism. First, late in the light period, when phytochrome (Pfr) levels are presumably saturated, the cab mRNA concentration drops in anticipation of darkness. Second, the *cab* transcript accumulates during the dark period 4 hours before illumination. Furthermore, this rhythm is continuous even after several days of darkness, when the Pfr is presumably depleted or in very low levels (Nagy *et al.*, 1988; Stayton *et al.*, 1989).

Studies carried out so far have determined the level of steady-state mRNA and do not distinguish between the two principal mechanistic alternatives to explain the circadian fluctuation in cab mRNA accumulation. The same result could be obtained by circadian fluctuation of mRNA stability or of the rates of transcription. Chimeric gene constructs in which the promoter and coding sequence are separate and linked to the corresponding parts of a constitutive gene are necessary to determine the role of transcription initiation and mRNA stability.

The fact that a wheat *cab-1* gene linked to the 35 S promoter is expressed without a circadian rhythm and transcription studies with pea and maize isolated nuclei suggest that the mechanism of control could be at the level of transcription (Taylor, 1989; Nagy *et al.*, 1988; Paulsen and Bogorad, 1988). If the circadian regulation acts at the level of transcription, the involvement of a repressor that is present or functional only during the dark phase of the rhythm could be postulated. This repressor would somehow block transcription and override the action of phytochrome, the determinant for the circadian rhythm being the activation or presence of this hypothetical repressor. However, because the effect of circadian rhythms cannot be detected after more than 4 days in darkness, it is probable that light, through the accumulation of Pfr, drives the activation of transcription, and when Pfr is depleted, the circadian rhythm can no longer be observed, even when the putative repressor is not active or present. A very simplistic model illustrating this hypothesis is shown in Fig. 1.

V. Effect of Nutritional Factors on Photosynthetic Genes

It is well known that nutritional deficiencies have a marked influence on the photosynthetic capacity of plants. The lack of sufficient nitrogen, iron, and many other nutrients leads to chlorotic plants in which the photosynthetic apparatus is severely damaged. Many biochemical and physiological studies have been carried out to study the effect of nutritional deficiencies on photosynthesis; however, little investigation has been done to determine the effect of nutrients on photosynthetic genes.

DIURNAL TRANSCRIPTION OF PHOTOSYNTHETIC GENES

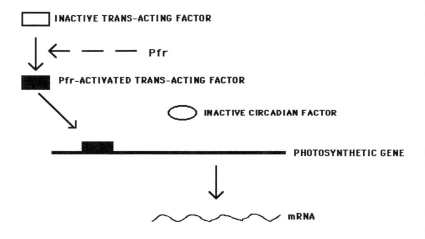

NOCTURAL INACTIVATION OF PHOTOSYNTHETIC GENES

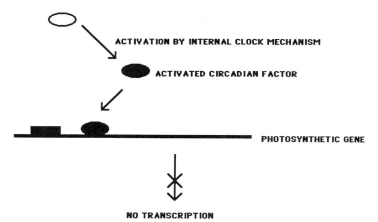

NO TRANSCRIPTION

FIG. 1. A putative repressor might be involved in the circadian regulation of photosynthetic genes. Circadian rhythms are dominant over the phytochrome-mediated light induction of the *cab* genes in all plant species analyzed to date. To explain this observation, a putative repressor has been postulated to bind the regulatory region of *cab* genes and to block transcription in the late part of the light phase despite the presence of active phytochrome (Pfr). Accumulation of *cab* transcripts in the late phase of the dark period could be explained by inactivation of the repressor, which in turn allows the remnant Pfr to activate transcription.

Nitrogen assimilation, photosynthetic electron transport, and CO_2 assimilation are interwoven processes limiting plant growth. Reducing equivalents generated from photosynthetic electron transport are required for nitrate reduction, and ATP-dependent NH_4^+ assimilation into amino acids occurs primarily within plastids. Nitrogen deficiency restricts photosynthetic efficiency and leads to chlorotic plants, having a specific effect on photosynthesis; the products of other metabolic pathways, such as lipids and carotenoids, accumulate normally under these conditions.

Using *Chlamydomonas* as a model system, Plumley and Schmidt (1989) have studied the effect of nitrogen deficiency on the expression of photosynthetic genes. It was found that when *Chlamydomonas* is grown under nitrogen-limiting conditions, there is a significant reduction in the level of photosynthetic proteins.

Both nuclear- and plastid-encoded proteins are reduced; cab-1 and cab-2 are reduced by 80%, being the most affected. Using *in vivo* pulse-labeling experiments, it was shown that protein turnover is not a major component in the reduction of photosynthetic proteins. Blot hybridizations demonstrated that plastid transcripts were present at normal levels, whereas nuclear-encoded transcripts were markedly reduced. Therefore, the reduced level of plastid-encoded photosynthetic proteins must be due to translational restraints; in contrast, the low levels of cytoplasmically synthesized proteins are explained on the basis of a reduction in the steady-state level of mRNAs. Whether the mechanism controlling nuclear-encoded transcripts is at the level of transcription initiation or mRNA stability remains to be determined. When nitrogen is added to starved cells within 4 hours there is a rapid accumulation of cab mRNAs. This latter finding implies that nitrogen, directly or through a rapid mechanism of transduction, activates transcription or stabilizes nuclear-encoded transcripts.

Iron is another nutrient that has a great influence on the photosynthetic capacity of plants. Iron-deficient pea and sugarbeet leaves contain plastids with few photosynthetic lamellae and only some rudimentary grana. Spiller *et al.* (1987) have shown that iron-deficient pea plants contain reduced levels of both plastid and nuclear-encoded photosynthetic transcripts. The cab, rbcS, psbA, and rbcL transcripts were found to be affected. When iron is added to this type of iron-deficient plant an increase in the level of the above-mentioned transcripts is observed. Transcripts start to accumulate 20 or more hours after iron is supplied to the plants. The long lag period preceding the iron-induced changes in the level of photosynthetic transcripts can in part be explained by the time required for the iron supplied to the roots to reach chlorotic leaves. However, there are several studies indicating that this process takes 3 hours or less (Branton and Jacobson, 1962). Therefore, iron must induce several biochemical events that are not reflected in transcript levels, before the transducing mechanism that activates transcription or alters mRNA stability is activated.

As we have already mentioned, the stage of development of the chloroplast is very important in the regulation of both plastid-encoded and nuclear-encoded photosynthetic genes. In the experiments involving nitrogen- and iron-deficient plants, we cannot distinguish if the effect of this deficiency directly affects gene regulation or whether the regulation is affected by damage to chloroplasts caused by iron or nitrogen deficiency.

VI. Effect of Growth Regulators on Photosynthetic Genes

It is also known that several environmental factors promote abscission and senescence, especially nitrogen deficiency and drought. One of the main characteristics of these physiological events is the loss of chlorophyll and a lowering of the photosynthetic capacity of the regions affected in the plants. Because both processes are controlled by growth regulators, we could consider that some environmental factors influence photosynthesis directly or indirectly, at least to some degree through the action of growth regulators. It is not known, however, how environmental factors could influence the synthesis activity, or translocation of growth regulators.

What it is well established is that cytokinins interact with light in triggering plastid development. The role of cytokinins in chloroplast development has been clearly demonstrated in experiments using cytokinin-deficient seedlings and cultured cell lines (Parthier, 1979; Feierabend, 1981; Axelos and Péaud-Lenoël, 1980).

To test whether cytokinins have a direct effect on the accumulation of photosynthetic transcripts, Flores and Tobin (1986) tested the effect of benzyladenine on the transcript level of dark-grown Lemna gibba plants. They found that cytokinin not only potentiates the effect of light but that by itself can stimulate the accumulation of cab and rbcS mRNAs in the dark, and that this induction is specific because the transcript level of a nonphotosynthetic gene, namely, β-actin, is not affected. It was proposed that the action of benzyladenine is at the level of transcript stability because this growth regulator has no effect on the level of cab nuclear transcripts. Similar results have been obtained in tobacco cell suspension cultures, a system in which kinetin is able to induce the accumulation of cab transcripts in dark-grown cells (Teyssendier de la Serve *et al.*, 1985). In both *Lemna* plants and tobacco cells the effect of cytokinins on photosynthetic genes takes more than 20 hours; these compounds therefore cannot directly be effectors of transcription or mRNA stability and must be involved in a transducing system that requires several processes to occur before it is activated. An interesting possibility is that the additive effect of cytokinins on the light-inducible mRNA accumulation of photosynthetic genes mediated by phytochrome action is not at the transcriptional level but rather at the level of mRNA stability; however, this requires more investigation to be demonstrated conclusively.

VII. Developmental Regulation of Photosynthetic Genes

Plant development is greatly influenced by environmental factors. During the different stages of plant development major differences in photosynthetic activity are observed; therefore, in one way or another, environmental factors must interact with the mechanisms regulating photosynthesis. One example of this interaction is observed during the basipetal differentiation gradient in maize leaves. Martineau and Taylor (1985) have shown that in the base of young maize leaves the mRNAs of *rbcS* and *rbcL* accumulate to high levels but no detectable amounts of the corresponding polypeptides are present. Because these mRNAs are associated with polysomes, these results suggest that although light is stimulating their transcription, no functional translation is occurring. This suggests that there is a developmental mechanism acting at the posttranscriptional level that is overriding the effect of light. This developmental control is associated with the differentiation of chloroplasts in bundle sheath and mesophyll cells.

The interaction of developmental factors with light regulatory effects on the expression of photosynthetic genes during fruit ripening in tomato has also been studied (Piechulla *et al.*, 1986). The chloroplast-encoded *psbA* and *rbcL* genes are expressed at much higher levels in fruit than are the light-inducible nuclear-encoded chlorophyll *a/b*-binding protein and ribulose bisphosphate carboxylase genes, and in the case of *psbA* for a much longer time, even after chromoplast development is well advanced. The *psbA* gene is expressed at 63% of the leaf expression in tomato pericarp 30 days after fruit development, and *rbcL* at 24% of leaf expression 14 days after fruit development, whereas the *cab* genes are expressed at 11% 7 days after fruit development and *rbcS* genes at 2% after 14 days of development. These results suggest that *rbcS* and *rbcL* are not so precisely coordinated in their regulation during development, as was previously assumed, and that the coordinate expression of *rbcS* and *rbcL* is affected by many factors, such as these developmental changes during fruit ripening.

VIII. Conclusions

As can be appreciated from the examples described previously, we are still at some distance from a clear view of the mechanisms by which environmental factors affect the expression of photosynthetic genes. Figure 2 summarizes the different levels at which environmental factors have been proved or suggested to act. Measuring mRNA transcription rates is one way to study gene expression in the plant of interest and is important in determining whether the regulatory control is at the level of transcription or posttranscription. Studying mRNA causes problems when, as in most cases, a family of genes is involved. The overall pattern of RNA may mask the expression of individual genes. Only in a few cases are specific probes available for detecting the mRNA from a single gene. To circumvent this, most studies have involved the use of chimeric gene constructs. These constructs typically comprise the 5' flanking sequences of the gene under study; the sequences are fused to an easily assayable marker gene, e.g., *cat, nptII,* or the *glucuronidase* gene. This has the great advantage of isolating the regulatory sequences from the other regions of the gene and the specific sequences involved in regulation can be delineated with confidence. In addition, there is no "noise" from the other members of the gene family. This technique, however, also brings with it its own special problems. In order to study the chimeric fusions *in vivo,* transformation experiments must be carried out. This means that the regulatory sequences are often studied in a heterologous plant background (it is now generally accepted that whole regenerated plants should be used rather than callus or suspension cultures). This raises the question as to whether the pattern of expression observed is that of the original plant from which the gene came or that of the new host plant (normally tobacco or petunia). This becomes more important when transferring genes between monocotyledons and dicotyledons, because it has been shown (Keith and Chua, 1986) that a wheat RbcS promoter failed to function in transgenic tobacco. With chimeric constructs it is impossible to know if the artificial fusion adequately reflects the natural gene structure or if some other element of the coding sequence or 3' sequences is also involved in the regulation. Another important point to be considered is

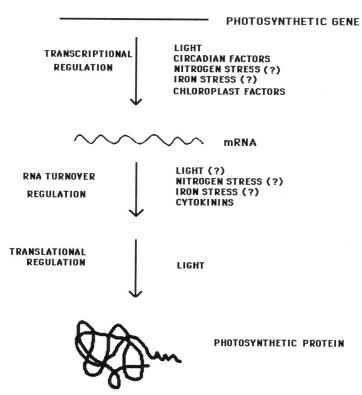

PHOTOSYNTHETIC GENE

TRANSCRIPTIONAL
REGULATION

LIGHT
CIRCADIAN FACTORS
NITROGEN STRESS (?)
IRON STRESS (?)
CHLOROPLAST FACTORS

mRNA

RNA TURNOVER
REGULATION

LIGHT (?)
NITROGEN STRESS (?)
IRON STRESS (?)
CYTOKININS

TRANSLATIONAL
REGULATION

LIGHT

PHOTOSYNTHETIC PROTEIN

FIG. 2. Environmental effects on photosynthetic gene expression. The expression of photosynthetic genes is influenced by many environmental factors. The mechanisms of regulation seem to be rather complex and to act at different molecular levels. Some factors have been shown to influence transcription initiation; others influence mRNA stability or mRNA translatability. The mechanism of regulation may vary from one species to another (see the text for a detailed explanation).

the effect of the position of insertion of the chimeric gene in the DNA of the transformed host. Surrounding host sequences may have a profound effect on the regulation of the foreign gene, usually in a quantitative manner as documented by Jones *et al.* (1985). This makes comparison of expression levels very difficult and the nature and expense of the reporter gene assays make the measurement of significant numbers of transgenic plants impossible. However, new and more sensitive histological techniques have been developed for certain reporter genes (Jefferson *et al.*, 1987; Teeri *et al.*, 1989), and this should make the screening of large numbers of transgenic plants more practical. The new histological techniques have also made it possible to examine the patterns of gene expression not just in whole organs, but in specific cell types. This has shown that the constructs normally used as constitutive controls, i.e., octopine synthase and the 35 S promoter fusions, do show a regulated pattern of expression at the cell-type-specific level (Jefferson *et al.*, 1987; Teeri *et al.*, 1989). In the future it will be necessary to choose and evaluate the "constitutive" controls carefully when studying regulatory patterns. One final point is that different groups measure regulatory patterns at different stages in development, and even taking samples at different times of the day could make a difference in the level of expression or pattern of expression observed. The conditions of assay should therefore always be reported.

Unravelling the mechanisms behind responses to environmental changes will not be easy. Detailed molecular and sequence analyses should eventually allow us to determine the chain of events, from direct stimulation by environmental fluctuations to response at the molecular level of gene expression. This should help us to understand photosynthesis and the mode of action of receptor molecules. We should also be able to determine the differential response of particular members of the same gene family and of the same genes in different plant species. All these aspects are currently under intense investigation for light-inducible responses; however, we should remember that although light is one of the most important stimuli, there are other environmental factors of equal importance.

Many groups have shown the importance of the developmental stage of the chloroplasts in the expression of photosynthesis-associated genes. This opens the question of coordinated chloroplast–nuclear gene expression and investigation into the factors mediating nuclear–chloroplast communication should prove very exciting. Another important point to remember is that the induction of mRNA synthesis by environmental factors is only half the story in the case of certain genes. In several cases a combination of transcriptional and posttranscriptional control is shown to occur, whereas in others control is primarily posttranscriptional.

Intensive studies are now being carried out on DNA-binding proteins and their involvement in gene regulation. However, these studies are in general carried out *in vitro* and often the DNA sequences involved are very short. It is necessary, therefore, to check that DNA sequences used do in fact have regulatory activity *in vivo* and that the bound proteins observed *in vitro* are conserved *on in vivo* binding. This will be facilitated by the development of *in vivo* footprinting techniques for plant species (Church *et al.*, 1985).

As a next step toward the understanding of the mechanisms of environmental signal transduction, several groups are pursuing the cloning of genes that encode the transacting factors that interact with the regulatory sequences of photosynthetic genes. Although this approach will generate much information on the way these proteins interact with their target DNA sequences and their involvement in regulating transcription, there is as yet no clear strategy to identify the earlier steps in the mechanism of transduction. If we envisage a cascade model of regulation, this approach presents several problems. First, in a regulatory sense it is difficult to imagine a long cascade having such rapid effects as light has on gene expression. Second, it will be very laborious to clone and characterize a whole series of genes. A third point is that recent results show that several proteins interact with each regulatory element, some of which may be unique and others constitutive binding factors. How do we know which ones to select for further study? Finally, most of these proteins are being selected on an *in vitro* basis. For these reasons, an old but very exciting strategy is beginning to be revived— the obtention of mutants that are altered in key steps in the transduction process. Chory (1989) has isolated *Arabidopsis* mutants that develop plastids and express photosynthetic genes in seedlings germinated in the dark. Because these types of mutants normally represent the loss of a function, this allows the postulation of the existence of a repressor that, in the dark, inhibits the expression of photosynthetic genes and hampers the development of plastids. Combining this classical approach with the use of transposon gene tagging, it should be possible not only to dissect the different steps of the transduction pathway but also to isolate the genes that encode the corresponding functions, with the sure knowledge that these are the key genes.

The elucidation of the molecular mechanism of response to environmental changes and how this response is modified by other factors will prove challenging, but, aided by these new strategies, should provide a wealth of insight into the regulatory mechanisms of plant cells.

REFERENCES

An, G. (1989). Integrated regulation of the photosynthetic gene family from Arabidopsis thaliana in transformed tobacco cells. *Mol. Gen. Genet.* **207,**210–216.

Apel, K., and Kloppstech, K. (1978). The plastid membranes of barley (*Hordeum vulgare*): Light-induced appearance of mRNA coding for the apoprotein of the light-harvesting chlorophyll a/b-protein. *Eur. J. Biochem.* **85,** 581–588.

Arntzen, C. (1978). Dynamic structural features of chloroplast lamellae. *Curr. Top. Bioenerg.* **8,** 111–160.

Axelos, M., and Péud-Lenoël, C. (1980). The apoprotein of the light harvesting chlorophyll a/b complex of tobacco cells as a molecular marker for cytokinin activity. *Plant Sci. Lett.* **19**, 33–41.

Bedbrook, J. R., Link, G., Coen, D. M., Bogorad, L., and Rich, A. (1978). Maize plastid gene expression during photoregulated development. *Proc. Natl. Acad. Sci. U.S.A.* **15**, 3060–3064.

Bennett, J., Jenkins, G., and Hartley, M. (1984). Differential regulation of the accumulation of the light harvesting chlorophyll a/b complex and ribulose bisphosphate/oxygenase in greening pea leaves. *J. Cell. Biochem.* **25**, 1–13.

Berry, O. J., Nikolav, B. J., Carr, J. P., and Klessig, D. (1985). Transcriptional and post-transcriptional regulation of ribulose bisphosphate carboxylase gene expression in light- and dark-grown amaranth cotyledons. *Mol. Cell. Biol.* **5**, 2238–2246.

Berry, O. J., Nikolav, B. J., Carr, J. P., and Klessig, D. (1986). Translational regulation of light-induced ribulose 1,5-bisphosphate carboxylase gene expression in amaranth. *Mol. Cell. Biol.* **6**, 2347–2353.

Berry, O., Carr, J. P., and Klessig, D. (1988). mRNAs encoding ribulose bisphosphate carboxylase remain bound to polysomes but are not translated in amaranth seedlings transferred to darkness. *Proc. Natl. Acad. Sci. U.S.A.* **85**, 4190–4194.

Berry-Lowe, S., and Meagher, R. (1985). Transcriptional regulation of a gene encoding the small subunit of the ribulose 1-5 bisphosphate carboxylase in soybean tissue is linked to the phytochrome response. *Mol. Cell. Biol.* **5**, 1910–1917.

Blowers, A. D., Bogorad, L., Shark, K. B., and Sanford, J. C. (1989). Studies on *Chlamydomonas* chloroplast transformation: Foreign DNA can be stably maintained in the chromosome. *Plant Cell* **1**, 123–132.

Boardman, N., Anderson, J., and Goodchild, D. J. (1978). Chorophyl–protein complexes and structure of mature and developing chloroplasts. *Curr. Top. Bioenerg.* **8**, 35–109.

Borner, T. (1986). Chloroplast control of nuclear gene function. *Endocyt. C. Res.* **3**, 265–274.

Boynton, J. E., Gillham, N. W., Harris, E. H., Hosler, J. P., Johnson, A. M., Jones, A. R., Randolph-Anderson, B. L., Robertson,D., Klein, T. M., Shark, K. B., and Sanford, J. C. (1988). Chloroplast transformation in *Chlamydomonas* with high velocity microprojectiles. *Science* **240**, 1534–1538.

Bradbeer, J., Atkinson, Y., Borner, T., and Hageman, R. (1979). Cytoplasmic synthesis of plastid polypeptides may be controlled by plastid synthesised RNA. *Nature (London)* **279**, 816–817.

Branton, D., and Jacobson, L. (1962). Iron transport in pea plants. *Plant Physiol.* **37**, 539–545.

Broglie, R., Coruzzi, G., Lamppa, G., Keith, B., and Chua, N.-H. (1983). Structural analysis of nuclear genes coding for the precursor to the small subunit of wheat ribulose 1-5 bisphosphate carboxylase. *Bio/Technology* **1**, 55–61.

Broglie, R., Coruzzi, G., Fraley, R., Rogers, S., and Horsch, R. (1984). Light regulated expression of pea ribulose 1-5 carboxylase small subunit gene in transformed plant cells. *Science* **224**, 838–843.

Burgess, D. G., and Taylor, W. C. (1987). Chloroplast photooxidation affects the accumulation of cytosolic mRNAs encoding chloroplast proteins in maize. *Planta* **170**, 520–527.

Burgess, D. G., and Taylor, W. C. (1988). The chloroplast affects the transcription of a nuclear gene family. *Mol. Gen. Genet.* **214**, 89–96.

Cashmore, A. R. (1983). Nuclear genes encoding the small subunit of ribulose 1-5 bisphosphate carboxylase. *In* "Genetic Engineering of Plants" T. Kosuge, C. P. Meredith, and A. Hollaender, eds.), pp. 29–38. Plenum, New York.

Castresana, C., Staneloni, R., Malik, V. S., and Cashmore, A. R. (1987). Molecular characterisation of two clusters of genes encoding the Type I CAB polypeptides of PSII in *Nicotiana plumbagenifolia. Plant Mol. Biol.* **10**, 117–126.

Castresana, C., Garcia-Luque, I., Alonso, E., Malik, V. S., and Cashmore, A. R. (1988). Both positive and negative regulatory elements mediate expression of a photoregulated *cab* gene from *Nicotiana plumbagenifolia. EMBO J.* **7**, 1929–1936.

Chory, J. (1989). Genetic analysis of photoreceptor action in *Arabidopsis, thaliana. J. Cell. Biochem., Suppl.* **13D**, 237.

Church, G. M., Ephrussi, A., Gilbert, W., and Tonegawa, S. (1985). Cell-type-specific contacts to immunoglobulin enhancers in nuclei. *Nature (London)* **313**, 798–801.

Corruzi, G., Broglie, R., Edwards, C., and Chua, N.-H. (1984). Tissue specific and light regulated expression of a pea nuclear gene encoding the small subunit of the ribulose-1,5-bisphosphate carboxylase. *EMBO J.* **3**, 1671–1679.

Dean, C., Van Den Elzen, P., Tamaki, S., Black, M., Dunsmuir, P., and Bedbrook, J. (1987). Molecular characterisation of the rbcS multigene family of Petunia (Mitchell). *Mol. Gen. Genet.* **206**, 465–474.

Deng, X.-W., and Gruissem, W. (1987). Control of plastid gene expression during development: The limited role of transcriptional regulation. *Cell* **49**, 379–387.

Deng, X.-W., Stern, D. B., Tonkyn, J. C., and Gruissem, W. (1987). Plastid run-on transcription—Application to determine the transcriptional regulation of spinach plastid genes. *J. Biol. Chem.* **262**, 9641–9648.

De Vries, S., De Vries, S., Springer, J., and Wessels, J. G. H. (1982). Diversity of abundant mRNA sequences and patterns of protein synthesis in etiolated and greened pea seedlings. *Planta* **156**, 129–135.

De Vries, S., De Vries, S., Springer, J., and Wessels, J. G. H. (1983). Sequence diversity of polysomal mRNAs in roots and shoots of etiolated and greened pea seedlings. *Planta* **158**, 42–50.

Dunsmuir, P. (1985). The petunia chlorophyll a/b binding protein genes. A comparison of *cab* genes from different gene families *Nucleic Acids Res.* **13**, 2502–2518.

Ellis, R. (1979). The most abundant protein in the world. *Trends Biochem. Sci. (Pers. Ed.)* **4**, 241–244.

Feierabend, J. (1981). Influence of cytokinins on plastid biogenesis in rye leaves. *In* "Metabolism and Activities of Cytokinins" J. Guern and C. Péaud-Lenoël, eds.), pp. 252–260. Springer-Verlag, Berlin and New York.

Flores, S., and Tobin, E.M. (1986). Benzyladenine modulation of the expression of two genes for nuclear-encoded chloroplast proteins in *Lemna Gibba*: Apparent post-transcriptional regulation. *Planta* **168**, 340–349.

Fluhr, R., and Chua, N.-H. (1986). Developmental regulation of two genes encoding ribulose bisphosphate carboxylase small subunit in pea and transgenic Petunia plants: Phytochrome and blue light induction. *Proc. Natl. Acad. Sci. U.S.A.* **83**, 2358–2362.

Fluhr, R., Kuhlemeier, C., Nagy, F., and Chua, N. (1986a). Organ specific and light induced expression of plant genes *Science* **282**, 1106–1112.

Fluhr, R., Moses, P., Morelli, G., Coruzzi, G., and Chua, N.-H. (1986b). Expression dynamics of the pea rbcS multigene family and organ distribution of the transcripts. *EMBO J.* **5**, 2063–2071.

Gallagher, T. F., and Ellis, R. (1982). Light-stimulated transcription of genes for two chloroplast polypeptides in isolated pea leaf nuclei. *EMBO J.* **1**, 1493–1498.

Gallagher, T., Jenkins, G., and Ellis, R. (1985). Rapid modulation of transcription of nuclear genes encoding chloroplast proteins by light. *FEBS Lett.* **186**, 241–245.

Giuliano, G., Pichersky, E., Malik, V. S., Timko, M. P., Scolnik, P. A., and Cashmore, A. R. (1988). An evolutionarily conserved protein binding sequence upstream of a plant light-regulated gene. *Proc. Natl. Acad. Sci. U.S.A.* **85**, 7089–7093.

Gollmer, I., and Apel, K. (1983). The phytochrome-controlled accumulation of mRNA sequences encoding the light-harvesting chlorophyll a/b-protein of barley (*Hordeum vulgare* L.). *Eur. J. Biochem.* **133**, 309–313.

Greeland, A., Thomas, M., and Walden, R. (1987). Expression of two nuclear genes encoding chloroplast proteins during early development of cucumber seedlings. *Planta* **170**, 99–110.

Green, P. J., Kay, S. A., and Chua, N.-H. (1987). Sequence-specific interactions of a pea nuclear factor with light-responsive elements upstream of the rbcS-3A gene. *EMBO J.* **6**, 2543–2549.

Green, P. J., Yong, M.-H., Cuozzo, M., Kano-Murakami, Y., Silverstein, P., and Chua, N.-H. (1988). Binding site requirements for pea nuclear protein factor GT-I correlate with sequences required for light-dependent transcriptional activation of the rbcS-3A gene. *EMBO J.* **7**, 4035–4044.

Herrera-Estrella, L., Van Den Broek, G., Maenhout, R., Van Montagu, M., Schell, J., Timko, M. P., and Cashmore, A. R. (1984). Light inducible and chloroplast associated expression of a chimaeric gene introduced into *Nicotiana tabacum* using a Ti-plasmid vector. *Nature (London)* **310**, 115–120.

Herrman, R. G., Westhoff, P., Alt, J., Tittgen, J., and Nelson, N. (1985). Thylakoid membrane proteins and their genes. *In* "Molecular Form and Function of the Plant Genome" (L. Van Vloten-Doting, G. S. P. Groot, and T. C. Hall, eds.), pp. 233–256. Plenum, New York.

Hoffman, N. E., Pichersky, E., Malik, V. S., Castresana, C., Kenton, K., Darr, S. C., and Cashmore, A. R. (1987). A cDNA clone encoding a photosystem I protein with homology to photosystem II chlorophyl a/b binding polypeptides. *Proc. Natl. Acad. Sci. U.S.A.* **84**, 8844–8848.

Inamine, G., Nash, B., Weissbach, H., and Nathan, B. (1985). Light regulation of the synthesis of the large subunit of ribulose-1-5-bisphosphate carboxylase in pea: Evidence for translational control. *Proc. Natl. Acad. Sci. U.S.A.* **82**, 5690–5694.

Jefferson, R. A., Kavanagh, T. A., and Bevan, M. (1987). GUS fusions: β-Glucuronidase as a sensitive and versatile gene fusion marker in higher plants. *EMBO J.* **6**, 3901–3907.

Jones, J. B., Dunsmuir, P., and Bedbrook, J. (1985). High level of expression of introduced chimaeric genes in regenerated transformed plants. *EMBO J.* **4**, 2411–2418.

Karlin, Neumann, G. A., Kohorn, B. D., Thornber, J. P., and Tobin, E. (1985). A chlorophyll a/b-protein encoded by a gene containing an intron with characteristics of a transposable element. *J. Mol. Appl. Genet.* **3**, 45–61.

Karlin-Neumann, G. A., Sun, L., and Tobin, E. M. (1988). Expression of light-harvesting chlorophyll a/b-protein genes is phytochrome regulated in etiolated *Arabidopsis thaliana* seedlings. *Plant Physiol.* **88**, 1323–1331.

Kaufman, L., Thompson, W., and Briggs, W. (1984). Different red light requirements for phytochrome-induced accumulation of cab RNA and rbcS RNA. *Science* **226**, 1447–1449.

Kaufman, L. S., Watson, J. C., Briggs, W. R., and Thompson, W. (1985). Photoregulation of nucleus-encoded transcripts: Blue-light regulation of specific transcript abundance. *In* "Molecular Biology of the Photosynthetic Apparatus" (K. Steinbeck, S. Bonitz, C. Arntzen, and L. Bogorad, eds.), pp. 367–372. Cold Spring Harbor Lab., Cold Spring Harbor, New York.

Keith, B., and Chua, N.-H. (1986). Monocot and dicot pre-mRNAs are processed with different efficiencies in transgenic tobacco. *EMBO J.* **5**, 2419–2425.

Klein, R. R., and Mullet, J. E. (1987a). Regulation of chloroplast encoded chlorophyl-binding protein translation during higher plant chloroplast biogenesis. *J. Biol. Chem.* **261**, 11138–11145.

Klein, R. R., and Mullet, J. E. (1987b). Control of gene expression during higher plant chloroplast biogenesis. *J. Biol. Chem.* **262**, 4341–4348.

Kloppstech, K. (1985). Diurnal and circadian rhythmicity in the expression of light-induced plant nuclear mRNAs. *Planta* **165**, 502–506.

Kuhlemeir, C., Green, P., and Chua, N.-H. (1987). Regulation of gene expression in higher plants. *Annu. Rev. Plant Physiol.* **38**, 221–257.

Kuhlemeir, C., Fluhr, R., and Chua, N.-H. (1988). Upstream sequences determine the difference in transcript abundance of pea rbcS genes. *Mol. Gen. Genet.* **212**, 405–411.

Leutwiler, L., Myerowitz, E., and Tobin, E. (1986). Structure and expression of three light-harvesting chlorophyl a/b binding protein genes in *Arabidopsis thaliana*. *Nucleic Acids Res.* **14**, 4051–4065.

Lorimer, G. (1981). The carboxylation and oxygenation of ribulose 1,5-bisphosphate: The primary events in photosynthesis and photorespiration. *Annu. Rev. Plant Physiol.* **32**, 349–383.

Martineau, B., and Taylor, W. C. (1985). Photosynthetic gene expression and cellular differentiation in developing maize leaves. *Plant Physiol.* **78**, 399–404.

Mayfield, P., and Taylor, W. C. (1984). Carotenoid-deficient maize seedlings fail to accumulate light-harvesting chlorophyll a/b binding protein (LHCP) mRNA. *Eur. J. Biochem.* **144**, 79–84.

Mayfield, S., Nelson, T., Taylor, W., and Malkin, R. (1986). Carotenoid synthesis and pleiotropic effects in carotenoid deficient seedlings of maize. *Planta* **169**, 23–32.

Meyer, H., Thienel, U., and Piechulla, B. (1989). Molecular characterization of the dirunal/circadian expression of the chlorophyll a/b binding proteins in leaves of tomato and other dicotyledonous and monocotyledonous plant species. *Planta* **180**, 5–15.

Mitra, A., Choi, H. K., and An, G. (1989). Structural and functional analysis of *Arabidopsis thaliana* chlorophyll a/b binding protein (cab) promoters. *Plant Mol. Biol.* **12**, 169–179.

Morelli, G., Nagy, F., Fraley, R., Rogers, S., and Chua, N.-H. (1985). A short conserved sequence is involved in the light-inducibility of a gene encoding ribulose 1,5-bisphosphate carboxylase small subunit of pea. *Nature (London)* **315**, 200–204.

Muller, M., Viro, M., Balke, C., and Kloppstech, K. (1980). Polyadenylated mRNA for the light-harvesting chlorophyll a/b protein. Its presence in green and absence in chloroplast-free plant cells. *Planta* **148**, 444–447.

Nagy, F., Fluhr, R., Kuhlemeier, C., Kay, S., Boutry, M., Green, P., Poulsen, C., and Chua, N.-H. (1986a). *CIS*-acting elements for selective expression of two photosynthetic genes in transgenic plants. *Philos. Trans. R. Soc. London, B* **314**, 493–500.

Nagy, F., Kay, S., Boutry, M., Hsu, M., and Chua, N.-H. (1986b). Phytochrome-controlled expression of a wheat Cab gene in transgenic tobacco seedlings. *EMBO J.* **5**, 1119–1124.

Nagy, F., Boutry, M., Hsu, M.-Y., Wong, M., and Chua, N.-H. (1987). The 5′ proximal region of the wheat Cab-1 gene contains a 268bp enhancer-like sequence for phytochrome response. *EMBO J.* **6**, 2537–2542.

Nagy, F., Kay, S. A., and Chua, N.-H. (1988). A circadian clock regulates transcription of the wheat cab-1 gene. *Genes Dev.* **2**, 376–382.

Nelson, T., Harpster, M., Mayfield, S., and Taylor, W. (1984). Light-regulated gene expression during maize leaf development. *J. Cell Biol.* **98**, 558–564.

Oelmuller, R., and Mohr, H. (1986). Photooxidative destruction of chloroplasts and its consequences for expression of nuclear genes. *Planta* **167**, 106–113.

Oelmuller, R., Dietrich, G., Link, G., and Mohr, H. (1986). Regulatory factors involved in gene expression (subunits of ribulose-1,5 bisphosphate carboxylase) in mustard (*Sinapis alba* L.) cotyledons. *Planta* **169**, 260–266.

Parthier, B. (1979). The role of phytohormones (cytokinins) in chloroplast development. *Biochem. Physiol. Pflanz.* **174**, 173–214.

Paulsen, H., and Bogorad, L. (1988). Diurnal and circadian rhythms in the accumulation and synthesis of mRNA for the light-harvesting chlorophyl a/b binding protein in tobacco. *Plant Physiol.* **88**, 1104–1109.

Piechulla, B. (1988). Plastid and nuclear mRNA fluctuations in tomato leaves: Diurnal and circadian rhythms during extended dark and light periods. *Plant Mol. Biol.* **11**, 345–353.

Piechulla, B. (1989). Changes of the diurnal and circadian (endogenous) mRNA oscillations of the chlorophyl a/b binding protein in tomato leaves during altered day–night (light/dark) regimes. *Plant Mol. Biol.* **12**, 317–327.

Piechulla, B., and Gruissem, W. (1987). Diurnal mRNA fluctuations of nuclear and plastid genes in developing tomatos fruits. *EMBO J.* **6**, 3595–3599.

Piechulla, B., Pichersky, E., Cashmore, A., and Gruissem, W. (1986). Expression of nuclear and plastid genes for photosynthesis-specific proteins during tomato fruit ripening. *Plant Mol. Biol.* **7**, 367–376.

Plumley, F. G., and Schmidt, G. W. (1989). Nitrogen-dependent regulation of photosynthetic gene expression. *Proc. Natl. Acad. Sci. U.S.A.* **86**, 2678–2682.

Rodermel, S., and Bogorad, L. (1985). Maize plastid photogenes: Mapping and photoregulation of transcript levels during light-induced development. *J. Cell Biol.* **100**, 463.

Sanford, J. C. (1988). The biolistic process. *Trends Biochem. Sci. (Pers. Ed.)* **6**, 299–302.

Sasaki, Y., Sakihama, T., Kamikubo, T., and Shinozaki, K. (1983). Phytochrome-mediated regulation of two mRNAs, encoded by nuclei and chloroplasts of ribulose 1,5 bisphosphate carboxylase/oxygenase. *Eur. J. Biochem.* **133**, 617–620.

Sheen, J. Y., and Bogorad, L. (1986). Differential expression of six light-harvesting chlorophyll a/b binding protein genes in maize leaf cell types. *Proc. Natl. Acad. Sci. U.S.A.* **83**, 7811–7815.

Sheen, J., and Bogorad, L. (1987). Regulation of nuclear transcripts for C4 photosynthesis in bundle sheath and mesophyl cells of maize leaves. *Plant Mol. Biol.* **8**, 227–238.

Silverthorne, J., and Tobin, E. (1984). Demonstration of transcriptional regulation of specific genes by phytochrome action. *Proc. Natl. Acad. Sci. U.S.A.* **81**, 112–116.

Simpson, J., Timko, M., Cashmore, A., Schell, J., Van Montagu, M., and Herrera-Estrella, L. (1985). Light-inducible and tissue specific expression of a chimaeric gene under control of the 5′ flanking sequence of a pea chlorophyll a/b binding protein gene. *EMBO J.* **4**, 2723–2729.

Simpson, J., Van Montagu, M., and Herrera-Estrella, L. (1986a). Photosynthesis associated gene families: Differences in response to tissue specific and environmental factors. *Science* **233**, 36–38.

Simpson, J., Schell, J., Van Montagu, M., and Herrera-Estrella, L. (1986b). The light-inducible and tissue specific expression of a pea LHCP gene involves an upstream element combining enhancer and silencer-like properties. *Nature (London)* **323**, 551–553.

Smith, S., and Ellis, R. (1978). Light-stimulated accumulation of transcripts of nuclear and chloroplast genes for ribulose-biphosphate carboxylase. *J. Mol. Appl. Genet.* **1**, 127–137.

Spiller, S. C., Kaufman, L. S., Thompson, W. F., and Briggs, W. R. (1987). Specific mRNA and rRNA levels in greening pea leaves during the recovery from iron-stress. *Plant Physiol.* **84**, 409–411.

Stayton, M. M., Brosio, P., and Dunsmuir, P. (1987). Characterization of a full length petunia cDNA encoding a polypeptide of the light harvesting complex associated with photosystem I. *Plant Mol. Biol.* **10**, 127–137.

Stayton, M. H., Brosio, P., and Dunsmuir, P. (1989). Photosynthetic genes of petunia (Mitchell) are differentially expressed during the diurnal cycle. *Plant Physiol.* **89**, 776–782.

Struhl, K. (1987). Promoters, activator proteins and the mechanism of transcriptional initiation in yeast. *Cell* **49**, 295–297.

Taylor, W. C. (1989). Transcriptional regulation by a circadian rhythm. *Plant Cell* **1**, 259–264.

Teeri, T. H., Lehvaslaiho, H., Franck, M., Votila, J., Heino, P., Palva, E. T., Van Montagu, M., and Herrera-Estrella, L. (1989). Gene fusions to *Lac Z* reveal new expression patterns of chimaeric genes in transgenic plants. *EMBO J.* **8**, 343–350.

Teyssendier de la Serve, B., Axelos, M., and Péaud-Lenoël, C. (1985). Cytokinins modulate the expression of genes encoding the protein of the light-harvesting chlorophyll a/b complex. *Plant Mol. Biol.* **5**, 155–163.

Thompson, W., and Cleland, R. (1972). Effect of light and gibberellin on RNA species of pea stem tissues as studied by DNA–RNA hybridization. *Plant Physiol.* **50**, 289–292.

Timko, M. P., and Cashmore, A. R. (1983). Nuclear genes encoding the constituent polypeptides of the light harvesting chlorophyl a/b protein complex from pea. *In* "Plant Molecular Biology" (R. Goldberg, ed.), pp. 403–412. Liss, New York.

Timko, M., Kausch, A., Castresana, C., Fassler, J., Herrera-Estrella, L., Van den Broeck, G., Van Montagu, M., Schell, J., and Cashmore, A. (1985a). Light regulation of plant gene expression by an upstream enhancer-like element. *Nature (London)* **318**, 579–582.

Timko, M., Kausch, A., Hand, J., Cashmore, A., Herrera-Estrella, L., Van den Broeck, G., and Van Montagu, M. (1985b). Structure and expression of nuclear genes encoding polypeptide of the photosynthetic apparatus. *In* "Molecular Biology of the Photosynthetic Apparatus" (K. E. Steinbeck, S. Bonitz, C. J. Arntzen, and L. Bogorad, eds.), pp. 381–396. Cold Spring Harbor Lab., Cold Spring Harbor, New York.

Tobin, E. (1981). White light effects on the mRNA for the light-harvesting chrorophyl a/b protein in *Lemna gibba* L.G.-3. *Plant Physiol.* **67**, 1073–1083.

Tobin, E., and Klein, R. R. (1975). A. Isolation and translation of plant messenger RNA. *Plant Physiol.* **56**, 88–92.

Tobin, E. M., and Silverthorne, J. (1985). Light regulation of gene expression in higher plants. *Annu. Rev. Plant Physiol.* **36**, 569–593.

Tobin, E., Wimpee, C., Silverthorne, J., Stiekema, W., Neumann, G., and Thornber, J. (1984). Phytochrome regulation of the expression of two nuclear coded chloroplast proteins. *UCLA Symp. Mol. Cell. Biol.* **14**, 325.

Vince-Prue, D. (1983). Photomorphogenesis and flowering. *In* "Encyclopedia of Plant Physiology" (W. Shropshire and H. Mohr, eds.), Vol. 16B, pp. 458–490. Springer-Verlag, Berlin.

ACTIVATION, STRUCTURE, AND ORGANIZATION OF GENES INVOLVED IN MICROBIAL DEFENSE IN PLANTS

Richard A. Dixon and Maria J. Harrison

Plant Biology Division, The Samuel Roberts Noble Foundation, Inc.,
Ardmore, Oklahoma 73402

165

ADVANCES IN GENETICS, Vol. 28

I. Introduction

A. SCOPE OF THE REVIEW

Over the last few years, increasing interest has been shown in the molecular biology of plant–pathogen interactions. Reasons for this include the realization that the powerful new tools of gene manipulation technology may profitably be applied to the design of plants with improved defense characteristics, as well as the attractiveness, from a more fundamental viewpoint, of biological systems that encompass, at the molecular level, recognition, signal transduction, and gene activation. The active nature of plant defense responses was first clearly documented in the mid-1970s, when it was shown that RNA and protein synthesis were often required for expression of resistance by the host, and that appearance of "molecular defensive barriers" required the *de novo* synthesis of specific enzymes (reviewed by Dixon *et al.*, 1983). At about the same time, molecules termed elicitors were attracting attention as potential microbial inducers of host defense (reviewed by Darvill and Albersheim, 1984). Since then, much progress has been made in our understanding of the biochemistry and molecular biology of induced host defenses, and the application of genetic transformation techniques to plant pathogenic bacteria, and more recently fungi, has led to the isolation of pathogen genes which may encode molecular determinants of avirulence or pathogenicity (reviewed by Dixon *et al.*, 1990).

The main aim of the present review is to discuss recent progress in the molecular genetics of activation of plant defenses in response to pathogen attack. Primary consideration is given to resistance to fungal and bacterial pathogens by members of the plant family Leguminosae; this bias simply reflects the authors' main research interests. The genes discussed in most detail are those whose activation contributes directly to potential resistance mechanisms. Such genes have been termed disease resistance response genes or defense response genes, and are distinct from resistance genes per se, which determine, by mechanisms as yet to be elucidated, whether or not resistance will be expressed in response to a particular pathogen. Several of the genes activated in response to fungal or bacterial infection are also inducible by viral infection, UV light, or physical wounding, and many are also involved in nondefensive roles within the plant, under the control of developmental cues. Reference to these aspects is included. Discussion of host defense response genes as a central theme requires a brief introduction to the biology and genetics of disease resistance, the nature of the microbial factors believed to be responsible for defense gene induction, and the signal transduction pathways directly leading to gene activation. Additionally, some comments on the potential of genetic transformation strategies for crop protection are given in Section VI.

A number of recent review articles can provide the reader with more detailed information on pathogenicity determinants in plant pathogenic bacteria (Daniels *et al.*, 1988), plant disease resistance genes (Pryor, 1987; Ellis *et al.*, 1988), and elicitors and signal transduction (Dixon 1986; Ryan, 1988; Templeton and Lamb, 1988; Lamb *et al.*, 1989; Dixon and Lamb, 1990b). Other reviews dealing more specifically with defense response gene activation include those by Dixon *et al.* (1986), Dixon and Bolwell (1986), Collinge and Slusarenko (1987), Ryan and An (1988), Dixon and Lamb (1990a), and Hahlbrock and Scheel (1989).

B. RESISTANCE AND SUSCEPTIBILITY IN PLANT–PATHOGEN INTERACTIONS

Many plant pathogens express a virulent phenotype (i.e., cause disease) on one or a limited number of host species. Host range is therefore an important genetically defined determinant associated with the ability to be a pathogen. Recent work on the molecular determinants of host range in both plant pathogenic and symbiotic bacteria has been reviewed (Keen and Staskawicz, 1988).

If a pathogen has the ability successfully to infect a host species, it may do so by one of two general strategies. In the terminology of Keen (1986), unspecialized "thugs" are necrotrophic, and damage host tissue through the production of enzymes and/or toxins. As they may induce a considerable response in the host, they are often able to avoid or inactivate host defenses. More specialized pathogens initially infect without causing damage to the host, and for some time grow biotrophically within plant tissue. Such "con artists" are usually susceptible to induced host defenses, the specificity of the interaction being dependent upon whether these responses are triggered. In an incompatible interaction (pathogen avirulent, host resistant), early molecular recognition events trigger the activation of host defense responses. In a compatible interaction (pathogen virulent, host susceptible), the pathogen either eludes or suppresses recognition and successful colonization ensues, leading to disease.

An example of this latter type of disease strategy is seen in the infection of bean *(Phaseolus vulgaris)* by the anthracnose fungus *Colletotrichum lindemuthianum*, the cytology of which has been studied in some detail (O'Connell *et al.*, 1985; O'Connell and Bailey, 1988). In susceptible tissue, after appressorium formation and penetration, intracellular infection vesicles form in epidermal cells, which remain alive during subsequent intracellular growth of primary hyphae. After a biotrophic phase of no more than 24 hours, the cytoplasm of the infected cell gradually degenerates and the cell dies; this cycle is repeated as cells successively become infected. In contrast, in resistant interactions, a rapid collapse of infected tissue [the hypersensitive response (HR)] limits the invading fungus to a single cell. The HR is a common feature of incompatible interactions between many fungal, bacterial, and viral plant pathogens.

Specific biochemical changes in the host accompany cell death in both compatible and incompatible interactions. In particular, the HR is often associated with the rapid accumulation, to high local concentrations around the lesion, of antimicrobial compounds termed phytoalexins. There is now good evidence that phytoalexins accumulate to sufficiently high concentrations in the immediate vicinity of the invading microorganism to enable them to act as effective resistance factors (Pierce and Essenberg, 1987), that inhibition of phytoalexin synthesis can result in loss of resistance (Moesta and Grisebach, 1982; Ralton *et al.*, 1988), and that, at least in some interactions, ability of the fungus to detoxify phytoalexins is an important virulence determinant (Kistler and VanEtten, 1984). Genes involved in phytoalexin synthesis are therefore important determinants of a plant's defensive capabilities.

Specific interactions between legumes and their fungal or bacterial pathogens often follow a pattern that operates in many plant–pathogen systems, wherein different host cultivars exhibit differential responses to distinct physiological races of the pathogen. Such race-specific resistance is believed to function via the interaction of dominant avirulence gene-encoded functions in the pathogen with dominant disease resis-

TABLE 1
The Quadratic Check[a]

Host cultivar	Pathogen race[b]			
	P1 P1 P2 P2	P1 P1 p2 p2	p1 p1 P2 P2	p1 p1 p2 p2
R1 R1 R2 R2	Res	Res	Res	Susc
R1 R1 r2 r2	Res	Res	Susc	Susc
r1 r1 R2 R2	Res	Susc	Res	Susc
r1 r1 r2 r2	Susc	Susc	Susc	Susc

[a] Pattern of interactions between a host and its pathogen wherein two resistance genes (R) are complemented by two avirulence genes (P). (After Ellingboe, 1981.)
[b] Res, Resistant (pathogen avirulent on the host cultivar); Susc, susceptible (pathogen virulent on the host cultivar).

TABLE 2
Race Specificity in Anthracnose Disease on
Bean Leaves

Bean cultivar	Race of *Colletotrichum lindemuthianum*[a]			
	α	β	γ	δ
Imuna	R	R	R	S
Kievit	R	R	R	R
Pinto	S	S	S	S
Dubbele Witte	S	R	R	S

[a] R, Hypersensitive resistance; S, susceptibility.

tance gene-encoded functions in the host. This gene-for-gene hypothesis (reviewed by Ellingboe, 1981), which was first described in 1947 for the interaction of flax with the rust fungus *Melampsora lini,* implies that resistance will be functionally dominant. Furthermore, if the plant and its pathogen each contain a number of complementary resistance and avirulence genes, respectively, any gene combination determining incompatibility will be epistatic on other complementary gene pairs conferring compatibility. This is illustrated in the so-called quadratic check (Table 1). An example of such race specificity in the interaction of bean with *Colletotrichum* is shown in Table 2. Similar differential patterns are observed in the host–pathogen interactions of other legumes, such as soybean (Staskawicz *et al.,* 1984) or cowpea (Ralton *et al.,* 1988). The genetics of resistance of cultivated legumes to many of their fungal, bacterial, and viral pathogens has been reviewed (Meiners, 1981).

C. Molecular Genetics of Virulence, Avirulence, and Resistance

Figure 1 is a simplified scheme outlining the classes of genes and their products that play a role in determining the outcome of a race-specific plant–pathogen interaction. A more detailed version of this scheme has been published (Lamb *et al.,* 1989). In the pathogen, pathogenicity genes encode a number of factors necessary for potentially successful infection of host tissues. These include genes for the synthesis and extracellular transport of hydrolytic enzymes, such as cutinase, protease, cellulase, pectin lyase, and polygalacturonase, which may be involved in host cell wall degradation and therefore entry into host

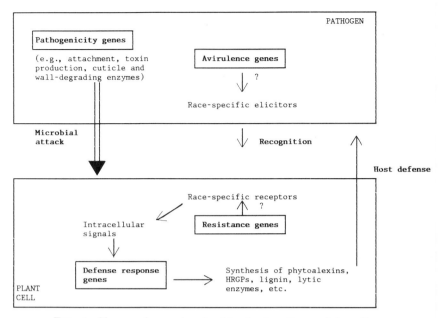

FIG. 1. Classes of genes involved in plant–pathogen interactions.

tissues; genes involved in attachment and differentiation of penetration structures; or genes for the biosynthesis of toxins or enzymes which may cause damage and death in advance of necrotrophic growth. Structural and regulatory genes thus involved in pathogenesis have now been cloned from a number of plant pathogenic bacteria and fungi. Strategies for cloning these genes have involved classical biochemistry, transposon mutagenesis, and mutant complementation by transformation. Much of this work has been reviewed (Mills, 1985; Daniels *et al.*, 1988; Dixon *et al.*, 1990; Lamb *et al.*, 1989).

The first cloned pathogen avirulence gene was obtained by transforming a race of the bacterial soybean pathogen *Pseudomonas syringae* pv. *glycinea*, which gave a susceptible interaction on a specific soybean cultivar, with a cosmid library of genomic DNA from a race (race 6) that induced the HR on the same cultivar (Staskawicz *et al.*, 1984). One particular transconjugant was altered to give an incompatible interaction on all soybean cultivars on which the DNA donor race 6 gave an incompatible interaction. The avirulence gene was further defined by transposon mutagenesis of the 27.2-kilobase (kb) insert in the cosmid, and an open reading frame encoding a 100-kDa polypeptide was identified as determining the race 6 phenotype (Napoli and Staskawicz, 1987). Similar strategies have been used recently for the cloning of further avirulence genes from phytopathogenic bacteria (Table 3). Two main conclusions can be drawn from these studies. First, in a number of cases, cloned avirulence genes appear to complement specific host resistance genes, thus providing molecular genetic support for the gene-for-gene hypothesis (Gabriel *et al.*, 1986; Staskawicz *et al.*, 1987). Second, sequence analysis has failed to show homology to known polypeptides and has not provided evidence for signal peptides for externalization of the avirulence gene products. The latter point has been confirmed by immunocytochemical studies. It is therefore unlikely that these avirulence gene products themselves act as race-specific elicitors that could interact directly with a host receptor to trigger resistance. Recently, it has been shown that overexpression of the *avrD* gene of *Pseudomonas syringae* pv. *tomato* in the nonpathogen *Escherichia coli* leads to induction of the hypersensitive response on tobacco by *E. coli* in the correct cultivar-specific manner (Stayton *et al.*, 1989). The transformed *E. coli* produce a low-molecular-weight metabolite which acts as a race-specific elicitor. The characterization of this metabolite is eagerly awaited, as is the confirmation of whether other avirulence gene products also function biosynthetically.

Although the preceding results confirm the basic gene-for-gene hypothesis, avirulence genes may themselves be under regulatory control. A cluster of genes (the *hrp* locus) has been shown to determine the ability of *P. syringae* pathovars to induce the hypersensitive response (Lindgren *et al.*, 1986, 1988). Furthermore, a number of bacterial avirulence genes are induced *in planta* (Table 3), but not under normal growth conditions in culture. This appears to reflect a requirement for somewhat nonspecific nutritional factors rather than for a specific class of inducer molecule, such as is required for the induction of the *vir* genes of *Agrobacterium* or the *nod* genes of *Rhizobium*. Further details of the molecular determinants of specificity in bacterial pathogen–plant interactions can be found in articles by Staskawicz *et al.* (1988) and by Keen and Staskawicz (1988).

Work on the genetic determinants of avirulence in plant pathogenic fungi has lagged behind that in bacteria because of the lack of defined genetic systems in many of these organisms, coupled with only recent development of suitable transformation vectors. However, rapid progress is now being made in transformation technology for several important plant pathogens, for example, the corn smut fungus *Ustilago maydis* (Wang *et al.*, 1988). Transformation of the saprophyte *Aspergillus nidulans* has recently been used for functional identification of a clone encoding a virulence gene responsible for phytoalexin detoxification by the pea pathogen *Nectria haematococca* (Weltring *et al.*, 1988).

TABLE 3

Cloned Avirulence Genes from Plant-Pathogenic Bacteria

Bacterium	Host	Gene	Open reading frame (kDa)	Induction in plant	Specificity	Reference
Pseudomonas syringae pv. glycinea race 6	Soybean	avrA	100	—	Determines race 6 phenotype	Staskawicz et al. (1984), Napoli and Staskawicz (1987)
Pseudomonas syringae pv. glycinea race 0	Soybean	$avrB_0$	36	Yes	Determines race 1 phenotype	Tamaki et al. (1988)
		avrC	39	Yes		Staskawicz et al. (1987)
Pseudomonas syringae pv. glycinea race 1	Soybean	$avrB_1$	—	—	Identical to $avrB_0$ from race 0. Interacts with soybean resistance locus Rgp1	Staskawicz et al. (1987)

Organism	Host	Avirulence gene(s)			Comments	Reference
Pseudomonas syringae pv. *tomato*	Tomato	*avrA*; *avrD*	34	—	Can confer cultivar specificity on soybean (*avrA* identical to *P. syringae* pv. *glycinea avrA*)	Kobayashi *et al.* (1988)
Pseudomonas syringae pv. *syringae*	Tobacco	—	—	—	Transfer to *Pseudomonas fluorescens* results in induction of HR	Huang *et al.* (1988)
Xanthomonas campestris pv. *vesicatoria*	Tomato and pepper	*avrRxv*	—	—	Determines hypersensitive resistance on nonhost species	Whalen *et al.* (1988)
		avrBs1	—	—		Kearney *et al.* (1988)
Xanthomonas campestris pv. *malvacearum*	Cotton	Five avirulence genes	—	—	Each interacts specifically with an individual resistance gene in congenic cotton lines	Gabriel *et al.* (1986)

173

A major phenotypic result of the action of host resistance genes is rapid responsiveness to avirulent pathogen races. At the cytological level, this is often associated with the localized cell death of the hypersensitive response. At the molecular level, defense response genes are activated for the production of antimicrobial barriers. In the scheme in Fig. 1, the simplest, most direct model is presented in which host resistance genes are predicted to encode, probably constitutively, receptors that bind microbial determinants of avirulence (race-specific elicitors). There is as yet no direct evidence to support this model; neither has any disease resistance gene been cloned to date. Shotgun cloning by function of resistance genes, using strategies analagous to those previously described for bacterial virulence and avirulence genes, is not yet feasible due to low plant transformation efficiencies and the large genome sizes of plants compared to bacteria.

The approaches currently being considered for the cloning of resistance genes include (1) the use of race-specific elicitors as probes for the cognate plant gene products; (2) the use of restriction fragment length polymorphisms (RFLPs) and other markers for linkage analysis, to be followed by cloning of large DNA fragments (separated by pulsed-field gel electrophoresis) for the construction of region-specific libraries for chromosome walking to the linked resistance locus; and (3) transposon tagging of resistance genes. All three approaches are not without considerable difficulties. In addition to the large amount of screening involved, transposon tagging of resistance genes is further complicated by either high copy number or low transposition rate of currently usable transposable elements, or, as in the case of the $Rp1$ rust-resistance locus in maize, high rates of spontaneous mutational events in genetic backgrounds lacking active transposable elements (Pryor, 1987; Bennetzen et al., 1988). The chromosome walking strategy is seriously complicated if the plant genome size is large and there is much repetitive DNA. Systems currently under investigation using this approach include lettuce (for resistance genes to the downy mildew fungus Bremia lactucae) (Michelmore et al., 1987), tomato (for which detailed RFLP maps are now available, which can be intensified around a number of disease resistance loci) (e.g., Young et al., 1988), and, as a model system, Arabidopsis thaliana. The latter is receiving particular attention in view of its very small genome size [7×10^4 kilobase pairs (kbp)], low amount of repetitive DNA, ability to be transformed by Agrobacterium tumefaciens, and availability of a detailed RFLP linkage map (Myerowitz and Pruitt, 1985; Chang et al., 1988; Myerowitz, 1989; Shields, 1989). The plant is, however, a weed of no economic importance, and its disease resistance characteristics still require further investigation.

Although the exact molecular nature of resistance genes is still unclear, genetic analysis has revealed that loci such as *Rp1* are complex functional units that may contain several closely linked specificity determinants. It is possible that rearrangements or recombinational events at these complex loci may lead to rapid reassortment to produce new specificities to match changes in the pathogen population (Pryor, 1987). The next few years should hopefully see major advances in our understanding of these genes and how they may act to regulate the expression of resistance response genes.

II. Biochemistry of Induced Resistance

A. PATHWAYS AND ENZYMOLOGY OF ISOFLAVONOID PHYTOALEXIN BIOSYNTHESIS

Phytoalexins have been defined as "low molecular weight compounds that are both synthesized by and accumulated in plant cells after exposure to microorganisms" (Paxton, 1981). For a compound to be properly classified as a phytoalexin it should also satisfy the criteria of *de novo* synthesis in response to infection and accumulation to antimicrobial concentrations in the area of infection. Evidence for the satisfaction of these latter criteria was listed in Section I,B. Most phytoalexins are broad-spectrum antibiotic compounds, which may also be induced in plants by a range of metabolic or physical stresses in addition to infection (Bailey, 1987). Phytoalexins have been isolated and characterized from members of a number of plant families, including the Leguminosae (in which they are isoflavonoid derivatives, or occasionally stilbenes), Solanaceae (sesquiterpenes and polyacetylenes), and Umbelliferae (furanocoumarins, isocoumarins, and chromones). Most phytoalexins are biogenetically related to known classes of secondary products which accumulate constitutively, or under developmental regulation, in the same plant species. A given plant may accumulate a small number of closely related phytoalexins, and these may sometimes prove useful as chemotaxonomic markers.

The pathway for the biosynthesis of isoflavonoid phytoalexins in legumes such as bean, soybean, and pea is outlined in Fig. 2. The early stages of this pathway are also involved in the synthesis of stilbene phytoalexins in peanuts and grapevine, and in the synthesis of wall-bound phenolics and lignin in a wide range of plant species. Flavonoids and wall-bound phenolics are normal constituents of healthy, uninfected plant tissues.

Over the last 10 years, nearly every enzyme involved in the biosynthesis of isoflavonoid phytoalexins in legumes has been character-

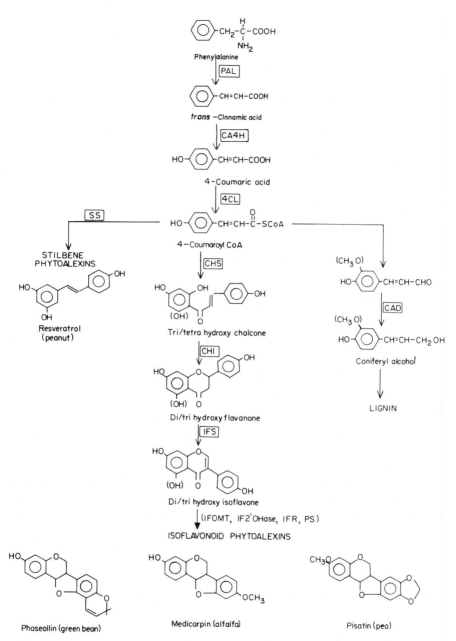

FIG. 2. Biosynthesis of phenylpropanoid defense metabolites in legumes. PAL, Phenylalanine ammonia-lyase; CA4H, cinnamic acid 4-hydroxylase; 4CL, 4-coumarate : CoA ligase; SS, stilbene synthase; CAD, coniferyl alcohol dehydrogenase; CHS, chalcone synthase; CHI, chalcone isomerase; IFS, isoflavone synthase; IFOMT, isoflavone *O*-methyl transferase; IF2'OHase, isoflavone 2' hydroxylase; IFR, isoflavone reductase; PS, pterocarpan synthase.

ized, the most recent examples being the latter enzymes specific for modifications of the basic isoflavonoid skeleton (Dixon *et al.*, 1983; Hagmann and Grisebach, 1984; Dixon and Bolwell, 1986; Biggs *et al.*, 1987; Tiemann *et al.*, 1987; Barz *et al.*, 1988; Bless and Barz, 1988; Welle and Grisebach, 1988a,b). These enzymes include cytochrome *P*-450-dependent activities (isoflavone synthase, isoflavone 2'-hydroxylase), and prenyl transferases, which have proved difficult to purify. Their molecular biology therefore remains to be investigated. However, in all systems studied so far, induction of isoflavonoid phytoalexin synthesis appears to be associated with increased activities of the relevant biosynthetic enzymes (Dixon *et al.*, 1983; Robbins *et al.*, 1985; Barz *et al.*, 1988; Daniel *et al.*, 1988; Dalkin *et al.*, 1990a) and this in turn, where investigated, has been shown to result from increased mRNA levels (see Section III). All the enzymes in Fig. 2 can therefore probably be described as products of defense response genes in legumes.

The early enzymes of the isoflavonoid pathway common to other areas of secondary metabolism have been extensively studied. In bean and alfalfa, L-phenylalanine ammonia-lyase exists as a number of isoforms with different K_m values, and the lower K_m forms are selectively induced in response to fungal elicitors and (in the case of bean) wounding or UV light (Bolwell *et al.*, 1985a; Liang *et al.*, 1989a; Jorrin and Dixon, 1990). This suggests a mechanism for channeling of phenylalanine into defense-related products specifically under stress conditions. Chalcone synthase, the key enzyme specific for flow into the flavonoid–isoflavonoid branch of phenylpropanoid biosynthesis, likewise exists as multiple isoforms (Hamdan and Dixon, 1987a), although their functional significance remains unclear. This complexity at the enzymatic level, arising from their being encoded by multigene families (Section IV,A), is consistent with the roles of these two enzymes as major control points in plant secondary metabolism. In contrast, later enzymes committed to formation of specific end-products, for example, chalcone isomerase (Robbins and Dixon, 1984), do not appear to exhibit such complex polymorphism.

B. CELL WALL MODIFICATIONS

1. Deposition of Phenolic Material

Rapid incorporation of phenolic material into plant cell walls, either by simple esterification of hydroxycinnamic acid derivatives or by peroxidase-catalyzed polymerization of phenylpropane alcohols to yield lignin, is often associated with induced resistance (Friend, 1976; Bolwell *et al.*, 1985c; Matern and Kneusel, 1988). Induced lignification may be an especially important defense response in certain monocots, in which its rapid induction appears to be controlled by major resistance genes such as the *Sr5* gene for stem rust resistance in wheat (Reisener *et al.*, 1986). Specific enzymes related to lignification include caffeic acid *O*-methyltransferase, coniferyl alcohol dehydrogenase, and peroxidase; elevated activities of these enzymes have been demonstrated in legume cells in response to infection or treatment with fungal elicitors (Robbins *et al.*, 1985; Grand *et al.*, 1987; Dalkin *et al.*, 1990a).

2. Cell Wall Proteins

Hydroxyproline-rich glycoproteins (HRGPs) accumulate in the cell walls of a number of plant species in response to microbial infection (Esquerré-Tugayé *et al.*, 1979; Bolwell *et al.*, 1985c; Roby *et al.*, 1985; Stermer and Hammerschmidt, 1987), and this is believed to be a defense mechanism. As such, they may act to strengthen cell walls, may be a matrix for the deposition of phenolic material, or may possibly agglutinate pathogenic bacteria. In addition to the predominance of hydroxyproline residues formed by the posttranslational hydroxylation of peptide proline, these proteins are also rich in serine, lysine, valine, and tyrosine, and are characterized by a repeating Ser-Hyp$_4$ unit. In bean, a number of HRGPs exist, differing in amino acid composition and induction pattern (Section IV,B). Most of the proline residues of HRGPs contain *O*-glycosidically attached oligoarabinosides, and protein arabinosyl transferase activity is coinduced with HRGP appearance in elicitor-treated bean cells (Bolwell *et al.*, 1985c). The mature proteins form extended helical rods that become immobilized in the cell wall, probably by cross-linking of tyrosine residues.

Posttranslational hydroxylation of proline residues of the HRGP apopolypeptide is catalyzed by the microsomal enzyme prolyl hydroylase during export through the endomembrane system. Prolyl hydroxylase is a very interesting enzyme. The β-subunit from humans and/or chicken has recently been shown to be identical to another enzyme, protein disulfide isomerase, and to exhibit extremely high sequence homology to thioredoxins, to a cellular thyroid hormone-binding protein, and to a polyphosphoinositide-specific phospholipase C (Pihlajaniemi *et al.*, 1987; Parkkonen *et al.*, 1988; Bennett *et al.*, 1988). The enzyme has been purified from bean (Bolwell *et al.*, 1985b) and is rapidly induced in response to fungal elicitors (Bolwell *et al.*, 1985c; Bolwell and Dixon, 1986).

C. Hydrolases and Pathogenesis-Related Proteins

Induction of chitinase and 1,3-β-D-glucanase activities appears to be closely related to induced defense in many plant species (Dixon *et al.*, 1990). These enzymes have the potential to hydrolyze polymers of the fungal cell wall (in fact, chitinase has no known substrate in plant tissues). Purified chitinase is antifungal, particularly against species such as *Trichoderma,* which have a high chitin content in their cell walls (Schlumbaum *et al.,* 1986); some chitinases can also exhibit lysozyme-like activity against bacteria, although a number of plant pathogenic bacteria appear to be resistant to this activity. Glucanase can act as a synergist to increase the antifungal activity of chitinase against a range of plant pathogenic fungi (Mauch *et al.,* 1988b). In addition to pathogen attack, chitinase and glucanase can also be induced in many plants by treatment with elicitors or ethylene, although pathogen-induced ethylene production does not appear to be causally related to hydrolase induction (Mauch *et al.,* 1984).

Chitinase and glucanase have been purified from a range of plant species (Young and Pegg, 1981; Boller *et al.,* 1983; Kurosaki *et al.,* 1987a; Mauch *et al.,* 1988a; Metraux *et al.,* 1988). Ethylene-induced chitinase in bean leaves appears to be located primarily in the central vacuole of the cell, whereas glucanase is also found in the cell wall space (Mauch and Staehelin, 1989). This has led to the proposal that vacuolar chitinase and glucanase represent a final line of defense in microbial infection, whereas wall-located glucanase could have a separate function in releasing elicitor-active molecules from invading pathogens and thereby activating "early" defense systems such as phytoalexin synthesis (Mauch and Staehelin, 1989). In other plants, such as cucumber, oats and carrot, chitinase may also be extracellular (Boller and Metraux, 1988; Fink *et al.,* 1988; Kurosaki *et al.,* 1987b); like glucanase, it could therefore also be involved in elicitor release, as chitin oligomers are active in eliciting phenolic synthesis in some systems (Kurosaki *et al.,* 1987a). In cucumber, induction of chitinase in response to viral, fungal, and bacterial plant pathogens, as well as abiotic stresses, occurs both locally and systemically (Metraux and Boller, 1986).

"Pathogenesis-related (PR) protein" is a term used to describe a class of low-molecular-weight soluble proteins that accumulate in stressed plant tissues, often in intercellular spaces. They were first detected in virus-infected tobacco leaves, from which 10 major acidic PR proteins have now been identified (Rigden and Coutts, 1988), as well as a number of related basic proteins. Two of the acidic PR proteins from tobacco and two of the basic ones are endochitinases (Legrand *et al.,* 1987), whereas three acidic PR proteins and one basic one have been shown to be β-glucanases (Kauffmann *et al.,* 1987). Similarly, in potato leaves, six distinct chitinases and two β-glucanases are induced in response to infection with the late blight fungus *Phytophthora infestans* (Kombrink *et al.,* 1988). The PR-1a -1b, -1c, and R classes of PR protein from tobacco have molecular weights of 17,000 and 13,000–15,000, respectively. They have no known enzymatic functions. The Pr-la–1c proteins are rich in acidic and aromatic amino acids, whereas the R protein is rich in cysteine and has extensive sequence homology to the sweet-tasting protein thaumatin.

D. PROTEINASE INHIBITORS

The proteinase inhibitors (PIs) are believed to act as antinutrient factors that can protect plants against insect attack by inhibiting digestive enzymes. They are induced both locally and systemically in leaves of solanaceous species in response to wounding or insect feeding (Ryan, 1984). They are included here because their induction shows many features in common with that of the antimicrobial factors discussed above, and good progress has been made in recent years in elucidating some of the mechanisms associated with signal transduction and gene activation for these systemic protectants (Ryan, 1988; Ryan and An, 1988). Some aspects of the regulation of PI genes are briefly discussed in Section V.

III. Gene Activation in Induced Defense

A. CHANGES IN mRNA POPULATIONS IN INFECTED OR ELICITED CELLS

Several research groups have indirectly assessed the extent of changes in gene expression in infected or elicitor-treated cells by looking at patterns of polypeptides on two-dimensional gels after translation *in vitro* from mRNA. Such an approach has usually shown that infection or elicitation results in dramatic increases in activities of certain mRNA species, whereas others remain unaffected or decline. The second stage of this approach is to identify the mRNAs whose behavior suggests a possible role in defense. In pea endocarp tissue infected with *Fusarium solani* f. sp. *phaseoli* (incompatible) or *Fusarium solani* f. sp. *pisi* (compatible), approximately 21 different mRNA species were induced in the early stage (6 hours) of infection with either pathogen, during which time growth of both fungi is suppressed; after approximately 18 hours, specificity is expressed and the compatible fungus continues growth. This phase was associated with a decline in a number of induced mRNA activities in the compatible interaction; these mRNAs were also induced by the incompatible *F. solani* f. sp. *pisi* (and remained at elevated levels) or by the elicitor chitosan (Hadwiger and Wagoner, 1983a; Loschke *et al.*, 1983), and their induction was inhibited by heat shock (Hadwiger and Wagoner, 1983b).

The same approach has been used to identify induced mRNA species in bean cell suspension cultures treated with elicitor from *C. lindemuthianum* (Cramer *et al.*, 1985b; Hamdan and Dixon, 1987a), in turnip leaves infected with *Xanthomonas campestris* (Collinge *et al.*, 1987), or in potato tuber disks treated with the elicitor arachidonic acid (Marineau *et al.*, 1987). Such experiments help define suitable times for the construction of cDNA libraries for the isolation of defense-related mRNAs by the strategies of either differential hybridization (for unknown mRNAs whose kinetics of induction are interesting) or screening with antibody or DNA probes for selection of clones to known enzymes/proteins. In the case of the pea–*Fusarium* system, a number of cDNA clones were identified on the basis of differential hybridization to cDNAs synthesized from RNA from infected and control pea tissues. The induction of the messages corresponding to these cDNAs in response to infection varied from 2-fold to 20-fold, and was also observed following infection of pea by the bacterial pathogen *P. syringae* pv. *pisi* (Riggleman *et al.*, 1985; Fristensky *et al.*, 1985; Daniels *et al.*, 1987). The sequences of these cDNAs so far do not suggest a function (Fristensky *et al.*, 1988); they most probably encode some type of PR protein. Similarly, among the 13 mRNA species shown to be induced by treatment of potato tuber disks with arachidonic acid, two of the most abundant may encode potato PR proteins (Marineau *et al.*, 1987).

A novel method recently used for the identification of plant genes whose expression changes in response to external stimuli has been to N-terminal microsequence polypeptides directly from two-dimensional gels, followed by synthesis of oligonucleotide probes based on the sequence. In this manner a manganese superoxide dismutase has been cloned from tobacco and was shown to be induced by a number of stress factors, including ethylene and infection with *P. syringae* (Bowler *et al.*, 1989).

B. Changes in Transcriptional and Translational Activity for Known Defense Response Enzymes–Proteins

cDNAs and genomic clones have now been obtained for a number of plant defense response genes encoding enzymes of phytoalexin and lignin biosynthesis, hydroxyproline-rich glycoproteins, hydrolases, and PR proteins. These are listed in Table 4, which is relatively complete as of January 1990. The list also includes references to sequences that have been cloned from plants in which the particular gene is not known

TABLE 4
Cloned Plant Defense Response Genes

Gene product	Source	Vector/clone	Comments	Reference
Phenylalanine ammonia-lyase	Ipomea batatas	pUC8/cDNA	One cDNA from cut roots	Tanaka et al. (1989)
	Oryza sativa L.	λgtll/cDNA λEMBL 3/genomic	One cDNA and one genomic sequence; small gene family	Minami et al. (1989)
	Phaseolus vulgaris	pAT153/cDNA	One cDNA species from elicitor-induced cell suspension cultures	Edwards et al. (1985)
	Phaseolus vulgaris	λgt WES, λ1059/genomic	Family of three divergent genes; differential regulation	Cramer et al. (1989a)
	Petroselinum crispum	λEMBL4/genomic	Family of at least four genes	Lois et al. (1989)
	Solanum tuberosum	λgt10/cDNA	From elicitor-induced cell suspension cultures	Fritzemeier et al. (1987)
4-Coumarate:CoA ligase	Petroselinum crispum	λEMBL4/genomic	Two genes, both induced by elicitor and UV light	Douglas et al. (1987)
	Solanum tuberosum	λgtll/cDNA	From elicitor-induced cell suspension cultures	Fritzemeier et al. (1987)
Chalcone synthase	Phaseolus vulgaris	pBR325/cDNA	From elicitor-induced cell suspension cultures	Ryder et al. (1984)
	Phaseolus vulgaris	λ1059/genomic	Family of six–eight genes, some linked; differential regulation	Ryder et al. (1987)

Enzyme	Species	Vector/type	Comments	Reference
	Petunia hybrida[a]	λEMBL3/genomic	Family of at least eight genes; one major expressed gene	Koes et al. (1986, 1987)
	Petroselinum crispum[a]	λEMBL4/genomic	Single gene but two alleles, one containing an insertional element; both induced by UV light	Herrmann et al. (1988)
	Arabidopsis thaliana[a]	λEMBL4/genomic	Single gene, induced by high-intensity light	Feinbaum and Ausubel (1988)
Chalcone isomerase	Phaseolus vulgaris	λgt11/cDNA	One cDNA species from elicitor-induced cell suspension cultures	Mehdy and Lamb (1987)
	Petunia hybrida[a]	λgt11/cDNA	One cDNA species from corolla tissue mRNA	van Tunen et al. (1988)
		λEMBL3/genomic	Two genes; differential expression (one UV induced)	van Tunen et al. (1989)
Stilbene synthase	Arachis hypogaea	NM1149/cDNA	From cell suspension cultures	Schroder et al. (1988)
		λEMBL3/genomic	Probably more than two genes; sequence similarity to chalcone synthase	
Cinnamyl alcohol dehydrogenase	Phaseolus vulgaris	λgt11/cDNA	Single cDNA from elicitor-induced cell suspension cultures; second gene involved in xylogenesis?	Walter et al. (1988)
Lignin peroxidase	Nicotiana tabacum	λgt11/cDNA	cDNA from leaf mRNA; four genes	Lagrimini et al. (1987)

(continued)

183

TABLE 4 (*Continued*)

Gene product	Source	Vector/clone	Comments	Reference
3-Hydroxymethyl-3-glutaryl coenzyme A reductase	*Lycopersicon esculentum*	λgt10/cDNA	cDNA from fruit; possibly more than one gene	Narita and Gruissem (1989)
	Solanum tuberosum	PCR[b]/cDNA	From elicitor-treated tuber tissue mRNA, amplified by PCR	Stermer *et al.*, (1990a)
	Arabidopsis thaliana	λcharon 35/genomic	Two genes	Caelles *et al.* (1989)
	Arabidopsis thaliana	λgt10/cDNA	cDNA characterized by complementation in yeast	Learned and Fink (1989)
Hydroxyproline-rich glycoprotein(s)	*Daucus carota*[a]	λcharon 4A/genomic	Wound induced; more than one gene	Chen and Varner (1985)
	Phaseolus vulgaris	pUC19/cDNA	Three different cDNAs corresponding to different length transcripts; three genes, differential induction	Corbin *et al.* (1987)
	Phaseolus vulgaris	pUC19/cDNA	Wound-induced cDNA, single gene (different from above)	Sauer *et al.* (1990)
	Nicotiana tabacum[a]	pUN121	cDNA for message induced by cytokinin	Memelink *et al.* (1987)
Chitinase	*Phaseolus vulgaris*	λgt11/cDNA	cDNAs from ethylene-treated leaves	Broglie *et al.* (1986)
		λEMBL4/genomic	Approximately four genes, at least two induced by ethylene	

	Phaseolus vulgaris	λgt11/cDNA	cDNA for message from elicitor-induced cell suspension cultures	Hedrick *et al.* (1988)
	Nicotiana tabacum	pBR322/cDNA	cDNA for hormonally regulated message	Shinshi *et al.* (1987)
	Nicotiana tabacum	pUN121/cDNA	cDNA for message induced by cytokinin	Memelink *et al.* (1987)
	Nicotiana tabacum	pUC9/cDNA	cDNAs to virus-induced acidic and basic chitinases; acidic and basic each encoded by at least two genes	Hooft van Huijsduijnen *et al.* (1987)
	Nicotiana tabacum	λongc/cDNA	Acidic endochitinases (PR P and Q)	Payne *et al.* (1990)
	Cucumis sativis	λZAP/cDNA	cDNA to extracellular, acidic chitinase; no homology to bean or tobacco chitinases; one gene	Metraux *et al.* (1989)
"Chitinase-like"	*Solanum tuberosum*	λEMBL4/genomic	Two tandemly linked, wound-induced genes homologous to chitin-binding proteins	Stanford *et al.* (1989)
1,3-β-D-glucanase	*Nicotiana tabacum*	pBR322/cDNA	cDNA for mRNA whose production is inhibited by both auxin and cytokinin	Mohnen *et al.* (1985)
	Nicotiana tabacum	pBR322/cDNA	cDNA sequences indicate at least four transcriptionally active genes	Shinshi *et al.* (1988)

(continued)

185

TABLE 4 (*Continued*)

Gene product	Source	Vector/clone	Comments	Reference
	Nicotiana plumbaginifolia	pUC18/cDNA	cDNA for hormonally regulated message	De Loose *et al.* (1988)
	Hordeum vulgare[a]	pUC9/cDNA	cDNA for germination-induced message (1,3 : 1,4 activity)	Fincher *et al.* (1986)
	Phaseolus vulgaris	pBR322/cDNA	Putative cDNA for ethylene-induced message	Vogeli *et al.* (1988)
	Phaseolus vulgaris	pUC19/cDNA	cDNA for message from elicitor-induced cell suspension cultures	Edington and Dixon (1990)
"Pathogenesis-related" proteins	*Nicotiana tabacum*	pUC9/cDNA	cDNAs encoding *PR-1a*, *PR-1b*, and *PR-1c* genes; virus-induced	Hooft van Huijsduijnen *et al.* (1985), Cornelissen *et al.* (1986)
	Nicotiana tabacum	pUC12/cDNA	cDNAs encoding *PR-1a*, *PR-1b*, and *PR-1c* genes; virus-induced	Pfitzner and Goodman (1987)
	Nicotiana tabacum	λEMBL3/genomic	*PR-1a* gene	Pfitzner *et al.* (1988)
	Nicotiana tabacum	pUC8/cDNA	*PR-1a* and *PR-1b* cDNAs; multigene families, at least two *PR-1a* and four *PR-1b* genes	Matsuoka *et al.* (1987)
	Nicotiana tabacum	pUN121	*PR-1* and "*PR-1*-like" cDNAs from message induced by cytokinin	Memelink *et al.* (1987)

	Nicotiana tabacum	λgt11/cDNA	*PR-1a, PR-1b,* and *PR-1c* cDNAs; induced by virus infection	Cutt *et al.* (1988)
	Nicotiana tabacum	pUC9/cDNA	cDNA for thaumatin-like protein	Hooft van Huijsduijnen *et al.* (1986)
	Petroselinum crispum	λEMBL4/genomic	At least two genes; rapid induction of *PR-1* by fungal infection; little homology to tobacco *PR-1* genes	Somssich *et al.* (1988)
"Disease resistance response" genes	*Pisum sativum*	pUC9/cDNA	Nine cDNA clones representing messages induced by fungal infection; at least two classes by sequence analysis	Riggleman *et al.* (1985), Fristensky *et al.* (1988)
Thionins	*Hordeum vulgare*	λcharon 35/genomic	Multigene family, 50–100 members per haploid genome; all on chromosome 6	Bohlmann *et al.* (1988)

[a] Genes are not necessarily involved in antimicrobial defense in the species from which they were isolated, but are involved in defense in other plant species.
[b] PCR, Polymerase chain reaction.

to function in a defense pathway, but which is involved in defense in other plant species. For example, the chalcone synthase gene is rapidly activated by fungus or elicitor in legume cells, but not in solanaceous plants (in which the gene is, however, induced in response to both UV light and developmental cues, as is also seen in legumes). This reflects the fact that different plant families use different biosynthetic pathways for the elaboration of defensive metabolites.

Phenylalanine ammonia-lyase (PAL) cDNAs have been cloned from bean (Edwards *et al.,* 1985), potato (Fritzemeier *et al.,* 1987), sweet potato (Tanaka *et al.,* 1989), rice (Minami *et al.,* 1989), and parsley (Lois *et al.,* 1989). Use of these clones in Northern blot analysis has shown that induction of PAL activity is preceded by a rapid increase in steady-state PAL mRNA level, usually followed by an equally rapid decline in the following systems: (1) elicitor-treated bean cell suspension cultures or cotyledons (Edwards *et al.,* 1985; Cramer *et al.,* 1985a; Ellis *et al.,* 1989; Tepper *et al.,* 1989); (2) *Colletotrichum*-infected bean hypocotyls (Bell *et al.,* 1986); (3) wounded leaves and roots or elicitor-treated cell suspension cultures of parsley (Lois *et al.,* 1989); (4) potato leaves infected with *P. infestans* (Fritzemeier *et al.,* 1987); and (5) soybean roots infected with the fungal pathogen *Phytophthora megasperma* f. sp. *glycinea* or soybean cell suspension cultures treated with elicitor from this fungus (Habereder *et al.,* 1989). These studies have confirmed earlier results in which the preceding treatments were shown transiently to increase PAL mRNA activities as assessed by immunoprecipitation, using monospecific anti-PAL sera, of PAL subunits synthesized *in vitro* from mRNA (Lawton *et al.,* 1983; Loschke *et al.,* 1983; Ebel *et al.,* 1984; Schmelzer *et al.,*1984).

Nuclear transcript runoff experiments have shown that the increase in PAL mRNA in response to elicitor treatment of bean or parsley cell cultures is due to rapid transcriptional activation (Lawton and Lamb, 1987; Bolwell *et al.,* 1988; Lois *et al.,* 1989). In bean, new PAL transcripts can be detected less than 5 minutes after exposure to elicitor. The rapidity of this response suggests that the signal transduction chain between interaction of elicitor with its putative receptor and gene activation is relatively short and comparable to some of the more direct pathways for gene activation in animal cells, e.g., the response to steroid hormones.

In bean, soybean, and alfalfa, similar rapid transcriptional activation and increased steady-state mRNA levels have been shown to underly elicitor-induced increases in the synthesis and activity of chalcone synthase (Lawton *et al.,* 1983; Bell *et al.,* 1984, 1986; Ebel *et al.,* 1984; Ryder *et al.,* 1984; Cramer *et al.,* 1985a; Grab *et al.,* 1985; Lawton and

Lamb, 1987; Dalkin *et al.*, 1990b) and chalcone isomerase (Robbins and Dixon, 1984; Mehdy and Lamb, 1987; Dixon *et al.*, 1988). In bean cell cultures, induced appearance of PAL and chalcone synthase (CHS) mRNAs is closely coordinated, suggesting similar mechanisms for transcriptional activation (Cramer *et al.*, 1985a); chalcone isomerase (CHI) mRNA is coordinately induced with PAL and CHS mRNAs in some bean cell culture lines, but not in others (Cramer *et al.*, 1985a; Bolwell *et al.*, 1988). Similar close coordination between the induction of PAL and 4-coumarate:CoA ligase (4CL) mRNAs is observed in parsley cell suspension cultures in response to fungal elicitor or UV light (Chappell and Hahlbrock, 1984; Kuhn *et al.*, 1984). Analysis of genomic sequences is now beginning to reveal the molecular basis for such coordinate induction (see Sections IV and V).

The induction of defense response gene transcripts in plant cell suspension cultures is usually transient. Cinnamic acid, the product of the PAL reaction, induces the appearance of a proteinaceous inhibitor of PAL activity (Bolwell *et al.*, 1986) and also inhibits the appearance of transcripts encoding phenylpropanoid biosynthetic enzymes (Bolwell *et al.*, 1988). Treatment of elicited bean cell suspensions with inhibitors of PAL activity results in increased appearance of PAL and CHS transcripts, suggesting that cinnamic acid, or some conjugate or metabolite of it, may act *in vivo* to down-regulate expression of the phenylpropanoid pathway (Dixon *et al.*, 1986; Bolwell *et al.*, 1988).

In *Colletotrichum*-infected bean hypocotyls, phytoalexin accumulation during the hypersensitive response is immediately preceded by increased activities and levels of mRNAs encoding PAL, CHS, and CHI, whereas in the compatible interaction at similar times postinoculation, no increase is observed. In the latter case, increases in these mRNAs are associated with the appearance of large necrotic lesions during symptom development, some 80 hours later than the HR in the incompatible interaction (Cramer *et al.*, 1985a; Dixon *et al.*, 1986). Presumably the signals for gene activation in the HR and compatible interaction are different. Similar differential timing in the accumulation of PAL and CHS transcripts has been observed in the interaction of compatible and incompatible races of *P. megasperma* f. sp. *glycinea* with soybean roots, although here there is only a 9-hour difference between the onset of increased levels in the two types of interaction (Habereder *et al.*, 1989).

Increased mRNA levels for other defense response genes, including HRGPs (Showalter *et al.*, 1985; Corbin *et al.*, 1987), chitinase (Hedrick *et al.*, 1988), 1,3,-β-D-glucanase (Edington and Dixon, 1990), and coniferyl alcohol dehydrogenase (CAD) (Walter *et al.*, 1988), are likewise induced in bean cells in response to fungal elicitors, infection, or wounding, and show differential timing in compatible and incompatible interactions. Particularly noteworthy is the very rapid induction of CAD in response to fungal elicitor (preceding the accumulation of PAL and CHS mRNAs); because PAL is the first enzyme in the synthesis of lignin precursors, and CAD is the last, this may suggest some alternative role for phenylpropanoid alcohols in the very early stages of the defense response. Dehydrodiconiferyl glucosides have recently been ascribed a role as plant cell division factors (Binns *et al.*, 1987); it is therefore possible to envisage some regulatory function for other related compounds. In contrast to the very early induction of CAD, rates of transcription of HRGP genes begin to increase in response to elicitor some 2 hours later than those of PAL and CHS, suggesting that a secondary signal may be responsible for the activation of these genes (Lawton and Lamb, 1987). The relative timings of appearance of defense response gene products and their transcripts in elicited bean cell suspension cultures are summarized in Table 5.

Infection-induced appearance of chitinase and/or glucanase transcripts has been measured in tobacco mosaic virus (TMV)-infected tobacco (Hooft van Huijsduijnen *et al.*, 1987; Vogeli-Lange *et al.*, 1988; Payne *et al.*, 1990), *Colletotrichum lagenarium*-infected melon (Roby and Gaynor, 1988), ethylene-treated bean leaves (Vogeli *et al.*, 1988), and *Pseudomonas syringae*-infected bean leaves (Voisey and Slusarenko, 1989). The induction of the mRNAs encoding the two hydrolases appears to be closely coordinated (Vogeli *et al.*, 1988). As well as being induced by ethylene, glucanase and chitinase mRNA expression, at least in tobacco cells, is controlled by levels of other growth regulators. In callus cultures, cytokinin and auxin appear to repress synthesis (Mohnen *et al.*, 1985; Shinshi *et al.*, 1987), whereas increased expression has been shown in shoots in response to high cytokinin levels (Memelink *et al.*, 1987). Similar regulation of phenylpropanoid biosynthetic enzyme activities by plant growth regulators has been reported (reviewed by Dixon *et al.*, 1983), but this requires further study at the transcriptional and translational levels. Such control is to be expected in view of the roles of these enzymes in hormonally regulated developmental processes such as xylogenesis or tissue-specific flavonoid accumulation.

cDNA clones encoding a number of related members of the PR1 protein family (see Section II,C) have been obtained from mRNA isolated from TMV-infected tobacco plants (Hooft van Huijsduijnen *et al.*, 1985, 1986; Cornelissen *et al.*, 1986; Pfitzner and Goodman, 1987; Cutt *et al.*, 1988), cytokinin-stressed tobacco shoots (Memelink *et al.*, 1987),

TABLE 5

Timing of Elicitor-Induced Gene Expression in Bean Cell Suspension Cultures[a]

Gene product	Time of attainment of maximum value following elicitation (hours)[a]						Reference
	Transcription rate	Steady-state mRNA level	Translatable mRNA activity	Rate of synthesis *in vivo*	Enzyme activity		
Phenylalanine ammonia-lyase (PAL)	1.3	2–4	2–4	4	6–8		Edwards *et al.* (1985), Bolwell *et al.* (1985a), Lawton and Lamb (1987)
Chalcone synthase (CHS)	1.3	2–4	2–4	4	6		Lawton and Lamb (1987), Ryder *et al.* (1984)
Chalcone isomerase (CHI)	—	2–4	6	6–10	16		Mehdy and Lamb (1987), Robbins and Dixon (1984), Dixon *et al.* (1988)
Cinnamyl alcohol dehydrogenase (CAD)	—	1.5	1.5–2.5	3	10		Grand *et al.* (1987), Walter *et al.* (1988)
Chitinase	~1.3	2	1–2	1–2	6		Hedrick *et al.* (1988)
Hydroxyproline-rich glycoproteins (HRGPs)	7	10–15	–	–	—		Corbin *et al.* (1987), Lawton and Lamb (1987)
Prolyl hydroxylase	—	—	2	2	6		Bolwell *et al.* (1985b), Bolwell and Dixon (1986)

[a] Values for PAL, CHS, and HRGPs do not distinguish between different members of the multigene families.

and elicitor-treated parsley cell suspension cultures (Somssich *et al.*, 1986). TMV infection of the lower leaves of tobacco plants results in systemic induction of mRNAs encoding the PR-1 proteins and the thaumatin-like PR protein (Hooft van Huijsduijnen *et al.*, 1986).

The preceding discussion of changes in mRNA levels has not addressed the question of differences associated with the expression of individual members of multigene families, such as those encoding PAL, CHS, and the HRGPs. This topic is included in Section IV, A.

C. Spatial Accumulation and Localization of Defense Response Gene Transcripts–Products

At the subcellular level, induced hydrolytic enzymes are often located in the vacuole, whereas a proportion of the glucanase and the bulk of the HRGPs are associated with the cell wall. A number of techniques, including immunocytochemistry, *in situ* hybridization, tissue dissection, and histochemical analysis of expression of reporter genes in transgenic plants, have recently revealed details of the spatial expression of defense response genes around the area of attempted infection. Such studies provide preliminary evidence for intercellular signaling pathways during the establishment of defensive barriers. For example, in an incompatible interaction of bean with *C. lindemuthianum*, PAL and CHS transcripts are localized predominantly in tissue directly adjacent to the site of infection, although slightly elevated transcript levels were observed in cells at uninfected sites approximately 0.5 cm from the area of application of fungal spores. In compatible interactions, a greater and more widespread response was observed in adjacent uninfected areas (Bell *et al.*, 1986). Immunogold histochemical labeling studies with a monospecific bean anti-PAL serum have shown that, in the above incompatible interaction, increased PAL protein is detectable in cells directly around the dead hypersensitive cell, but does not extend more than two cells inward from the epidermis (R. J. O'Connell, J. A. Bailey, and R. A. Dixon, unpublished observations). Similarly, use of *in situ* hybridization has shown the rapid accumulation of PAL mRNA in a halo of apparently healthy cells around penetration sites of an incompatible race of *P. infestans* on young potato leaf tissue (Cuypers *et al.*, 1988). A combination of enzyme activity measurements and *in situ* hybridization has also been used to define the localization of enzymes of flavonoid and furanocoumarin biosynthesis in uninfected seedlings of parsley in relation to the role of such compounds as preformed antimicrobial compounds and UV protectants (Jahnen and Hahlbrock, 1988; Schmelzer *et al.*, 1988). In parsley leaves infected with *P. megasperma*

f. sp. *glycinea,* there is a rapid and massive accumulation of transcripts encoding the pathogenesis-related protein PR-1 around hypersensitive sites (Somssich *et al.,* 1988).

Analysis of the expression of a bean chalcone synthase gene promoter linked to the *E. coli* β-glucuronidase (GUS) reporter gene in transgenic tobacco has confirmed the basic pattern of expression observed for the gene in infected bean cells. In the hypersensitive response induced by the bacterial pathogen *P. syringae,* highest levels of GUS were observed in healthy cells directly aroung the collapsed hypersensitive tissue, whereas lower but still significantly elevated levels were seen throughout the leaf, decreasing with distance from the infection site (Stermer *et al.,* 1990). This system may be of value in the future for the assay of potential intercellular signal transduction molecules.

IV. Structure and Organization of Defense Response Genes

A.   ENZYMES OF ISOFLAVONOID PHYTOALEXIN BIOSYNTHESIS

A major goal in studying the structure of defense response genes is to elucidate the nature of the cis-acting sequences involved in gene activation in response to infection (or, where appropriate, other environmental or developmental stimuli). This functional analysis is considered in Section V,C. Studies on cloned defense response genes are also giving new insights into plant genome organization, processing, gene evolution, functional polymorphisms, and differential regulation, and these topics are dealt with in the following sections. The reader should refer back to Table 4 for a summary of plant defense response genes cloned to date.

1. Phenylalanine Ammonia-Lyase

The polymorphic nature of bean PAL at the enzymatic level (Section II, A) is now known to result from expression of a small family of three divergent classes of the PAL gene (Cramer *et al.,* 1989a). There is no evidence for clustering of the PAL gene family within the bean genome. The nucleotide sequences of two complete genes (gPALs 2 and 3) have been determined. These genes contain open reading frames encoding polypeptides of 712 and 710 amino acids, respectively, consistent with the molecular weight of 77,000 for the native PAL subunit. Single introns in the two genes (1720 bp in gPAL 2, 447 bp in gPAL 3) occur in identical positions. At the nucleotide level, gPAL 2 and gPAL 3 show only 59% sequence similarity in exon I, 74% similarity in exon II, and extensive divergence in the intron, 5', and 3' flanking sequences. The full sequence of the 5' half of the PAL 1 gene, the gene which corresponds to the cDNA originally characterized from elicitor-induced cell suspension cultures (Edwards *et al.*, 1985), has yet to be determined. Use of gene-specific oligonucleotides or subclones has shown that, in cell suspensions of bean cultivar Canadian Wonder, gPAL 1 and gPAL 2 are rapidly activated by elicitor, whereas all three genes are activated by wounding of hypocotyls (Cramer *et al.*, 1989a; Liang *et al.*, 1989a). In suspension cultures of the bean cultivar Imuna, all three PAL genes are elicitor inducible (Ellis *et al.*, 1989). The 5' flanking regions of gPALs 2 and 3 contain classical TATA and CAAT boxes, and transcription start sites are located at 99 and 35 bp upstream from the initiator ATG codon respectively. Both genes also contain 5' elements resembling the SV40 viral enhancer core.

The coding sequence divergence between the bean PAL genes suggests possible functional and regulatory differences. Use of gene-specific probes has enabled a relation to be defined among PAL genes, subunits (as characterized by gene-specific hybrid arrest translation followed by analysis of *in vitro* translated, immunoprecipitated polypeptides on two-dimensional gels), and intact PAL tetramers (as characterized by chromatofocussing) (Liang *et al.*, 1989a). However, posttranslational modifications may yield a higher level of polymorphism at the subunit and tetramer level (Bolwell *et al.*, 1985a). Differences in K_m values for the polymorphic forms of the active enzyme (Bolwell *et al.*, 1985a) and tissue-specific differences in transcript levels for the three PAL genes (Liang *et al.*, 1989a) (Table 6) are consistent with, but do not yet prove, different functional roles for the individual PAL genes. Similar polymorphisms at the active enzyme or subunit levels suggest that PAL may be encoded by similar multigene families in alfalfa (Jorrin and Dixon, 1990) and potato (Fritzemeier *et al.*, 1987).

Genomic sequences have also been reported for PAL from yeast (Vaslet *et al.*, 1988; Filpula *et al.*, 1988), rice (Minami *et al.*, 1989), and parsley (Lois *et al.*, 1989). In the latter case, the enzyme is encoded by a family of at least four genes, of which three appear to be activated by UV light, fungal elicitors, or wounding. The parsley PAL genes appear to be less divergent than the bean PAL genes. Parsley PAL 1 has an open reading frame encoding a 716-amino acid polypeptide subunit, with a single intron of approximately 800 bp flanked, as in the bean PAL genes, with AG/GT consensus sequences (Lois *et al.*, 1989).

TABLE 6
Organ-Specific Expression of PAL Transcripts
in Bean[a]

Organ	Specific transcript level (pg/μg total cellular RNA)		
	PAL 1	PAL 2	PAL 3
Stem	14	40	0
Wounded hypocotyls	294	542	88
Leaves	10	0	0
Sepals	14	8	0
Petals	7	441	0
Root	22	55	17

[a] From Liang *et al.* (1989a).

2. 4-Coumarate: CoA Ligase

Two homologous 4CL genes have been described in parsley (Douglas *et al.*, 1987), and both appear to be activated, at identical transcriptional start sites, by treatment of cell cultures with UV light or fungal elicitors. The 5′ flanking regions of the two genes are nearly identical for approximately 400 bp, and then retain significant regions of homology up to − 1100 bp upstream of the transcription initiation site.

3. Chalcone Synthase

Chalcone synthase has been studied extensively at the molecular genetic level. This reflects its multiple roles in plant defense (isoflavonoid synthesis) in legumes, and in flavonoid synthesis both during development and, as a protection mechanism, in response to UV irradiation. The structure and regulation of CHS genes have recently been reviewed in detail (Dangl *et al.*, 1989). It would appear that the requirement for flexibility of expression of CHS in relation to its multiple metabolic roles has been met in different plant species by different gene organizational and regulatory strategies. To list three extremes, at least six distinct genes in bean are differentially regulated in response to different environmental stimuli (Ryder *et al.*, 1987); in *Petunia,* multiple genes exist, although the UV response and floral specific expression are largely mediated by a single gene (Koes *et al.*, 1986, 1987); in parsley, the enzyme is encoded by a single gene, although two alleles exist, differing in the insertion of a 927-bp transposon-like element in the 5' upstream region of one allele (Herrmann *et al.*, 1988).

Clustering of chalcone synthase genes has been reported in bean (Ryder *et al.*, 1987), soybean (see Dangl *et al.*, 1989), and *Petunia* (Koes *et al.*, 1987). In bean, three different classes of genomic clone have been identified that each carry two CHS genes. Sequence analysis of one clone has shown that two genes, CHS 14 and CHS 17, are arranged 12 kb apart in opposite transcriptional orientations (Ryder *et al.*, 1987). Further linkage is suggested. The two most closely related subclasses of CHS genes in *Petunia hybrida* are linked, and their chromosomal location has been defined (Koes *et al.*, 1987); however, current evidence suggests that they are not expressed, although only one of the eight complete genes in *Petunia* appears to be a pseudogene.

cDNA sequences for chalcone synthases of nonlegume origin have been used for the construction of a phylogenetic tree (Niesbach-Klosgen *et al.*, 1987), and this has revealed the interesting possibility that the first exon of the parsley gene may have evolved differently from this exon in other species, i.e., that the whole gene is not necessarily the evolutionary unit. The apparent amplification of CHS genes in legumes may reflect the evolution of the flavonoid–isoflavonoid pathway for the extra role of providing antimicrobial defense compounds in this family. If so, it would appear based on present evidence that this has led to extra gene-regulatory flexibility, perhaps coupled to increased expression capacity, rather than to the capacity to synthesize enzymatic forms with different properties; the individual CHS genes in bean contain very similar open reading frames flanked by divergent 3' and 5' sequences.

Peanut (*Arachis hypogaea*) produces stilbene phytoalexins by a reaction involving the same substrates as CHS (Fig. 2). cDNA and genomic clones for stilbene synthase not surprisingly show considerable (70–75%) homology to CHS sequences, and the intron splits the same codon in stilbene synthase and CHS (Schroder *et al.*, 1988). There appear to be several stilbene synthase genes in the peanut genome.

4. Chalcone Isomerase

In contrast to the complexity of the CHS gene family, the next enzyme in the flavonoid/isoflavonoid pathway, CHI, is encoded by a single gene in bean (Mehdy and Lamb, 1987). In *P. hybrida,* in which CHI is not involved in defense against pathogens, two genes are found (van Tunen *et al.*, 1988). Interestingly, two different promoters appear to exist upstream of the coding region of the *Petunia* CHI A gene, one being active in flower corolla and tube tissue, the other, expression from which results in a 437-bp extended transcript, being active in pollen grains and in the late stages of anther development (van Tunen *et al.*, 1989). The open reading frames of the CHIs from bean and *Petunia* are approximately 50% homologous at the nucleotide and amino acid sequence levels (Blyden *et al.*, 1989). These enzymes have different substrate specificities; the single gene product in legumes catalyzes the isomerization of chalcones both with and without a 6' hydroxyl group (most antimicrobial isoflavonoids lack a hydroxyl group at the 6' position, chalcone numbering), whereas the enzymes from *Petunia* require a 6' hydroxyl group for activity (Dixon *et al.*, 1983). Sequence homology comparisons suggest that it may be possible to understand the different substrate specificities for the two different types of CHIs by molecular modeling and site-directed mutagenesis studies (Blyden *et al.*, 1989).

B. Enzymes and Proteins Involved in Cell Wall Modification

1. Enzymes of Lignin Biosynthesis

Our understanding of genes specifically involved in lignin biosynthesis is still at a rudimentary stage. Increased interest in this area is likely as a basis for attempts to modify lignin composition by genetic engineering. This will require cloning the enzymes that determine the substitution pattern (hydroxylation and methoxylation) of the aromatic rings of lignin precursors. cDNAs encoding coniferyl alcohol dehydrogenase (CAD, Fig. 2) and lignin peroxidase have been cloned from bean and tobacco, respectively (Walter *et al.*, 1988; Lagrimini *et al.*, 1987). The bean CAD cDNA encodes a 65-kDa polypeptide with several features that are conserved in alcohol dehydrogenases; a single gene appears to be rapidly activated in response to a fungal elicitor. A second divergent bean CAD gene may be involved in lignification during xylogenesis (Walter *et al.*, 1988).

2. Hydroxyproline-Rich Glycoproteins

The first HRGP gene to be cloned was that encoding the cell wall extensin from carrot (Chen and Varner, 1985). Use of the carrot cDNA as a hybridization probe led to the identification of a differentially regulated HRGP gene family in bean (Corbin *et al.*, 1987). Proteins encoded by transcripts from two distinct genes contained a proline-rich domain involving tandem repeats of the 16-amino-acid unit Tyr_3 Lys Ser Pro_4 Ser Pro Ser Pro_4. The gene *Hyp 2.13* contains 12 copies of this sequence, immediately upstream of which is an in-frame 327-bp open reading frame containing a non-proline-rich domain. *Hyp 3.6* contains eight tandemly reiterated copies of the 16-amino-acid sequence. Interestingly, *Hyp 2.13* is expressed to much higher levels in the compatible as compared to incompatible interaction of bean with *C. lindemuthianum*, whereas *Hyp 3.6* is equally expressed in the two types of interaction. A third transcript, corresponding to the gene *Hyp 4.1*, is expressed in infected bean hypocotyls in a manner similar to that of *Hyp 3.6;* this transcript encodes a protein containing a hydrophobic 5' leader sequence (for a putative transit peptide) followed by a 200-amino-acid proline-rich domain containing the sequence Tyr_3 His Ser Pro_4 Lys His Ser Pro_4 tandemly reiterated 12 times. *Hyp 4.1* is expressed earlier than *Hyp 3.6* in the incompatible interaction.

All three classes of HRGP gene are induced by wounding, but with different kinetics. Furthermore, a *Hyp 4.1* transcript approximately 300 bp smaller than that induced in infected tissues was observed in wounded hypocotyls (Corbin *et al.,* 1987). A further distinct HRGP transcript, encoded by a single gene (*Hyp 2.11*) was found to disappear rapidly when bean cell cultures were exposed to elicitor, or during infection by compatible or incompatible races of *C. lindemuthianum.* This gene was specifically wound inducible and encoded a protein of a sequence somewhat different from that of the other bean HRGP genes. Its open reading frame contained 39 Ser Pro_x repeats (where $x = 1$ to 6), of which 16 are Ser Pro_4. These repeats are concentrated in the C-terminal portion of the polypeptide. The proline-rich region is immediately preceded by three closely linked repeats of the sequence Cys Lys Ser Phe (Sauer *et al.,* 1990). Interesting differences in codon usage are observed among the members of the bean HRGP family.

The different induction patterns and amino acid sequences of these HRGPs suggest but do not yet explain possible functional differences. Clearly the large number of hydroxyl groups, not only on proline residues, indicates a high capacity for substitution by carbohydrates, linking of phenolics, and oxidative cross-linking to other cellular components.

C. HYDROLASES AND PATHOGENESIS-RELATED PROTEINS

cDNA sequences have been published for the chitinases from bean (Broglie *et al.,* 1986; Hedrick *et al.,* 1988), tobacco (Shinshi *et al.,* 1987; Payne *et al.,* 1990), and cucumber (Metraux *et al.,* 1989). In bean, there appear to be at least three chitinase genes, of which at least two are expressed on treatment of leaves with ethylene (Broglie *et al.,* 1986). One full-length bean chitinase cDNA encoded a polypeptide of M_r

35,400, with a 27-amino-acid N-terminal leader sequence characteristic of other eukaryotic signal peptides. This signal peptide, which suggests that the chitinase is synthesized on membrane-bound ribosomes, is similar to the signal sequence for wound-induced, vacuolar-located proteinase inhibitors. The first 32 acids of the mature bean chitinase show strong homology (particularly in the placing of conserved cysteine residues) to a repeated region within the N-acetyl glucosamine-specific lectin wheat germ agglutinin, indicating that the chitin-binding side of the enzyme is located at the N-terminus (Lucas *et al.*, 1985).

Two wound-induced genes from potato, *win-1* and *win-2*, encode proteins with 25-amino-acid leader sequences followed by a region of extensive homology to chitinase, wheat germ agglutinin, and rice and stinging nettle lectins (Stanford *et al.*, 1989); it is not yet known whether these proteins exhibit chitinase activity. Genes *win-1* and *win-2* are organized in close tandem array and are found in the potato genome as members of a small multigene family. Although their coding sequences are 88% homologous, the two genes exhibit differential organ-specific expression after wounding.

The mature basic chitinase from tobacco shows 73% homology, at the amino acid level, to the mature bean chitinase (Shinshi *et al.*, 1987), and acidic tobacco chitinases are similarly homologous to their basic counterparts (Hooft van Huijsduijnen *et al.*, 1987). In contrast, the cDNA sequence of an extracellular acidic chitinase encoded by a single gene in cucumber shows no homology to the bean or tobacco enzymes. It is, however, closely related to a bifunctional lysozyme–chitinase from *Parthenocissus quinquifolia* (Metraux *et al.*, 1989). The leader sequence of this chitinase is presumably involved in targeting to the extracellular space. The homologies between various chitinase sequences and wheat germ agglutinin are summarized in Fig. 3.

The 1, 3-β-D-glucanase of *Nicotiana tabacum* is encoded by a family of at least four genes (Shinshi *et al.*, 1988). The amino acid sequence shows regions of homology to the 1,3:1,4-β-D-glucanase involved in endosperm wall degradation during the germination of barley grains (Fincher *et al.*, 1986). Close homology is also seen between the *N. tabacum* glucanase and the 1,3-β-D-glucanases of *Nicotiana plumbaginifolia* (De Loose *et al.*, 1988) and barley (Hoj *et al.*, 1988). Use of oligonucleotides based on two areas of conserved sequence among tobacco 1,3-β-D-glucanase and barley 1,3: 1,4-β-D-glucanase has enabled a bean glucanase cDNA to be cloned by use of the polymerase chain reaction (Edington and Dixon, 1990). This glucanase diverges from the barley and tobacco sequences in exactly the areas where they diverge from one another.

A. Leader sequences

Bean chitinase　　M K K N R M M M M I W S V G V V W M L L L V G G S Y G

Tobacco chitinase (partial sequence)　　　　　　　S L L L L S A S A

Cucumber chitinase　　M A A H K I T T T L S I F F L L S S I F R S S D A

Potato *win-1*　　　M V K L I S N S T I L L S L F L * F S I A A I A N A

B. N-Terminal mature protein sequences

Wheat germ　　　Q R C G E Q G S M M E C P N N L C C S Q Y G Y C G M G G D Y
agglutinin

Bean chitinase　　E Q C G R Q A G G A L C P G G N C C S Q F G W C G S T T D Y

Tobacco　　　　E Q C G S Q A G G A R C A S G L C C S K F G W C G N T N D Y
chitinase

Potato *win-1*　　Q Q C G R Q K G G A L C S G N L C C S Q F G W C G S T P E F

Cucumber　　　A G I A I Y W G Q N G N E G S L A S T C A T G N Y E F V N I
chitinase

FIG. 3. Sequence homology between chitinases, potato wound-induced genes, and wheat germ agglutinin.

A number of workers have reported cDNA sequences for the PR-1 family of tobacco PR proteins (Cornelissen *et al.,* 1986; Matsuoka *et al.,* 1987; Pfitzner and Goodman, 1987; Cutt *et al.,* 1988; Pfitzner *et al.,* 1988). The PR-1a, -1b, and -1c open reading frames are closely homologous, and are preceded by signal peptides of approximately 30 amino acids. In contrast, there is little homology to the PR-1 gene sequences from parsley (Somssich *et al.,* 1988). Hydropathy and secondary structure comparisons of tobacco and tomato PR proteins have led to the suggestion that the C-terminal half of the protein may be of functional importance (Matsuoka *et al.,* 1987). Tobacco contains several PR-1 genes. The 5' upstream region of one intronless PR-1a gene contains a number of interesting potential regulatory elements, including two copies of an 11-bp imperfect repeat and a sequence, C--GAA---TTC--G, homologous to the heat-shock regulatory element (Pfitzner *et al.,* 1988). This latter sequence is also found in a parsley PR-1 gene (Somssich *et al.,* 1988). Although both are of similar size, there is no coding sequence homology between small heat-shock proteins and PR proteins, and PR proteins are not induced by elevated temperatures. The functional significance of the heat-shock element in the PR promoter therefore remains to be determined.

D. Other Defense Response Genes

cDNAs for two enzymes involved in the biosynthesis of terpenoid phytoalexins have recently been cloned. Hydroxymethylglutaryl-CoA reductase (HMGR) is a key regulatory enzyme in the early stages of isoprenoid biosynthesis. The extensive homology in the catalytic site region among HMGRs from organisms as divergent as human and yeast has allowed the use of oligonucleotides or heterologous clones for the detection and cloning of HMGR sequences by the direct screening of cDNA or genomic libraries from tomato (Cramer *et al.,* 1989b; Narita and Gruissem, 1989) and *Arabidopsis* (Caelles *et al.,* 1989; Learned and Fink, 1989) or by use of the polymerase chain reaction with potato cDNA/mRNA (Stermer *et al.,* 1990a). These studies have suggested the presence of at least two HMGR genes in solanaceous species and *Arabidopsis*. Tomato and potato HMGR transcripts are rapidly induced by elicitors (Cramer *et al.,* 1989b). Casbene synthase catalyzes the formation of the cyclic diterpene phytoalexin casbene directly from the central isoprene pathway intermediate geranylgeranyl pyrophosphate in castor bean, and its mRNA is induced on infection of castor bean seedlings with the fungus *Rhizopus stolonifer* (Moesta and West, 1985). cDNA clones have been obtained and induction shown to result from increased transcriptional activation (Lois and West, 1990). It will be interesting to see whether transfer of the casbene synthase gene to other plant species results in controlled production of the phytoalexin.

A novel class of cysteine-rich, pathogen-inducible antifungal proteins, the thionins, has recently been detected in the cell walls of barley leaves. They are also present in dicots, and their structure is somewhat variable. In barley, they are encoded by a large family of 50–100 members per haploid genome, all localized on chromosome six. They are synthesized as high-molecular-weight precursors containing a leader sequence, thionin sequence, and 3' acidic protein sequence (Bohlmann *et al.,* 1988). It remains to be determined whether their main function is a protective one.

V. Signal Transduction Pathways for Activation of Defense Response Genes

A. Microbial Elicitors and Their Receptors

The concept that pathogen avirulence genes may somehow encode race-specific elicitors of defense responses was outlined in Section I,C. To date, avirulence genes have been cloned from plant pathogenic bacteria (Table 3) but not from fungi. In contrast, most of the biochemical studies on elicitors have involved fungal pathogens (reviewed by Dixon, 1986; Dixon *et al.*, 1990). There is therefore no genetic confirmation that any of the molecules so far proposed to act as inducers of defense in plant–fungal pathogen interactions are primary determinants of avirulence.

Table 7 lists the best known examples of fungal elicitors isolated to date. It should be noted that reports of race specificity are few, the best documented example being for the elicitor of necrosis from the tomato pathogen *Cladosporium fulvum,* which is only effective on tomato cultivars carrying the *A9* resistance gene (De Wit *et al.*, 1985). The role of non-race-specific elicitors in plant–pathogen interactions is still not clear. However, they are clearly potent inducers of defense gene activation and, whatever their physiological relevance, can be envisaged in model systems to act as the primary signals in the transduction pathway leading to induced defense. The accepted, but by no means proved, model for elicitor action, at least in the case of protein–polysaccharide elicitors, involves binding of the elicitor to a receptor assumed to be localized in the plant plasma membrane. It is perhaps surprising that the only reported attempts to isolate or characterize elicitor receptors have been limited to soybean and the fungus *P. megasperma* f. sp. *glycinea* (Schmidt and Ebel, 1987; Cosio and Ebel, 1988; Cosio *et al.*, 1988). These studies have identified high-affinity binding sites for a purified oligoglucoside from the fungal cell walls, and some degree of purification has been achieved. The binding appears to be associated with the plasma membrane, and appears to be competed with by elicitor-active but not elicitor-inactive derivatives of the oligoglucoside.

The elicitor/receptor area is a complex one. Not only are a wide variety of different chemical structures from plant pathogens able to act as elicitors, but some pathogens also appear to produce molecules that act as suppressors of elicitation (Kessmann and Barz, 1986). The site at which such suppressors bind is not yet clear (Dixon, 1986). Plants themselves also contain molecules with elicitor activity. The best defined are oligogalacturonides, which may be released from cell wall pectic polymers during infection, or perhaps by host processes in response to tissue damage (Lee and West, 1981; Davis and Hahlbrock, 1987). It is possible that other as yet unidentified plant components, possibly including secondary metabolites, may also have elicitor activity (Dixon *et al.*, 1989b; R. Edwards and A. Harris, unpublished observations). Pectic polysaccharides are believed to play an important role in the systemic induction of proteinase inhibitors in plants in response to wounding (Ryan, 1988). Oligogalacturonides, which are not usually

TABLE 7

Elicitors from Plant Pathogenic Fungi

Source	Structure and/or composition	Biological activity	Race specificity	Reference
Cladosporium fulvum	Extracellular glycopeptides	Phytoalexin induction in tomato	No	De Wit and Kodde (1981)
Cladosporium fulvum	Low-molecular-weight cysteine-rich peptide	Necrosis in tomato with A9 resistance gene	Yes	De Wit *et al.* (1985)
Colletotrichum lindemuthianum	3- and 4-linked glucan	Browning of bean hypocotyls	No	Anderson-Prouty and Albersheim (1975)
Colletotrichum lindemuthianum	Galactose/mannose-rich polysaccharides	Phytoalexin induction in bean cotyledons	Yes	Tepper and Anderson (1986)
Colletotrichum lindemuthianum	Galactose/mannose-rich polysaccharides	Phytoalexin induction in bean cell cultures	No	Hamdan and Dixon (1987b)
Fusarium solani	Chitosan (poly-1,4-β-D-glucosamine)	Phytoalexin induction in pea	No	Hadwiger and Beckman (1980)
Phytophthora infestans	Fatty acids (arachidonic and eicosapentaenoic)	Phytoalexin induction in potato tuber	No	Bostock *et al.* (1981)
Phytophthora megasperma f. sp. glycinea	3-Linked glucan/associated with mannose-containing components and protein	Phytoalexin induction in soybean cotyledons	No	Ayers *et al.* (1976a,b)
Phytophthora megasperma f. sp. glycinea	1,3 : 1,6-Linked hepta-β-D-glucopyranoside	Phytoalexin induction in soybean cotyledons	No	Sharp *et al.* (1984)
Phytophthora megasperma f. sp. glycinea	Cell surface glycoproteins	Phytoalexin induction in soybean cotyledons	Yes	Keen and Legrand (1980)
Phytophthora megasperma f. sp. glycinea	Cell wall glucomannan	Phytoalexin induction in soybean cotyledons	No	Keen *et al.* (1983)
Rhizopus stolonifer	Enzyme (polygalacturonase)	Phytoalexin induction in castor bean	No	Lee and West (1981)

potent elicitors on a weight basis, may act as synergists to increase the potency of low concentrations of fungal elicitor (Davis *et al.,* 1986).

B. Inter- and Intracellular Signaling Pathways

It is helpful to divide potential signaling pathways for defense gene activation into three categories: intracellular signals, short-distance intercellular signals, and systemic signals. The first category would be analogous to the range of factors such as cyclic nucleotides, calcium ions, and phosphoinositides, which are important signal transducers in animal cells. Short-distance intercellular signals could account for the transcriptional activation observed in tissue adjacent to hypersensitively responding cells, as seen, for example, with CHS induction in bean infected with *Colletotrichum* (Bell *et al.,* 1986). Systemic signals are required for the induction of proteinase inhibitors and some pathogenesis-related proteins, particularly in solanaceous species in response to wounding or insect or virus attack.

There is evidence in the literature both for and against the involvement of calcium as an intracellular signal for the activation of defense responses. In soybean and carrot cell suspension cultures, experiments involving addition or chelation of extracellular calcium, use of calcium ionophores, and use of calcium or calmodulin antagonists have suggested a requirement for increased cellular calcium associated with elicitation (Stab and Ebel, 1987; Kurosaki *et al.,* 1987c). They have not, however, directly demonstrated changes in calcium fluxes as an initial event in elicitation. In contrast, calcium and calmodulin were discounted as factors in the induction of phytoalexins and resistance to *F. solani* in pea (Kendra and Hadwiger, 1987). Cellular components necessary for potential calcium-mediated signal transduction in plants are now being better defined; a number of protein kinases, some calcium dependent, have been purified (Elliott and Kokke, 1987; Bogre *et al.,* 1988; Olah *et al.,* 1989), and probing of libraries with oligonucleotides homologous to conserved sequences in animal protein kinases has led to the first cloning of plant protein kinases (Lawton *et al.,* 1989b). Elucidating the function of these kinases will be a challenging but exciting area for future research.

A role for cyclic AMP in elicitation seems unlikely, as its concentration appears to be very low in plants and does not appear to alter in a manner consistent with a regulatory function during infection (Hahn and Grisebach 1983). Similar considerations have ruled against a role for polyphosphoinositides (Strasser *et al.*, 1986), although improved methods of detection for these molecules in plant cells may reveal meaningful changes. Calcium-stimulated polyphosphoinositide phospholipase Cs have been identified in plant plasma membranes (Melin *et al.*, 1987; Tate *et al.*, 1989), and a phospholipid similar to mammalian platelet activating factor, which stimulates protein kinase activity, has recently been proposed as a new type of plant regulatory molecule (Scherer *et al.*, 1988). It would appear that phosphoinositides and protein kinases have the potential to be important factors in plant gene regulation, but whether they are directly involved in defense gene activation remains to be established. Potential candidates as substrates for protein kinases in plants could be gene-regulatory trans-acting DNA-binding proteins (see later).

Treatment with glutathione selectively induces high levels of expression of defense response genes in bean cell suspension cultures (Wingate *et al.*, 1988), and activates the bean CHS promoter linked to a reporter gene in electroporated soybean or alfalfa protoplasts (Dron *et al.*, 1988; Choudhary *et al.*, 1990). The effects appear to be specific for the glutathione tripeptide as opposed to other thiol compounds or γ-glutamyl peptides. Glutathione could potentially act as a signal molecule for defense gene activation, particularly in those legumes that use homoglutathione for maintaining thiol homeostasis. Currently, however, there is no direct evidence available for a causal involvement of glutathione in microbial elicitor-mediated processes.

Very little is known about the nature of possible short-distance intercellular signaling molecules in plant defense; after or during hypersensitive cell death, it is possible that intermediates of the cell necrotizing reactions could act as diffusible signals. In contrast, pectic fragments have been strongly implicated in the systemic induction of proteinase inhibitor genes. Proteinase inhibitor inducing factor (PIIF) has been shown to be a pectic component of the tomato cell wall, although PIIF does not appear to move in the plant (Ryan, 1988). More recently, an unidentified low-molecular-weight compound ("super-PIIF") has been shown to be an extremely potent inducer of systemic proteinase inhibitor synthesis. Both PIIF and super-PIIF stimulate the selective phosphorylation of a subset of plasma membrane proteins (Farmer *et al.*, 1989), and this may be related to their ability to mediate systemic gene activation. Abscisic acid has also been implicated as a causal factor in systemic induction of proteinase inhibitor genes (Peña-Cortés *et al.*, 1989).

C. FUNCTIONAL ANALYSIS OF CIS-ACTING REGULATORY SEQUENCES
AND ISOLATION OF TRANS-ACTING DNA-BINDING PROTEINS

Numerous studies in animal systems and yeast have defined cis-acting sequences necessary for regulated gene expression, and a range of DNA-binding trans-acting factors that recognize these sequences have been purified and/or cloned. Many of the DNA-binding factors are "modular" proteins with separate defined regions responsible for the DNA-binding interaction and transcriptional activation. Domain swapping experiments have shown that the domains often have the ability to function alone or in heterologous chimeric constructs. Four main structural motifs have been described for DNA-binding proteins: the zinc finger, leucine zipper, helix-turn helix, and POU protein (Evans and Hollenberg, 1988; Landschulz et al., 1988; Robertson, 1988). Functional assays using in vitro transcription systems have allowed the binding of factors to be correlated with the activation or repression of the cognate gene. The possible interactions between cis-acting elements and trans-acting factors in the regulation of transcription in animal cells have been described (Muller et al., 1988; Wasylyk, 1988; Schaffner, 1989).

In contrast, our understanding of plant transcription factors is still in its infancy. As described previously, a number of important developmentally and environmentally regulated plant genes have been cloned. However, there is as yet no efficient in vitro transcription system for plant genes. Functional analysis has therefore had to rely on measurements of expression of promoter constructs in transgenic plants, which is time-consuming, variable (due to position effects), and, until recently, only routinely possible for homologous studies with a few solanaceous plant species, such as tobacco, tomato, Petunia, or potato. Transient expression of electroporated constructs in isolated protoplasts has also been studied, but results require care in extrapolation to possible regulatory functionality in the intact plant.

In spite of these difficulties, cis-acting elements have been functionally identified in the upstream regions of a number of plant genes, including defense response genes such as chalcone synthase, proteinase inhibitor II, and the potato wound-inducible gene wun-1. These are summarized in Table 8. So far, the most detailed analysis of plant gene promoter elements and their potential trans-acting factors has been accomplished for the pea ribulose bisphosphate carboxylase small subunit 3a gene (see references in Table 8), and this illustrates the potential complexity when a gene is regulated by multiple control circuits, as will presumably occur for many defense response genes. The promoter contains a series of overlapping positive and negative regulatory elements with a second set of similar elements exhibiting conditional redundancy in the upstream region of the promoter. A region capable of conferring light-dependent and tissue-specific expression to a reporter gene in transgenic plants has been defined, and a factor, GT-1, has been shown to bind to two elements within this region. GT-1 is thought to be a transcriptional activator, and mutations in the promoter that decrease binding *in vitro* eliminate expression in transgenic plants. GT-1 has also been shown to bind to a conserved site in the 5' upstream region of the rice phytochrome gene (Kay *et al.*, 1989). So far, however, no transcription factors have been purified to homogeneity in analyzable quantities from plants. Recently, a factor from tobacco recognizing sequences within the cauliflower mosaic virus 35 S promoter has been cloned by screening a phage expression library with an oligonucleotide complementary to the sequences to which the factor binds (Lam *et al.*, 1989); a second factor, ASF-2 binds to both the 35 S promoter and a conserved GATA motif in plant *cab* genes (Lam and Chua, 1989). The cloned factor binding to the 35 S promoter and a recently cloned factor which binds upstream of a wheat histone gene (Tabata *et al.*, 1989) contain leucine zipper motifs typical of certain mammalian trans factors.

Genetic analysis and transposon tagging have led to the cloning of two regulatory loci from *Zea mays, c1* and *opaque-2* (Paz-Ares *et al.*, 1986, 1987; Schmidt *et al.*, 1989). The *c1* locus is regulatory for anthocyanin biosynthesis via the pathway shown in Fig. 2, and encodes a protein with defined acidic and basic domains, of which the basic domains show 40% homology to the animal *myb* protooncogene. Therefore, *c1* probably encodes a DNA-binding regulatory protein (Paz-Ares *et al.*, 1987). These results suggest that plant trans-acting factors may be modular-type proteins similar to many of the mammalian and yeast factors already described.

Transcription of the bean CHS gene λ*15* is rapidly activated in response to fungal elicitor or glutathione in cultured bean cells (see Section III,B). The 5' upstream sequence of this gene is shown in Fig. 4. The 336 bp upstream of the transcription start site is sufficient to confer tissue-specific and wound/elicitor responsiveness to the β-glucuronidase (GUS) reporter gene in transgenic tobacco (Stermer *et al.*, 1990b; P. Doerner, J. Schmidt, R. A. Dixon, and C. J. Lamb, unpublished observations) and, when fused to the bacterial *CAT* gene as shown in Fig. 4, drive *CAT* expression in an elicitor-inducible manner on electroporation of the *CHS-CAT-NOS* 3' construct into protoplasts of soybean, tobacco, or alfalfa (Dron *et al.*, 1988; Choudhary *et al.*, 1990).

TABLE 8

Plant Cis-Acting Elements and Corresponding Trans-Acting Factors

Gene	Cis-acting elements	Identification of binding activities	Reference
Pea *rbcS-3a*	Three elements, boxes I, II, and III, described in a region required for light-dependent organ-specific expression	Binding factor GT-1 binds to boxes II and III	Aoyagi *et al.* (1988), Kuhlemeier *et al.* (1988), Green *et al.* (1987, 1988)
Tomato *rbcS-3a*	Light inducibility and organ specificity conferred by sequences within −374-bp proximal part of the promoter	Binding factor GBF recognizes a sequence distinct from GT-1 motif, which has also been identified in *Arabidopsis rbcS*	Ueda *et al.* (1989), Giuliano *et al.* (1988)
Arabidopsis *cab-1*	At least three cis-acting elements involved in light-dependent developmental expression are contained within 1120 bp of upstream sequence	—	Ha and An (1988)
Wheat *cab-1*	The 286-bp enhancer for the phytochrome response	From −89 to −357	Nagy *et al.* (1987)
Rice phytochrome gene	—	GT-1 binds to two core sequences (competition for binding with *rbcS-3a* promoter)	Kay *et al.* (1989)
Tomato *E8* fruit ripening gene	A 2-kb promoter sequence contains elements responsible for ethylene induction and developmental regulation	Binding activity increases during fruit ripening	Deikman and Fischer (1988)
Rice α-amylase	—	Factor from aleurone layer; inducible by gibberellin	Ou-Lee *et al.* (1988)
Maize *zein* gene	—	Four protein binding sites, two of which show tissue-specific binding; this includes a	Maier *et al.* (1987, 1988)

Gene/promoter	Sequence/description	Comment	References
Potato *wun-1*	1.2 kb of promoter sequence capable of conferring wound inducibility	22-bp sequence present in a number of other cereal storage protein gene promoters	Logemann *et al.* (1989)
Potato proteinase inhibitor-II	100 bp of 3′ flanking sequence shown to be important for expression	Binding activity to 5′ sequence identified; activity is wound inducible	An *et al.* (1989), Palm and Ryan (1988), Palm *et al.* (1990)
Antirrhinum CHS gene	UV-responsive elements and general enhancer defined within −661 bp	Three light-dependent binding activities as defined by *in vivo* footprinting	Kaulen *et al.* (1986), Lipphardt *et al.* (1988), Schulze-Lefert *et al.* (1989)
Soybean lectin gene	—	A 60-kDa binding protein	Jofuku *et al.* (1987)
Soybean leghemoglobin gene *lbc₃*	The 1200 bp promoter contains elements capable of conferring nodule-specific expression	Binding activity nodule specific; recognizes two AT-rich elements in this region	Stougaard *et al.* (1987), Jensen *et al.* (1988)
Soybean *β*-conglycinin	A sequence between −143 and −257 bp in the upstream promoter sequence confers organ-specific enhancement on reporter gene	A number of factors identified binding to this and other regions of the promoter	Chen *et al.* (1988)
Cauliflower mosaic virus 35 S promoter	Two protein binding sites between −59 and −107	Two factors isolated from tobacco bind to these sites; a λgt11 clone has been obtained for one of these factors	Lam *et al.* (1989), Lam and Chua (1989)
Carrot extension	A 245bp region 200 bp upstream of the transcription start involved in ethylene induction	Nuclear factor EGBF-1 involved in ethylene induction; factor is inhibited on wounding	Holdsworth and Laties (1989a,b)
Wheat histone H3	ACGTCA motif upstream of TATA box essential for transcription	Leucine-zipper-containing factor binds to hexamer motif	Tabata *et al.* (1989)

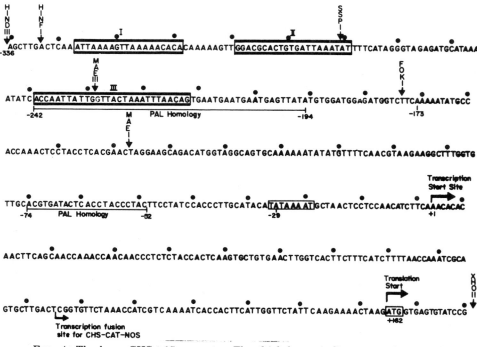

FIG. 4. The bean CHS λ15 promoter. The thick boxes indicate putative protein-binding sites in the functionally defined silencer element. Areas of homology to the bean *PAL 2* gene are underlined.

A longer promoter from the bean CHS8 gene is also inducible by bacterial infection of transgenic tobacco plants harboring a CHS8-GUS chimeric gene (Stermer *et al.*, 1990b).

The sequence CCTACC(N)₇CT is found at positions −61 to −47, −154 to −140, and in the opposite orientation on the opposite strand, −122 to −136, in the CHS λ15 promoter. The same sequence occurs five times in the promoter of the bean *gPAL 2* gene, which is coinduced with CHS in response to elicitor. Furthermore, regions from −74 to −52, and −242 to −194, show extensive homology to regions in the *gPAL 2* promoter. Interestingly, a sequence overlapping the 5' end of the −74 to −52 region has extensive homology to the so-called G-box, a sequence identified as a potential protein-binding regulatory site in a number of light-induced plant genes (Giuliano *et al.*, 1988) (see Table 8). Furthermore, a sequence within the −74 to −52 region occurs in the parsley

PAL promoter and has been identified as a potential regulatory site (Lois *et al.*, 1989; see below). The relationship between these sequences is shown in Fig. 5. They may also be significant for the regulation of the bean *CHS* promoter, as sequences between −130 and the TATA box have been shown to be essential for elicitor regulation (see later), and gel retardation assays have revealed specific binding of bean nuclear proteins to this region (L. Yu, unpublished observations). The above functionally active sequences correlate in part with DNase I hypersensitive regions in chromatin from elicitor-treated bean cells (Lawton *et al.*, 1990a).

A number of deletions of the CHS λ15 promoter with 5′ end points of −326, −173, −130, −72, and −19 were fused to the bacterial *CAT* gene and introduced into soybean protoplasts by electroporation. The −326 construct was weakly expressed in unelicited protoplasts, but was inducible by *Colletotrichum* elicitor or glutathione (Dron *et al.*, 1988). Deletion to −173 resulted in a small increase in basal expression and a 2.5-fold increase in expression in response to glutathione. It is therefore likely that the −326 to −173 region contains a negative regulatory silencer element that is functionally active in soybean protoplasts. When these experiments were repeated in alfalfa protoplasts, deletion to −173 resulted in decreased expression (A. D. Choudhary and R. A. Dixon, unpublished observations), suggesting that the region may contain both positive and negative elements whose operation depends on the cell-type environment. In soybean protoplasts, coelectroporation of the −326 to −173 region alone on a separate plasmid along with the full-length *CHS-CAT-NOS* 3′ construct resulted in increased *CAT* expression, presumably by competition *in trans* for putative repressor factors (Lawton *et al.*, 1990b). Deletion of the λ15 promoter to −72 almost totally abolished *CAT* expression, whereas the −130 deletion gave expression similar to that of the −326 construct (Dron *et al.*, 1988). These results therefore define the region between −130 and the TATA box as containing sequences essential for transcriptional activation, as predicted from homology comparisons. The second region of homology to the bean *PAL 2* gene lies within the region defined as a silencer element in soybean cells.

Migration of the −326 to −141 (*Mae1-1*) fragment of the bean *CHS* promoter (Fig. 4), containing the above-defined silencer region, was retarded in polyacrylamide gels to a single position when the labeled fragment was mixed with a protein extract from nuclei from elicited or unelicited bean cells. This binding was reversible, and could be competed with by excess unlabeled fragment or with the bean *PAL 2* promoter, which contains the region homologous to positions −242 to

−194. It was not competed with by bean CHS coding sequences (Lawton *et al.*, 1988, 1990b; Dixon *et al.*, 1989a). The binding activity was destroyed by preincubation of the bean nuclear extract with proteinase K. Retardation of the fragment to an identical position was observed by a nuclear extract from cultured alfalfa cells (Harrison, *et al.*, 1990).

DNase-1 footprinting *in vitro* showed that components from bean nuclei appear to bind to three distinct regions within the functionally defined silencer fragment, sites I, II, and III (Fig. 4) (Lawton *et al.*, 1990b; Dixon *et al.*, 1989a). Site III lies within the region of homology to the *PAL 2* promoter. A ligated multimer of a 33-bp double-stranded oligonucleotide homologous to site III totally competed with binding of bean nuclear extract to the complete silencer region in gel retardation assays and was itself retarded to a single position. Migration of the unligated oligonucleotide was similarly retarded. UV cross-linking of bean nuclear proteins to the site III binding sequence revealed a protein–DNA complex of M_r approximately 85,000 on SDS–PAGE of an irradiated mixture of oligonucleotide and bean nuclear extract. This may reflect the size of the binding protein subunit. Nuclear extracts from alfalfa cell cultures also reversibly and specifically bound the site III oligonucleotide in gel retardation assays (Harrison, *et al.*, 1990).

Use of *in vivo* genomic footprinting techniques (Saluz and Jost, 1989a,b) has revealed a number of constitutive and inducible potential protein-binding sites in the parsley *CHS* and *PAL* promoters (Schulze-Lefert *et al.*, 1989; Lois *et al.*, 1989). In the *PAL* gene, two constitutive footprints with similar sequence elements, TCTCCAC and TGTCCACGT, map in the region between −100 and −275 with respect to the transcription start site. An overlapping element, TCCACGTGGC, is found in the *CHS* promoter and is believed to be important for UV induction of transcription. This latter element is found in a number of plant gene promoters (Schulze-Lefert *et al.*, 1989). Three inducible *in vivo* footprints, two responsive to both fungal elicitor and UV, the third to elicitor alone, map between the transcription start site and the constitutive *in vivo* footprints in the parsley *PAL* gene (Lois *et al.*, 1989). The fact that the sequence of one of these elicitor-inducible footprints is found in the −52 to −74 region of *PAL* homology in the elicitor-induced bean *CHS* promoter (Fig. 5), and that parsley *PAL in vivo* footprinted sequences also occur in the coordinately induced parsley 4-coumarate : CoA ligase gene (Lois *et al.*, 1989) (although in this gene the sequences are in the intron), suggest that common cis elements may exist in plants as putative binding sites for trans-acting factors involved in UV- and elicitor-induced gene activation.

Other plant defense response genes whose potential regulatory sequences have recently been investigated by functional analysis include

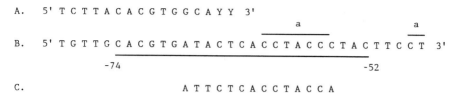

A.   5' T C T T A C A C G T G G C A Y Y 3'

B.   5' T G T T G C A C G T G A T A C T C A C C T A C C C T A C T T C C T 3'

C.   A T T C T C A C C T A C C A

FIG. 5. Putative regulatory sequences within the bean CHS λ15 promoter (B). The region between −74 to −52 is homologous to a region within the bean *gPAL2* promoter. The CCTACC(N)$_7$CT motif is marked "a." (A) The "G-box" present in light-induced gene promoters (Giuliano *et al.*, 1988). (C) A sequence from the promoter of a parsley PAL gene identified as a potential protein-binding site by *in vivo* footprinting (Lois *et al.*, 1989).

bean chitinase, in which an ethylene-responsive element has been localized to within a 228-bp region upstream of the transcription start site (Broglie *et al.*, 1989a,b); the wound-inducible potato proteinase inhibitor II gene, whose 5' *and* 3' regions appear to be necessary for full regulation (Thornburg *et al.*, 1987; An *et al.*, 1989; Keil *et al.*, 1990) and for which proteins binding to 5' upstream sequences are currently under investigation (Palm and Ryan, 1988; Palm *et al.*, 1990); and tobacco *PR-1a* genes (Montoya *et al.*, 1989).

A very different model for defense gene activation, which does not appear to invoke specific trans-acting protein factors, has been proposed based on the observations that a wide variety of different chemical structures can elicit phytoalexin induction and that some of these molecules have the ability to interact with DNA, and on the belief that they may do so *in vivo* (Kendra *et al.*, 1987; Hadwiger, 1988). The model proposes that resistance response genes are chromosomally located near nuclear scaffold attachment regions, and that elicitors may directly influence loop structures within these regions (Hadwiger, 1988). There is at present no direct evidence in support of this hypothesis.

In the absence of a good *in vitro* transcription system, useful new techniques for the analysis of plant defense gene regulatory elements and their corresponding binding factors will include the development of better protoplast systems for transient expression assays, the use of *in vivo* genomic footprinting to identify potential protein-binding regions in defense gene promoters, and the use of expression library screening with recognition sequences for the cloning of trans-acting factors. It is essential in all these studies to relate potential binding sites to functional behavior. In this respect, transient assays provide a temptingly rapid screen of functionality, but particular problems may be encountered with defense response genes over and above those associated with possible loss of tissue-specific regulation. The act of producing pro-toplasts by enzymatic digestion of cell walls releases elicitor-active molecules that may elicit the protoplasts only transiently (Dron *et al.*, 1988) or in such a way as to give unacceptably high basal expression throughout the experiment (Meith *et al.*, 1986). In alfalfa cell culture protoplasts, the extent of protoplasting-induced elicitation is variable, but the relative inducibility above background levels depends on the age (number of passages) of the suspension stock from which the pro-toplasts were isolated (Choudhary *et al.*, 1990). Parsley protoplasts appear not to be elicited during protoplasting, and retain excellent responsiveness to UV light or fungal elicitor (Dangl *et al.*, 1987).

In vivo genomic footprinting can detect binding sites inducible by elicitors, or possibly even infection, providing that sufficient re-sponding cells can be harvested at a given time point, but such sites will still require functional confirmation. The technique also requires the analysis of large amounts of genomic DNA. Amplification of the se-quence ladder with *Taq* polymerase in place of Southern hybridization in the final stages should increase the sensitivity of the technique (Saluz and Jost, 1989a).

As mentioned previously, recognition-site screening of expression libraries for plant DNA-binding proteins is just beginning to yield successful results (Lam *et al.*, 1989; Tabata *et al.*, 1989). The various protocols have been described in detail (Singh *et al.*, 1989), and there is every reason to expect reports of the cloning of defense gene transcrip-tion factors by this technique in the near future, although factors which are posttranslationally modified or bind as heterodimers are not amenable to cloning by this approach. Screening of a bean λgt11 library with the *CHS* promoter site III sequence yielded a single positive clone out of 250,000 recombinants screened, but this proved to be an artifact even though an extract from the lysogen of the clone appeared to retard specifically the *CHS* silencer region in gel-shift assays (M. J. Harrison, unpublished observations). It would appear on the basis of the limited information available that mRNAs encoding plant transcription fac-tors may be of low abundance.

VI. Expression of Defense Response Genes in Transgenic Plants

A. REGULATION AND CONSEQUENCES

Currently available data on the functional analysis of plant defense response gene promoters in transgenic plants, usually tobacco, appear to suggest that these genes are regulated in a manner approximating that observed in the plant of origin. For example, developmental expression from bean *PAL* or *CHS* promoters does not appear to be abnormal in transgenic tobacco (Liang *et al.*, 1989b; Bevan *et al.*, 1989; Schmid *et al.*, 1990); and the *CHS* promoter retains elicitor, wound, and infection inducibility (see Section V,C). It is therefore likely that attempts to express defense response genes in new plant species will not be seriously hampered by lack of compatibility of basic signal transduction pathways.

The expression of defense response gene coding sequences in transgenic plants will provide information in three areas: (1) the potential for improving resistance by manipulating response pathways, (2) defining the role of individual response components in active defense, and (3) examining novel phenotypes associated with altered expression resulting in perturbations in hitherto unrecognized functions. Currently these possibilities are just beginning to be explored.

Two main strategies are available for examining roles in disease resistance. In the first, overexpression of certain response genes under the control of a constitutive promoter may predictably result in increased resistance. Two examples of this approach to date have failed to yield the expected results. Expression of a bacterial chitinase under the control of photosynthetic gene promoters in tobacco leads to elevated chitinase levels (Jones *et al.*, 1988), but this does not appear to result in increased plant protection, possibly in view of the weaker antifungal activity of bacterial chitinases compared to their plant counterparts (Roberts and Selitrennikoff, 1988). Overexpression of the tobacco *PR-1b* gene in transgenic tobacco resulted in elevated levels of the PR protein in the extracellular spaces of leaves, with apparently no effect on the onset or severity of symptoms of systemic TMV infection (Cutt *et al.*, 1989). The second strategy, whose use has not yet been reported for the manipulation of defense functions, involves elimination of expression by antisense RNA (van der Krol *et al.*, 1988b) or sequence-specific ribozymes (Herget *et al.*, 1989). The potential effectiveness of this approach is seen in the startling changes in floral pigmentation patterns in transgenic *Petunia* expressing an antisense *CHS* gene (van der Krol *et al.*, 1988a). However, the multicomponent nature of many plant's defense responses may make interpretation difficult; the approach may work well for factors such as glucanase and chitinase, whose localization and activity is different from that of other antimicrobial barriers, such as phytoalexins or wall depositions.

Overexpression of the bean *PAL 2* gene in tobacco by incorporation of 35S enhancer sequences into its own promoter leads to severe symptoms, including reduced xylem development, interveinal lesions on leaves, and drastically reduced extractable PAL activity (Y. Elkind, unpublished observations). Analysis of these plants suggests that this might reflect perturbations of endogenous signaling pathways involving the product of the PAL reaction, *trans*-cinnamic acid (Y. Elkind, R. Edwards, M. Mavandad, R. A. Dixon, and C. J. Lamb, unpublished observations).

B. PROSPECTS FOR ENGINEERING IMPROVED DISEASE RESISTANCE

Plants and their specific pathogens coevolve, and it is likely that a wide and subtle variety of induced defense strategies have arisen throughout the plant kingdom. However, these may involve a limited number of basic components, such as phytoalexins, hydrolases, etc. This suggests that there will be scope for introducing new genetic combinations, which may not have been selected for during evolution, in order to improve plant defense characteristics. Potential examples include the following alterations.

1. The modification of phytoalexin biosynthetic pathways to either increase antimicrobial spectrum or activity, or to inhibit the ability of the pathogen to detoxify the phytoalexin. In the case of isoflavonoid phytoalexins, the genes required for this could be involved in substitution (methylation or prenylation), or in determining stereochemistry. Alteration of secondary metabolite profiles by gene transfer has already been successfully achieved in the case of flower pigmentation in *Petunia* (Meyer *et al.*, 1987).

2. Changes in temporal or spatial expression of endogenous defense response genes under the control of different inducible promoters or leader sequences. By these means it should be possible readily to target enzymes and other proteins into the intercellular space to provide a front line of attack against pathogens.

3. The introduction of novel genes under the control of defense response promotors or constitutive promoters. Potential examples may include antimicrobial lectins or proteins, which can inhibit fungal cell wall degrading enzymes (Cervone *et al.*, 1987; Degra *et al.*, 1988).

The techniques for plant transformation are now routine for a number of species (Weising *et al.*, 1988), and new technologies such as particle bombardment (Klein *et al.*, 1988) should further increase the number of species that can be engineered. Targeting of plant genes to specific locations in the genome, thereby overcoming variability of expression due to position effects, is now becoming feasible (Paszkowski *et al.*, 1988). The next few years should see exciting new advances in our understanding of the molecular basis of plant defense, and the resulting development of novel defense strategies.

ACKNOWLEDGMENTS

We thank Scotty McGill for preparation of the manuscript. We are grateful to the following persons for communicating the results of their work prior to publication: J. Dangl, D. Dudits, R. Edwards, Y. Elkind, K. Hahlbrock, N. T. Keen, C. J. Lamb, M. A. Lawton, J. N. M. Mol, J. Schmidt, and A. J. van Tunen.

REFERENCES

An, G., Mitra, A., Choi, H. K., Costa, M. A., An, K., Thornburg, R. W., and Ryan, C. A. (1989). Functional analysis of the 3' control region of the potato wound-inducible proteinase inhibitor II gene. *Plant Cell* 1, 115–122.

Anderson-Prouty, A. J., and Albersheim P. (1975). Host–pathogen interactions. VIII. Isolation of a pathogen-synthesized fraction rich in glucan that elicits a defense response in the pathogen's host. *Plant Physiol.* 56, 286–291.

Aoyagi, K., Kuhlemeier, C., and Chua, N.-H.(1988). The pea rbcS-3A enhancer-like element directs cell specific expression in transgenic tobacco. *Mol. Gen. Genet.* 213, 179–185.

Ayers, A. R., Ebel, J., Valent, B., and Albersheim, P. (1976a). Host–pathogen interactions. X. Fractionation and biological activity of an elicitor isolated from the mycelial walls of *Phytophthora megasperma* var. *sojae. Plant Physiol.* 57, 760–765.

Ayers, A. R., Valent, B., Ebel, J., and Albersheim, P. (1976b). Host–pathogen interactions. XI. Composition and structure of wall-released elicitor fractions. *Plant Physiol.* 57,766–774.

Bailey, J. A. (1987). Phytoalexins: A genetic view of their significance. *In* "Genetics and Plant Pathogenesis" (P. R. Day and G. J. Jellis, eds.), pp. 233–244. Blackwell, Oxford.

Barz, W., Daniel, S., Hinderer, W., Jaques, U., Kessmann, H., Koster, J., Otto, C., and Tiemann, K. (1988). Elicitation and metabolism of phytoalexins in plant cell cultures. *Ciba Found. Symp.* 137, 178–198.

Bell, J. N., Dixon, R. A., Bailey, J. A., Rowell, P. M., and Lamb, C. J. (1984). Differential induction of chalcone synthase mRNA activity at the onset of phytoalexin accumulation in compatible and incompatible plant–pathogen interactions. *Proc. Natl. Acad. Sci. U.S.A.* 81, 3384–3388.

Bell, J. N., Ryder, T. B., Wingate, V. P. M., Bailey, J. A., and Lamb, C. J. (1986). Differential accumulation of plant defense gene transcripts in a compatible and incompatible plant–pathogen interaction. *Mol. Cell. Biol.* 6, 1615–1623.

Bennett, C. F., Balcarek, J. M., Varrichio, A., and Crooke, S. T. (1988). Molecular cloning and complete amino-acid sequence of form-I phosphoinositide-specific phospholipase C. *Nature (London)* 334, 268–270.

Bennetzen, J. L., Qin, M. M., Ingels, S., and Ellingboe, A. H. (1988). Allele-specific and mutator-associated instability at the *Rp1* disease-resistance locus of maize. *Nature (London)* 332, 369–370.

Bevan, M., Shufflebottom, D., Edwards, K., Jefferson, R., and Schuch, W. (1989). Tissue- and cell-specific activity of a phenylalanine ammonia-lyase promoter in transgenic plants. *EMBO J.* 8, 1899–1906.

Biggs, D. R., Welle, R., Visser, F. R., and Grisebach, H. (1987). Dimethylallylpyrophosphate: 3,9-dihydroxypterocarpan 10-dimethylallyl transferase from *Phaseolus vulgaris. FEBS Lett.* 220, 223–226.

Binns, A. N., Chen, R. H., Wood, H. N., and Lynn, D. G. (1987). Cell division promoting activity of naturally occurring dehydrodiconiferyl glucosides: Do cell wall components control cell division? *Proc. Natl. Acad. Sci. U.S.A.*, **84**, 980–984.

Bless, W., and Barz, W. (1988). Isolation of pterocarpan synthase, the terminal enzyme of pterocarpan phytoalexin biosynthesis in cell suspension cultures of *Cicer arietinum*, *FEBS Lett.* **235**, 47–50.

Blyden, E. R., Dixon, R. A., Lamb, C. J., Vriglandt, E., van Tunen, A. J., and Mol, J. N. M. (1989). Comparative molecular studies of chalcone isomerases. *J. Cell. Biochem.*, *Suppl.* **13d**, 274.

Bogre, L., Olah, Z., and Dudits, D. (1988). Ca^{2+}-dependent protein kinase from alfalfa (*Medicago varia*): Partial purification and autophosphorylation. *Plant Sci.* **58**, 135–144.

Bohlmann, H., Clausen, S., Behnke, S., Giese, H., Hiller, C., Reimann-Phillip, U., Schrader, G., Barkholt, V., and Apel, K. (1988). Leaf-specific thionins of barley—A novel class of cell wall proteins toxic to plant-pathogenic fungi and possibly involved in the defense mechanism of plants. *EMBO J.* **7**, 1559–1565.

Boller, T., and Metraux, J. P. (1988). Extracellular localization of chitinase in cucumber. *Physiol. Mol. Plant Pathol.* **33**, 11–16.

Boller, T., Gehri, A., Mauch, F., and Vogeli, U. (1983). Chitinase in bean leaves: Induction by ethylene, purification, properties and possible function. *Planta* **157**, 22–31.

Bolwell, G. P., and Dixon, R. A. (1986). Membrane bound hydroxylases in elicitor-treated bean cells. Rapid induction of the synthesis of prolyl hydroxylase and a putative cytochrome P450. *Eur. J. Biochem.* **159**, 163–169.

Bolwell, G. P., Bell, J. N., Cramer, C. L., Schuch, W., Lamb, C. J., and Dixon, R. A. (1985a). L-Phenylalanine ammonia-lyase from *Phaseolus vulgaris*: Characterization and differential induction of multiple forms from elicitor-treated cell suspension cultures. *Eur. J. Biochem.* **149**, 411–419.

Bolwell, G. P., Robbins, M. P., and Dixon, R. A. (1985b). Elicitor-induced prolyl hydroxy-lase from *Phaseolus vulgaris*: Localization, purification and properties. *Biochem. J.* **229**, 693–699.

Bolwell, G. P., Robbins, M. P., and Dixon, R. A. (1985c). Metabolic changes in elicitor-treated bean cells. Enzymic responses in relation to rapid changes in cell wall composition. *Eur. J. Biochem.* **148**, 571–578.

Bolwell, G. P., Cramer, C. L., Lamb, C. J., Schuch, W., and Dixon, R. A. (1986). L-Phenylalanine ammonia-lyase from *Phaseolus vulgaris*. Modulation of the levels of active enzyme by *trans*-cinnamic acid. *Planta* **169**, 97–107.

Bolwell, G. P., Mavandad, M., Millar, D. J., Edwards, K. H., Schuch, W., and Dixon, R. A. (1988). Inhibition of mRNA levels and activities by *trans*-cinnamic acid in elicitor-induced bean cells. *Phytochemistry* **27**, 2109–2117.

Bostock, R. M., Kuć, J., and Laine, R. A. (1981). Eicosapentaenoic acid and arachidonic acids from *Phytophthora infestans* elicit fungitoxic sesquiterpenes in potato. *Science* **212**, 67–69.

Bowler, C., Alliotte, T., De Loose, M., van Montagu, M., and Inzé, D. (1989). The induction of manganese superoxide dismutase in response to stress in *Nicotiana plumbaginifo-lia*. *EMBO J.* **8**, 31–38.

Broglie, K. E., Gaynor, J. J., and Broglie, R. M. (1986). Ethylene-regulated gene expression: Molecular cloning of the genes encoding an endochitinase from *Phaseolus vulgaris*. *Proc. Natl. Acad. Sci. U.S.A.* **83**, 6820–6824.

Broglie, R., Broglie, K., Roby, D., and Gaynor, J. (1989a). Functional analysis of DNA sequences responsible for ethylene and elicitor regulation of the defense related gene chitinase. *J. Cell. Biochem.*, *Suppl.* **13d**, 247.

Broglie, K. E., Biddle, P., Cressman, R., and Broglie, R. (1989b). Functional analysis of

DNA sequences responsible for ethylene regulation of a bean chitinase gene in transgenic tobacco. *Plant Cell* 1, 599–607.

Caelles, C., Ferrer, A., Balcella, L., Hegardt, F. G., and Boronat, A. (1989). Isolation and structural characterization of a cDNA encoding *Arabidopsis thaliana* 3-hydroxy-3-methylglutaryl coenzyme A reductase. *Plant Mol. Biol.* 13, 627–638.

Cervone, F., De Lorenzo, G., Degra, L., Salvi, G., and Bergami, M. (1987). Purification and characterization of a polygalacturonase-inhibiting protein from *Phaseolus vulgaris* L. *Plant Physiol* 85, 631–637.

Chang, C., Bowman, J. L., DeJohn, A. W., Lander, E. S., and Myerowitz, E. M. (1988). Restriction fragment length polymorphism map for *Arabidopsis thaliana*. *Proc. Natl. Acad. Sci. U.S.A.* 85, 6856–6860.

Chappell, J., and Hahlbrock, K. (1984). Transcription of plant defense genes in response to UV light or fungal elicitor. *Nature (London)* 311, 76–78.

Chen, J., and Varner, J. E. (1985). An extracellular matrix protein in plants: Characterization of a genomic clone for carrot extensin. *EMBO J.* 4, 2145–2151.

Chen, Z.-L., Pan, N.-S., and Beachy, R. N. (1988). A DNA sequence element that confers seed specific enhancement to a constitutive promoter. *EMBO J.* 7, 297–302.

Choudhary, A. D., Kessmann, H., Lamb, C. J., and Dixon, R. A. (1990). Stress responses in alfalfa (*Medicago sativa L.*) IV. Expression of defense gene constructs in electroporated suspension cell protoplasts. *Plant Cell Rep.,* in press.

Collinge, D. B., and Slusarenko, A.J. (1987). Plant gene expression in response to pathogens. *Plant Mol. Biol.* 9, 389–410.

Collinge, D. B., Milligan, D. E., Dow, J. M. Scofield, G., and Daniels, M. (1987). Gene expression in *Brassica campestris* showing a hypersensitive response to the incompatible pathogen *Xanthomonas campestris* p.v. *vitians. Plant Mol. Biol.* 8, 405–414.

Corbin, D. R., Sauer, N., and Lamb, C. J. (1987). Differential regulation of a hydroxyproline-rich glycoprotein gene family in wounded and infected plants. *Mol. Cell. Biol.* 7, 6337–6344.

Cornelissen, B. J. C., Hooft van Huijsduijnen, R. A. M., van Loon, L. C., and Bol, J. F. (1986). Molecular characterization of messenger RNAs for "pathogenesis-related" proteins 1a, 1b and 1c, induced by TMV infection of tobacco. *EMBO J.* 5, 37–40.

Cosio, E. G., and Ebel, J. (1988). Solubilization of fungal oligoglucan binding proteins from soybean cell membranes. *Plant Physiol., Suppl.* 86, 22.

Cosio, E. G., Popperl, H., Schmidt, W., and Ebel, J. (1988). High-affinity binding of fungal β-glucan fragments to soybean (*Glycine max L.*) microsomal fractions and protoplasts. *Eur. J. Biochem.* 175, 309–315.

Cramer, C. L., Bell, J. N., Ryder, T. B., Bailey, J. A., Schuch, W., Bolwell, G. P., Robbins, M. P., Dixon, R. A., and Lamb, C. J. (1985a). Co-ordinated synthesis of phytoalexin biosynthetic enzymes in biologically-stressed cells of bean (*Phaseolus vulgaris L.*). *EMBO J.* 4, 285–289.

Cramer, C. L., Ryder, T. B., Bell, J. N., and Lamb, C. J. (1985b). Rapid switching of plant gene expression by fungal elicitor. *Science* 227, 1240–1243.

Cramer, C. L., Edwards, K., Dron, M., Liang, X., Dildine, S. L., Bolwell, G. P., Dixon, R. A., Lamb, C. J., and Schuch, W. (1989a). Phenylalanine ammonia-lyase gene organization and structure. *Plant Mol. Biol.* 12, 367–383.

Cramer, C. L., Park, H. S., Denbow, C. J., Yang, Z., and Lacy, G. H. (1989b). Molecular cloning and defense-related expression of a tomato HMG CoA reductase gene. *J. Cell. Biochem., Suppl.* 13d, 316.

Cutt, J. R., Dixon, D. C., Carr, J. P., and Klessig, D. F. (1988). Isolation and nucleotide sequence of cDNA clones for the pathogenesis-related proteins PR1a, PR1b and PR1c of *Nicotiana tabacum* cv. *Xanthi nc* induced by TMV infection. *Nucleic Acids Res.* **16,** 9861.

Cutt, J., Harpster, M., Carr, J., Dixon, D., Dunsmuir, P., and Klessig, D. (1989). Is the PR1 family of "pathogenesis-related" proteins in tobacco a component of viral disease resistance? *J. Cell. Biochem., Suppl.* **13d,** 333.

Cuypers, B., Schmelzer, E., and Hahlbrock, K. (1988). *In situ* localization of rapidly accumulated phenylalanine ammonia-lyase mRNA around penetration sites of *Phytophthora infestans* in potato leaves. *Mol. Plant–Microbe Interact.* **1,** 157–160.

Dalkin, K., Edwards, R., Edington, B., and Dixon, R. A. (1990a). Defense responses in alfalfa (*Medicago sativa* L.). I. Elicitor induction of phenylpropanoid biosynthesis and hydrolytic enzymes in cell suspension cultures. *Plant Physiol.* **92,** 440–446.

Dalkin, K., Jorrin, J., and Dixon, R. A. (1990b). Stress responses in alfalfa (*Medicago sativa* L.). VIII. Induction of mRNA activities in elicitor-treated cell suspension cultures. *Physiol. Mol. Plant. Pathol.* (submitted).

Dangl, J. L., Hauffe, K. D., Lipphardt, S., Hahlbrock, K., and Scheel, D. (1987). Parsley protoplasts retain differential responsiveness to U.V. light and fungal elicitor. *EMBO J.* **6,** 2551–2556.

Dangl, J. L., Hahlbrock, K., and Schell, J. (1989). Regulation and structure of chalcone synthase genes. *In* "Cell Culture and Somatic Cell Genetics of Plants" (I. K. Vasil and J. Schell, eds.) Vol. 6, pp. 155–174. Academic Press, San Diego, California.

Daniel, S., Hinderer, W., and Barz, W. (1988). Elicitor-induced changes of enzyme activities related to isoflavone and pterocarpan accumulation in chickpea (*Cicer arietinum* L.) cell suspension cultures. *Z. Naturforsch., C: Biosci.* **43c,** 536–544.

Daniels, C. H., Fristensky, B., Wagoner, W., and Hadwiger, L. A. (1987). Pea genes associated with non-host disease resistance to *Fusarium* are also active in race-specific disease resistance to *Pseudomonas. Plant Mol. Biol.* **8,** 309–316.

Daniels, M. J., Dow, J. M., and Osbourn, A. E., (1988). Molecular genetics of pathogenicity in phytopathogenic bacteria. *Annu. Rev. Phytopathol.* **26,** 285–312.

Darvill, A. G., and Albersheim, P. (1984). Phytoalexins and their elicitors—A defense against microbial infection in plants. *Annu. Rev. Plant Physiol.* **35,** 243–298.

Davis, K. R., and Hahlbrock, K. (1987). Induction of defense responses in cultured parsley cells by plant cell wall fragments. *Plant Physiol.* **85,** 1286–1290.

Davis, K. R., Darvill, A. G., and Albersheim, P. (1986). Host–pathogen interactions. XXXI. Several biotic and abiotic elicitors act synergistically in the induction of phytoalexin accumulation in soybean. *Plant Mol. Biol.* **6,** 23–32.

Degra, L., Salvi, G., Mariotti, D., De Lorenzo, G., and Cervone, F. (1988). A polygalacturonase-inhibiting protein in alfalfa callus cultures. *J. Plant Physiol.***133,** 364–366.

Deikman, J., and Fischer, R. L. (1988). Interaction of a DNA binding factor with the 5′ flanking region of an ethylene responsive fruit ripening gene from tomato. *EMBO J.* **7,** 3315–3320.

De Loose, M., Alliotte, T., Gheysen, G., Genetello, C., Gielen, J., Soetaert, P., van Montagu, M., and Inzé, D. (1988). Primary structure of a hormonally-regulated β-glucanase of *Nicotiana plumbaginifolia. Gene* **70,** 13–23.

De Wit, P. J. G. M., and Kodde, E. (1981). Further characterization and cultivar-specificity of glycoprotein elicitors from culture filtrates and cell walls of *Cladosporium fulvum* (syn. *Fulvia fulva*). *Physiol. Plant Pathol.* **18,** 297–314.

De Wit, P. J. G. M., Hofman, A. E., Velthuis, G. C. M., and Kúc, J. A. (1985). Isolation and characterization of an elicitor of necrosis isolated from intercellular fluids of compatible interactions of *Cladosporium fulvum* (syn. *Fulvia fulva*) and tomato. *Plant Physiol.* **77,** 642–647.

Dixon, R. A. (1986). The phytoalexin response: Elicitation, signalling and the control of host gene expression. *Biol. Rev.* **61**, 239–291.

Dixon, R. A., and Bolwell, G. P. (1986). Modulation of the phenylpropanoid pathway in bean (*Phaseoulus vulgaris*) cell suspension cultures. *In* "Secondary Metabolism in Plant Cell Cultures" (P. Morris, A. H. Scragg, A. Stafford, and M. W. Fowler, eds.), pp. 89–102. Cambridge Univ. Press, London and New York.

Dixon, R. A., and Lamb, C. J. (1990a). Regulation of the secondary metabolism at the biochemical and genetic levels. *In* "Secondary Products from Plant Tissue Culture" (B. V. Charlwood, ed.), pp. 101–116. Oxford University Press, Oxford.

Dixon, R. A., and Lamb, C. J. (1990b). Molecular communication in interactions between plants and microbial pathogens. *Annu. Rev. Plant Physiol. Plant Mol. Biol.* **41**, 339–367.

Dixon, R. A., Dey, P. M., and Lamb, C. J. (1983). Phytoalexins: Enzymology and molecular biology. *Adv. Enzymol.* **55**, 1–136.

Dixon, R. A., Bailey, J. A., Bell, J. N., Bolwell, G. P., Cramer, C. L., Edwards, K., Hamdan, M. A. M. S., Lamb, C. J., Robbins, M. P., Ryder, T. B., and Schuch, W. (1986). Rapid changes in gene expression in response to microbial elicitors. *Philos. Trans. R. Soc. London, B* **314**,411–426.

Dixon, R. A., Bylden, E. R., Robbins, M. P., van Tunen, A. J., and Mol, J. N. M. (1988). Comparative biochemistry of chalcone isomerases. *Phytochemistry* **27**, 2801–2808.

Dixon, R. A., Harrison, M. J., Lawton, M. A., Jenkins, S., and Lamb, C. J. (1989a). Defense gene transcription factors. *J.Cell. Biochem., Suppl.* **13d**, 247.

Dixon, R. A., Jennings, A. C., Davies, L. A., Gerrish, C., and Murphy, D. L. (1989b). Elicitor-active components from French bean hypocotyls. *Physiol. Mol. Plant Pathol.* **34**, 99–115.

Dixon, R. A., Blyden, E. R., and Ellis, J. A. (1990). Biochemistry and molecular genetics of plant–pathogen systems. *In* "Biochemical Aspects of Crop Improvement" (K. R. Khanna, ed.). CRC Press, Boca Raton, Florida (in press).

Douglas, C., Hoffmann, H., Schulz, W., and Hahlbrock, K. (1987). Structure and elicitor or U.V.-light-stimulated expression of two 4-coumarate: CoA ligase genes in parsley. *EMBO J.* **6**, 1189–1195.

Dron, M., Clouse, S. D., Lawton, M. A., Dixon, R. A., and Lamb, C. J. (1988). Glutathione and fungal elicitor regulation of a plant defense gene promoter in electroporated protoplasts. *Proc. Natl. Acad. Sci. U.S.A.* **85**, 6738–6742.

Ebel, J., Schmidt, W. E., and Loyal, R. (1984). Phytoalexin synthesis in soybean cells: Elicitor induction of phenylalanine ammonia-lyase and chalcone synthase mRNAs and correlation with phytoalexin accumulation. *Arch. Biochem. Biophys.* **232**, 240–248.

Edington, B., and Dixon, R. A. (1990). cDNA cloning of an elicitor-induced β-glucanase transcript from bean cell suspension cultures. *Plant Mol. Biol.* (submitted).

Edwards, K., Cramer, C. L., Bolwell, G. P., Dixon, R. A., Schuch, W., and Lamb, C. J. (1985). Rapid transient induction of phenylalanine ammonia-lyase mRNA in elicitor-treated bean cells. *Proc. Natl. Acad. Sci. U.S.A.* **82**, 6731–6735.

Elliott, D. C., and Kokke, Y. S. (1987). Partial purification and properties of a protein kinase C type enzyme from plants. *Phytochemistry* **26**, 2929–2935.

Ellingboe, A. H. (1981). Changing concepts in host–pathogen genetics. *Annu. Rev. Phytopathol.* **19**, 125–143.

Ellis, J. G., Lawrence, G. J., Peacock, W. J., and Pryor, A. J. (1988). Approaches to cloning plant genes conferring resistance to fungal pathogens. *Annu. Rev. Phytopathol.* **26**, 245–263.

Ellis, J. S., Jennings, A. C., Edwards, L. A., Lamb, C. J., and Dixon, R. A. (1989). Defense gene expression in elicitor-treated cell suspension cultures of French bean cv. Imuna. *Plant Cell Rep.* **8,** 504–507.

Esquerré -Tugayé, M.-T., Lafitte, C., Mazau, D., Toppan, A., and Touzé, A. (1979). Cell surfaces in plant–microorganism interactions. II. Evidence for the accumulation of hydroxyproline-rich glycoproteins in the cell walls of diseased plants as a defense mechanism. *Plant Physiol.* **64,** 320–326.

Evans, R. M., and Hollenberg, S. M. (1988). Zinc fingers: Gilt by association. *Cell* **52,** 1–3.

Farmer, E. E., Pearce, G., and Ryan, C. A. (1989). In vitro phosphorylation of plant plasma membrane proteins in response to proteinase inhibitor inducing factor. *Proc. Natl. Acad. Sci. U.S.A.* **86,** 1539–1542.

Feinbaum, R. L., and Ausubel, F. M. (1988). Transcriptional regulation of the *Arabidopsis thaliana* chalcone synthase gene. *Mol. Cell. Biol.* **8,** 1985–1992.

Filpula, D., Vaslet, C. A., Levy, A., Sykes, A., and Strausberg, R. L. (1988). Nucleotide sequence of the gene for phenylalanine ammonia-lyase from *Rhodotorula rubra*. *Nucleic Acids Res.* **16,** 11381.

Fincher, G. B., Lock, P. A., Morgan, M. M., Lingelback, K., Wettenhall, R. E. H., Mercer, J. F. B., Brandt, A., and Thomsen, K. K. (1986). Primary structure of the (1-3,1-4)-β-D-glucan 4-glucohydrolase from barley aleurone. *Proc. Natl. Acad. Sci. U.S.A.* **83,** 2081–2085.

Fink, W., Liefland, M., and Mendgen, K. (1988). Chitinases and β-1,3-glucanases in the apoplastic compartment of oat leaves (*Avena sativa* L.). *Plant Physiol.* **88,** 270–275.

Friend, J. (1976). Lignification in infected tissues. In "Biochemical Aspects of Plant–Parasite Interactions" (J. Friend and D. R. Threlfall, eds.), pp. 291–303. Academic Press, London and New York.

Fristensky, B., Riggleman, R. C., Wagoner, W., and Hadwiger, L. A. (1985). Gene expression in susceptible and disease resistant interactions of peas induced with *Fusarium solani* pathogens and chitosan. *Physiol. Plant Pathol.* **27,** 15–28.

Fristensky, B., Horovitz, D., and Hadwiger, L. A. (1988). cDNA sequences for pea disease resistance response genes. *Plant Mol. Biol.* **11,** 713–715.

Fritzemeier, K.-H., Cretin, C., Kombrink, E., Rohwer, F., Taylor, J., Scheel, D., and Hahlbrock, K. (1987). Transient induction of phenylalanine ammonia-lyase and 4-coumarate:CoA ligase mRNAs in potato leaves infected with virulent or avirulent races of *Phytophthora infestans*. *Plant Physiol.* **85,** 34–41.

Gabriel, D., Burges, A., and Lazo, G. (1986). Gene-for-gene interactions of five cloned avirulence genes from *Xanthomonas campestris* pv. *malvacearum* with specific resistance genes in cotton. *Proc. Natl. Acad. Sci. U.S.A.* **83,** 6415–6419.

Giuliano, G., Pichersky, E., Makik, V. S., Timko, M. P., Scilnik, P. A., and Cashmore, A. R. (1988). An evolutionarily conserved protein binding sequence upstream of a plant light-regulated gene. *Proc. Natl. Acad. Sci. U.S.A.* **85,** 7089–7093.

Grab, D., Loyal, R., and Ebel, J. (1985). Elicitor-induced phytoalexin synthesis in soybean cells: Changes in the activity of chalcone synthase mRNA and the total population of translatable mRNA. *Arch. Biochem. Biophys.* **243,** 523–529.

Grand, C., Sarni, F., and Lamb, C. J. (1987). Rapid induction by fungal elicitor of the synthesis of cinnamyl alcohol dehydrogenase, a specific enzyme of lignin synthesis. *Eur. J. Biochem.* **169,** 73–77.

Green, P. J., Kay, S. A., and Chua, N.-H., (1987). Sequence specific interactions of a pea nuclear factor with light responsive elements upstream of the *rbcS-3A* gene. *EMBO J.* **6,** 2543–2549.

Green, P. J., Yong, M.-H., Cuozzo, M., Kano-Murakami, Y., Silverstein, P., and Chua, N.-H. (1988). Binding site requirements for pea nuclear protein factor GT-1 correlate with sequences required for light-dependent transcriptional activation of the *rbcS-3A* gene. *EMBO J.* **7**, 4035–4044.

Ha, S. B., and An, G. (1988). Identification of upstream regulatory elements involved in the developmental expression of the *Arabidopsis thaliana cab 1* gene. *Proc. Natl. Acad. Sci. U.S.A.* **85**, 8017–8021.

Habereder, H., Schroder, G., and Ebel, J. (1989). Rapid induction of phenylalanine ammonia-lyase and chalcone synthase mRNAs during fungus infection of soybean (*Glycine max* L.) roots or elicitor treatment of soybean cell cultures at the onset of phytoalexin synthesis. *Planta* **177**, 58–65.

Hadwiger, L. A. (1988). Possible role of nuclear structure in disease resistance of plants. *Phytopathology* **78**, 1009–1014.

Hadwiger, L. A., and Beckman, J. M. (1980). Chitosan as a component of pea–*Fusarium solani* interactions. *Plant Physiol.* **66**, 205–211.

Hadwiger, L. A., and Wagoner, W. (1983a). Electrophoretic patterns of pea and *Fusarium solani* proteins synthesized *in vitro* or *in vivo* which characterize the compatible and incompatible interactions. *Physiol. Plant Pathol.* **23**, 153–162.

Hadwiger, L. A., and Wagoner, W. (1983b). Effect of heat-shock on the mRNA-directed disease resistance response of peas. *Plant Physiol.* **72**, 553–556.

Hagmann, M., and Grisebach, H. (1984). Enzymatic rearrangement of flavanone to isoflavone. *FEBS Lett.* **175**, 199–202.

Hahlbrock, K., and Scheel, D. (1989). Physiology and molecular biology of phenylpropanoid metabolism. *Annu. Rev. Plant Physiol. Plant Mol. Biol.* **40**, 347–369.

Hahn, M. G., and Grisebach, H. (1983). Cyclic AMP is not involved as a second messenger in the response of soybean to infection by *Phytophthora megasperma* f. sp. *glycinea*. *Z. Naturforsch., C: Biosci.* **38C**, 578–582.

Hamdan, M. A. M. S., and Dixon, R. A. (1987a). Differential patterns of protein synthesis in bean cells exposed to elicitor fractions from *Colletotrichum lindemuthianum*. *Physiol. Mol.Plant Pathol.* **31**, 105–121.

Hamdan, M. A. M. S., and Dixon, R. A. (1987b). Fractionation and properties of elicitors of the phenylpropanoid pathway from culture filtrates of *Colletotrichum lindemuthianum*. *Physiol. Mol. Plant Pathol.* **31**, 91–103.

Harrison, M. J., Choudhary, A. D., Kooter, J., Lamb, C. J., Dixon, R. A. (1990). *Cis*-elements and *trans*-acting factors for the quantitative expression of a bean chalcone synthase gene promoter in electroporated alfalfa protoplasts. *Plant Mol. Biol.* (submitted).

Hedrick, S. A., Bell, J. N., Boller, T., and Lamb, C. J. (1988). Chitinase cDNA cloning and mRNA induction by fungal elicitor, wounding and infection. *Plant Physiol.* **86**, 182–186.

Herget, T., Schell, J., and Schreier, P. H. (1989). Designed ribozymes to study and manipulate gene expression in plants. *J. Cell. Biochem., Suppl.* **13d**, 303.

Herrmann, A., Schulz, W., and Hahlbrock, K. (1988). Two alleles of the single-copy chalcone synthase gene in parsley differ by a transposon-like element. *Mol. Gen. Genet.* **212**, 93–98.

Hoj, P. B., Slade, A. M., Wettenhall, R. E. H., and Fincher, G. B. (1988). Isolation and characterization of a (1-3)-β-glucan endohydrolase from germinating barley (*Hordeum vulgare*): Amino acid sequence similarity with barley (1-3,1-4)-β-glucanases. *FEBS Lett.* **230**, 67–71.

Holdsworth, M. J., and Laties, G. G. (1989a). Site-specific binding of a nuclear factor to the carrot extension gene is infuenced by both ethylene and wounding. *Planta* **179**, 17–23.

Holdsworth, M. J., and Laties, G. G. (1989b). Identification of a wound-induced inhibitor of a nuclear factor that binds the carrot extension gene. *Planta* **80,** 74–81.

Hooft van Huijsduijnen, R. A. M., Cornelissen, B. J. C., van Loon, L. C., van Boom, J. H., and Bol, J. F. (1985). Virus-induced synthesis of messenger RNAs for precursors of pathogenesis related proteins in tobacco. *EMBO J.* **4,** 2167–2171.

Hooft van Huijsduijnen, R. A. M., van Loon, L. C., and Bol, J. F. (1986). cDNA cloning of six mRNAs induced by TMV infection of tobacco and a characterization of their translation products. *EMBO J.* **5,** 2057–2061.

Hooft van Huijsduijnen, R. A. M., Kauffmann, S., Brederode, F. T., Cornelissen, B. J. C., Legrand, M., Fritig, B., and Bol, J. F. (1987). Homology between chitinases that are induced by TMV infection of tobacco. *Plant Mol. Biol.* **9,** 411–420.

Huang, H.-C., Schuurink, R., Denny, T. P., Atkinson, M. M., Baker, C. J., Yucel, I., Hutcheson, S. W., and Collmer, A. (1988). Molecular cloning of a *Pseudomonas syringae* pv. *syringae* gene cluster that enables *Pseudomonas fluorescens* to elicit the hypersensitive response in tobacco plants. *J. Bacteriol.* **170,** 4748–4756.

Jahnen, W., and Hahlbrock, K. (1988). Differential regulation and tissue-specific distribution of enzymes of phenylpropanoid pathways in developing parsley seedlings. *Planta* **173,** 453–458.

Jensen, E. O., Marcker, K. A., Schell, J., and de Bruijn, F. J. (1988). Interaction of a nodule specific transacting factor with distinct DNA elements in soybean leghaemoglobin lbc$_3$ 5′ upstream region. *EMBO J.* **7,** 1265–1271.

Jofuku, K. D., Okamuro, J. K., and Goldberg, R. B. (1987). Interaction of an embryo DNA binding protein with a soybean lectin gene upstream region. *Nature (London)* **320,** 734–737.

Jones, J. D. G., Dean, C., Godoni, D., Gilbert, D., Bond-Nutter, D., Lee, R., Bedbrook, J., and Dunsmuir, P. (1988). Expression of bacterial chitinase protein in tobacco leaves using two photosynthetic gene promoters. *Mol. Gen Genet.* **212,** 536–542.

Jorrin, J., and Dixon, R. A. (1990). Defense responses in alfalfa (*Medicago sativa* L.). II. Purification, characterization and induction of phenylalanine ammonia-lyase isoforms from elicitor-treated cell suspensions. *Plant Physiol.* **92,** 447–455.

Kauffmann, S., Legrand, M., Geoffroy, P., and Fritig, B. (1987). Biological function of pathogenesis-related proteins: Four PR proteins of tobacco have 1,3-β-glucanase activity. *EMBO J.* **6,** 3209–3212.

Kaulen, H., Schell, J., and Kreuzaler, F. (1986). Light induced expression of chimaeric chalcone synthase–NPTII gene in tobacco cells. *EMBO J.* **5,** 1–8.

Kay, S. A., Keith, B., Shinozaki, K., Chye, M. L., and Chua, N.-H., (1989). The rice phytochrome gene: Structure, autoregulated expression and binding of GT-1 to a conserved site in the 5′ upstream region. *Plant Cell* **1,** 351–360.

Kearney, B., Ronald, P. C., Dahlbeck, D., and Staskawicz, B. J. (1988). Molecular basis for the evasion of plant host defense in bacterial spot disease of pepper. *Nature (London)* **332,** 541–543.

Keen, N. T. (1986). Pathogenic strategies for fungi. *In* "Recognition in Microbe–Plant Symbiotic and Pathogenic Interactions" (B. Lugtenberg, ed.), NATO–ASI Ser. H, Vol. 4, pp. 171–188. Springer-Verlag, Berlin and New York.

Keen, N. T., and Legrand, M. (1980). Surface glycoproteins: Evidence that they may function as the race-specific phytoalexin elicitors of *Phytophthora megasperma* f. sp. *glycinea. Physiol. Plant Pathol.* **17,** 175–192.

Keen, N. T., and Staskawicz, B. (1988). Host range determinants in plant pathogens and symbionts. *Annu. Rev. Microbiol.* **42,** 421–440.

Keen, N. T., Yoshikawa, M., and Wang, M. C. (1983). Phytoalexin elicitor activity of carbohydrates from *Phytophthora megasperma* f. sp. *glycinea* and other sources. *Plant Physiol.* **71,** 466–471.

Keil, M., Sánchez-Serrano, J., Schell, J., and Willmitzer, L. (1990). Localization of elements important for wound-inducible expression of a chimeric potato proteinase inhibitor II-CAT gene in transgenic tobacco plants. *Plant Cell* **2**, 61–70.

Kendra, D. F., and Hadwiger, L. A. (1987). Calcium and calmodulin may not regulate the disease resistance and pisatin formation responses of *Pisum sativum* to chitosan or *Fusarium solani. Physiol. Mol. Plant Pathol.* **31**, 337–348.

Kendra, D. F., Fristensky, B., Daniels, C. H., and Hadwiger, L. A. (1987). Disease resistance response genes in plants: Expression and proposed mechanisms of induction. *In* "Molecular Strategies for Crop Protection." pp. 13–24. Liss, New York.

Kessmann, H., and Barz, W. (1986). Elicitation and suppression of phytoalexin and isoflavone accumulation in cotyledons of *Cicer arietinum* L. as caused by wounding and by polymeric components from the fungus *Ascochyta rabiei. J. Phytopathol.* **117**, 321–335.

Kistler, K. C., and VanEtten, H. D. (1984). Regulation of pisatin demethylation in *Nectria haematococca* and its influence on pisatin tolerance and virulence. *J. Gen. Microbiol.* **130**, 2605–2613.

Klein, T. M., Harper, E. C., Svab, Z., Sanford, J. C., Fromm, M. E., and Maliga, P. (1988). Stable genetic transformation of intact *Nicotiana* cells by the particle bombardment process. *Proc. Natl. Acad. Sci. U.S.A.* **85**, 8502–8505.

Kobayashi, D. Y., Tamaki, S. J., and Keen, N. T. (1988). Cloned avirulence genes from the tomato pathogen *Pseudomonas syringae* pv. *tomato* confer cultivar specificity on soybean. *Proc. Natl. Acad. Sci. U.S.A.* **85**, 157–161.

Koes, R. E., Spelt, C. E., Reif, H. J., van den Elzen, P. J. M., Veltkamp, E., and Mol, J. M. N. (1986). Floral tissue of *Petunia hybrida* (V30) expresses only one member of the chalcone synthase multigene family. *Nucleic Acids Res.* **14**, 5229–5239.

Koes, R. E., Spelt, C. E., Mol, J. N. M., and Gerats, A. G. M. (1987). The chalcone synthase multigene family of *Petunia hybrida*: Sequence homology, chromosomal localization and evolutionary aspects. *Plant Mol. Biol.* **10**, 159–169.

Kombrink, E., Schroder, M., and Hahlbrock, K. (1988). Several pathogenesis-related proteins in potato are 1,3-β-D-glucanases and chitinases. *Proc. Natl. Acad. Sci. U.S.A.* **85**, 782–786.

Kuhlemeier, C., Couzzo, M., Green, P. J., Goyvaerts, E., Ward, K., and Chua, N.-H. (1988). Localization and conditional redundancy of regulatory elements in *rbcS-3A*, a pea gene encoding the small subunit of ribulose-bisphosphate carboxylase. *Proc. Natl. Acad. Sci. U.S.A.* **85**, 4662–4666.

Kuhn, D. N., Chappell, J., Boudet, A., and Hahlbrock, K. (1984). Induction of phenylalanine ammonia-lyase and 4-coumarate : CoA ligase mRNAs in cultured plant cells by UV light or fungal elicitor. *Proc. Natl. Acad. Sci. U.S.A.* **81**, 1102–1106.

Kurosaki, F., Tashiro, N., and Nishi, A. (1987a). Induction, purification and possible function of chitinase in cultured carrot cells. *Physiol. Mol. Plant Pathol.* **31**, 201–210.

Kurosaki, F., Tashiro, N., and Nishi, A. (1987b). Secretion of chitinase from cultured carrot cells treated with fungal mycelial walls. *Physiol. Mol. Plant Pathol.* **31**, 211–216.

Kurosaki, F., Tsurusawa, Y., and Nishi, A. (1987c). The elicitation of phytoalexins by Ca^{2+} and cyclic AMP in carrot cells. *Phytochemistry* **26**, 1919–1923.

Lagrimini, L. M., Burkhart, W., Moyer, M., and Rothstein, S. (1987). Molecular cloning of complementary DNA encoding the lignin-forming peroxidase from tobacco: Molecular analysis and tissue-specific expression. *Proc. Natl. Acad. Sci. U.S.A.* **84**, 7542–7546.

Lam, E., and Chua, N.-H. (1989). ASF-2: A factor that binds to the cauliflower mosaic virus 35S promoter and a conserved GATA motif in *Cab* promoters. *Plant Cell* **1**, 1147–1156.

Lam, E., Benfrey, P. N., Gilmartin, P. M., Katagiri, F., and Chua, N.-H. (1989). *In vitro* and *in vivo* characterization of trans-acting factors which bind to the CaMV 35S promoter. *J. Cell. Biochem., Suppl.* **13d**, 236.

Lamb, C. J., Lawton, M. A., Dron, M., and Dixon, R. A. (1989). Signals and transduction mechanisms for activation of plant defenses against microbial attack. *Cell* **56**, 215–224.

Landschulz, W. H., Johnson, P. F., and McKnight, S. L. (1988). The leucine zipper: A hypothetical structure common to a new class of DNA binding protein. *Science* **240**, 1759–1764.

Lawton, M. A., and Lamb, C. J. (1987). Transcriptional activation of plant defense genes by fungal elicitor, wounding and infection. *Mol. Cell. Biol.* **7**, 335–341.

Lawton, M. A., Dixon, R. A., Hahlbrock, K., and Lamb, C. J. (1983). Elicitor-induction of mRNA activity: Rapid effects of elicitor on phenylalanine ammonia-lyase and chalcone synthase mRNA activities in bean cells. *Eur. J. Biochem.* **130**, 131–139.

Lawton, M. A., Yamamoto, R. T., Hanks, S. K., and Lamb, C. J. (1989). Molecular cloning of plant transcripts encoding protein kinase homologs. *Proc. Natl. Acad. Sci. U.S.A.* **86**, 3140–3144.

Lawton, M. A., Clouse, S. D., Lamb, C. J. (1990a). Glutathione-elicited changes in chromatin structure within the promoter of the defense gene chalcone synthase. *Plant Cell Rep.* **8**, 561–564.

Lawton, M. A., Jenkins, S. M., Dron, M., Kooter, J. M., Kragh, K. M., Harrison, M. J., Yu, L. Tanguay, L., Dixon, R. A., and Lamb, C. J. (1990b). Silencer region of a chalcone synthase promoter contains multiple binding sites for GT-1. *EMBO J.* (submitted).

Learned, R. M., and Fink, G. R., (1989). 3-Hydroxy-3-methylglutaryl-coenzyme A reductase from *Arabidopsis thaliana* is structurally distinct from the yeast and animal enzymes. *Proc. Natl. Acad. Sci. U.S.A.* **86**, 2779–2783.

Lee, S.-C., and West, C. A. (1981). Polygalacturonase from *Rhizopus stolonifer*, an elicitor of casbene synthetase activity in castor bean (*Ricinus communis* L.) seedlings. *Plant Physiol.* **67**, 633–639.

Legrand, M., Kauffman, S., Geoffroy, P., and Fritig, B. (1987). Biological function of pathogenesis-related proteins: Four tobacco pathogenesis-related proteins are chitinases. *Proc. Natl. Acad. Sci. U.S.A.* **84**, 6750–6754.

Liang, X., Dron, M., Cramer, C. L., Dixon, R. A., and Lamb, C. J. (1989a). Differential regulation of phenylalanine ammonia-lyase genes during plant development and by environmental cues. *J. Biol. Chem.* **264**, 14486–14492.

Liang, X., Dron, M., Schmid, J., Dixon, R. A., and Lamb, C. J. (1989b). Developmental and environmental regulation of a phenylalanine ammonia-lyase-β-glucuronidase gene fusion in transgenic tobacco plants. *Proc. Natl. Acad. Sci. U.S.A.* **86**, 9284–9288.

Lindgren, P., Peet, R., and Panopoulos, N. (1986). Gene cluster of *Pseudomonas syringae* pv. *phaseolicola* controls pathogenicity on bean plants and hypersensitivity on non-host plants. *J. Bacteriol.* **168**, 512–522.

Lindgren, P., Panopoulos, N., Staskawicz, B. J., and Dahlbeck, D. (1988). Genes required for pathogenicity and hypersensitivity are conserved and interchangeable among pathovars of *Pseudomonas syringae*. *Mol. Gen. Genet.* **211**, 489–506.

Lipphardt, S., Brettschneider, R., Kreuzaler, F., Schell, J., and Dangl, J. L. (1988). UV-inducible transient expression in parsley protoplasts identifies regulatory *cis*-elements of a chimeric *Antirrhinum majus* chalcone synthase gene. *EMBO J.* **7**, 4027–4033.

Logemann, J., Lipphardt, S., Lorz, H., Hauser, I., Willmitzer, L., and Schell, J. (1989). 5′ Upstream sequences from *Wun 1* gene are responsible for gene activation by wounding in transgenic plants. *Plant Cell* 1, 151–158.

Lois, A. F., and West, C. A. (1990). Regulation of expression of the casbene synthetase gene during elicitation of castor bean seedlings with pectic fragments. *Arch. Biochem. Biophys.* 276, 270–277.

Lois, R., Dietrich, A., and Hahlbrock, K. (1989). A phenylalanine ammonia-lyase gene from parsley: Structure, regulation and identification of elicitor and light responsive cis-acting elements. *EMBO J.* 8, 1641–1648.

Loschke, D. C., Hadwiger, L. A., and Wagoner, W. (1983). Comparison of mRNA populations coding for phenylalanine ammonia-lyase and other peptides from pea tissue treated with biotic and abiotic phytoalexin inducers. *Physiol. Plant Pathol.* 23, 163–173.

Lucas, J., Henschen, A., Lottspeich, F., Vogeli, U., and Boller, T. (1985). Amino-terminal sequence of ethylene-induced bean chitinase reveals similarities to sugar-binding domains of wheat germ agglutinin. *FEBS Lett.* 193, 208–210.

Maier, U.-G., Brown, J. W. S., Toloczyki, C., and Feix, G. (1987). Binding of a nuclear factor to a consensus sequence in the 5′ flanking region of zein genes from maize. *EMBO J.* 6, 17–22.

Maier, U.-G., Brown, J. W. S., Schmitz, L., Schwall, M., Dietrich, G., and Feix, G. (1988). Mapping of tissue dependent and independent protein binding sites to the 5′ upstream region of a zein gene. *Mol. Gen. Genet.* 212, 241–245.

Marineau, C., Matton, D. P., and Brisson, N. (1987). Differential accumulation of potato tuber mRNAs during the hypersensitive response induced by arachidonic acid. *Plant Mol. Biol.* 9, 335–342.

Matern, U., and Kneusel, R. E. (1988). Phenolic compounds in plant disease resistance. *Phytoparasitica* 16, 153–170.

Matsuoka, M., Yamamoto, N., Kano-Murakami, Y., Tanaka, Y., Ozeki, Y., Hirano, H., Kagawa, H., Oshima, M., and Ohashi, Y. (1987). Classification and structural comparison of full-length cDNAs for pathogenesis-related proteins. *Plant Physiol.* 85, 942–946.

Mauch, F., and Staehelin, L. A. (1989). Functional implications of the subcellular localization of ethylene-induced chitinase and β-1,3-glucanase in bean leaves. *Plant Cell* 1, 447–457.

Mauch, F., Hadwiger, L. A., and Boller, T. (1984). Ethylene: Symptom, not signal for the induction of chitinase and β-1,3-glucanase in pea pods by pathogens and elicitors. *Plant Physiol.* 76, 607–611.

Mauch, F., Hadwiger, L. A., and Boller, T. (1988a). Antifungal hydrolases in pea tissue. I. Purification and characterization of two chitinases and two β-glucanases differentially regulated during development and in response to fungal infection. *Plant Physiol.* 87, 325–333.

Mauch, F., Mauch-Mani, B., and Boller, T. (1988b). Antifungal hydrolases in pea tissue. II. Inhibition of fungal growth by combinations of chitinase and β-1,3-glucanase. *Plant Physiol.* 88, 936–942.

Mehdy, M., and Lamb, C. J. (1987). Chalcone isomerase cDNA cloning and mRNA induction by fungal elicitor, wounding and infection. *EMBO J.* 6, 1527–1533.

Meiners, J. P. (1981). Genetics of disease resistance in edible legumes. *Annu. Rev. Phytopathol.* 21, 189–209.

Meith, H., Speth, V., and Ebel, J. (1986). Phytoalexin production by isolated soybean protoplasts. *Z. Naturforsch., C: Biosci.* 41C, 193–201.

Melin, P.-M., Sommarin, M., Sandelius, A. S., and Jergil, B. (1987). Identification of

Ca^{2+}-stimulated polyphosphoinositide phospholipase C in isolated plant plasma membranes. *FEBS Lett.* **223,** 87–91.

Memelink, J., Hoge, J. H. C., and Schilperoort, R. A. (1987). Cytokinin stress changes the developmental regulation of several defense-related genes in tobacco. *EMBO J.* **6,** 3579–3584.

Metraux, J. P., and Boller, T. (1986). Local and systemic induction of chitinase in cucumber plants in response to viral, bacterial and fungal infections. *Physiol. Mol. Plant Pathol.* **28,** 161–169.

Metraux, J. P., Strett, L., and Staub, T. (1988). A pathogenesis-related protein in cucumber is a chitinase. *Physiol. Mol. Plant Pathol.* **33,** 1–9.

Metraux, J. P., Burkhart, W., Moyer, M., Dincher, S., Middlesteadt, W., Williams, S., Payne, G., Carnes, M., and Ryals, J. (1989). Isolation of a complementary DNA encoding a chitinase with structural homology to a bifunctional lysozyme/chitinase. *Proc. Natl. Acad. Sci. U.S.A.* **86,** 896–900.

Meyer, P., Heidmann, I., Forkmann, G., and Saedler, H. (1987). A new petunia flower colour generated by transformaton of a mutant with a maize gene. *Nature (London)* **330,** 677–678.

Michelmore, R. W., Hulbert, S. H., Landry, B. S., and Leung, H. (1987). Towards a molecular understanding of lettuce downy mildew. *In* "Genetics and Plant Pathogenesis" (P. R. Day and G. J. Jellis, eds.), pp. 220–231. Blackwell, Oxford.

Mills, D. (1985). Transposon mutagenesis and its potential for studying virulence genes in plant pathogens. *Annu. Rev. Phytopathol.* **23,** 297–320.

Minami, E., Ozeki, Y., Matsuoka, M., Koizuka, N., and Tanaka, Y. (1989). Structure and some characterization of the gene for phenylalanine ammonia-lyase from rice plants. *Eur. J. Biochem.* **185,** 19–25.

Moesta, P., and Grisebach, H. (1982). L-2-Aminooxy-3-phenylpropionic acid inhibits phytoalexin accumulation in soybean with concomitant loss of resistance against *Phytophthora megasperma* f. sp. *glycinea. Physiol. Plant Pathol.* **21,** 65–70.

Moesta, P., and West, C. A. (1985). Casbene synthetase: Regulation of phytoalexin biosynthesis in *Ricinus communis* L. seedlings. *Arch. Biochem. Biophys.* **238,** 325–333.

Mohnen, D., Shinshi, H., Felix, G., and Meins, F., Jr. (1985). Hormonal regulation of β-1,3-glucanase messenger RNA levels in cultured tobacco tissues. *EMBO J.* **4,** 1631–1635.

Montoya, A. L., Bethards, L., Buchanan, L. A., Dincher, S., Howe, G., Thompson-Talor, H., Desai, N., and Ryals, J. (1989). Functional analysis of the PR-1a promoter from *Nicotiana tabacum* cv. Xanthi: Identification of *cis*-acting sequences sufficient for induction by salicyclic acid and tobacco mosaic virus. *J. Cell. Biochem., Suppl.* **13d,** 308.

Muller, M. M., Gerster, T., and Schaffner, W. (1988). Enhancer sequences and the regulation of gene transcription. *Eur. J. Biochem.* **176,** 485–495.

Myerowitz, E. M. (1989). *Arabidopsis,* a useful weed. *Cell* **56,** 263–269.

Myerowitz, E. M., and Pruitt, R. E. (1985). *Arabidopsis thaliana* and plant molecular genetics. *Science* **229,** 1214–1218.

Nagy, F., Bantry, M., Hsu, M. Y., Wang, M., and Chua, N.-H. (1987). The 5′ proximal region of the wheat *Cab-1* gene promoter contains a 286 bp enhancer-like sequence for the phytochrome response. *EMBO J.* **6,** 2537–2542.

Napoli, C., and Staskawicz, B. J. (1987). Molecular characterization and nucleic acid sequence of an avirulence gene from race 6 of *Pseudomonas syringae* pv. *glycinea. J. Bacteriol.* **169,** 572–578.

Narita, J. O., and Gruissem, W. (1989). Tomato hydroxymethylglutaryl-CoA reductase is required early in fruit development but not during ripening. *Plant Cell* **1**, 181–190.

Niesbach-Klosgen, U., Barzen, E., Bernhardt, J., Rohde, W., Schwarz-Sommer, Z., Reif, H. J., Wienand, U., and Saedler, H. (1987). Chalcone synthase genes in plants: A tool to study evolutionary relationships. *J. Mol. Evol.* **26**, 213–235.

O'Connell, R. J., and Bailey, J. A. (1988). Differences in the extent of fungal development, host cell necrosis and symptom expression during race–cultivar interactions between *Phaseolus vulgaris* and *Colletotrichum lindemuthianum. Plant Pathol.* **37**, 351–362.

O'Connell, R. J., Bailey, J. A., and Richmond, D. V. (1985). Cytology and physiology of infection of *Phaseolus vulgaris* by *Colletotrichum lindemuthianum. Physiol. Plant Pathol.* **27**, 75–98.

Olah, Z., Bogre, L., Lehel, C., Farago, A., Seprodi, J., and Dudits, D. (1989). The phosphorylation site of Ca^{2+}-dependent protein kinase from alfalfa. *Plant Mol. Biol.* **12**, 453–461.

Ou-Lee, T.-M., Turgeon, R., and Wu, R. (1988). Interaction of a gibberellin-induced factor with the upstream region of an α-amylase gene in rice aleurone tissue. *Proc. Natl. Acad. Sci. U.S.A.* **85**, 6366–6369.

Palm, C. J., and Ryan, C. A. (1988). Identification and partial purification of a protein that binds the promoter region of a wound-inducible proteinase inhibitor II gene. *Plant Physiol., Suppl.* **86**, 20.

Palm, C. J., Costa, M. A., An, G., and Ryan, C. A. (1990). Wound-inducible nuclear protein binds DNA fragments that regulate a proteinase inhibitor II gene from potato. *Proc. Natl. Acad. Sci. U.S.A.* **87**, 603–607.

Parkkonen, T., Kivirikko, K. O., and Pihlajaniemi, T. (1988). Molecular cloning of a multifunctional chicken protein acting as the proly-hydroxylase β-subunit, protein disulphide-isomerase and a cellular thyroid-hormone-binding protein. *Biochem. J.* **256**, 1005–1011.

Paszkowski, J., Baur, M., Bogucki, A., and Potrykus, I. (1988). Gene targetting in plants. *EMBO J.* **7**, 4021–4026.

Paxton, J. D. (1981). Phytoalexins—A working redefinition. *Phytopathol. Z.* **101**, 106–109.

Payne, G., Ahl, P., Moyer, M., Maspen, A., Beck, J., Meins, F., Jr., and Ryals, J. (1990). Isolation of complementary DNA clones encoding pathogenesis-related proteins P and Q, two acidic chitinases from tobacco. *Proc. Natl. Acad. Sci. U.S.A.* **87**, 98–102.

Paz-Ares, J., Wienand, U., Peterson, P. A., and Saedler, H. (1986). Molecular cloning of the *c* locus of *Zea mays*: A locus regulating the anthocyanin pathway. *EMBO J.* **5**, 829–833.

Paz-Ares, J., Ghosal, D., Wienand, U., Peterson, P., and Saedler, H. (1987). The regulatory *c1* locus of *Zea mays* encodes a protein with homology to *myb* proto-oncogene products and with structural similarities to transcriptional activators. *EMBO J.* **6**, 3553–3558.

Peña-Cortés, M., Sánchez-Serrano, J., Mertens, R., and Willmitzer, L. (1989). Abscisic acid is involved in the wound-induced expression of the proteinase inhibitor II gene in potato and tomato. *Proc. Natl. Acad. Sci. U.S.A.* **86**, 9851–9855.

Pfitzner, U. M., and Goodman, H. M. (1987). Isolation and characterization of cDNA clones encoding pathogenesis-related proteins from tobacco mosaic virus infected tobacco plants. *Nucleic Acids Res.* **15**, 4449–4465.

Pfitzner, U. M., Pfitzner, A. J. P., and Goodman, H. M. (1988). DNA sequence analysis of a *PR-1a* gene from tobacco: Molecular relationship of heat shock and pathogen responses in tobacco. *Mol. Gen. Genet.* **211**, 290–295.

Pierce, M., and Essenberg, M. (1987). Localization of phytoalexins in fluorescent meso-phyll cells isolated from bacterial blight-infected cotton cotyledons and separated from other cells by fluorescence-activated cell sorting. *Physiol. Mol. Plant Pathol.* **31**, 273–290.

Pihlajaniemi, T., Helaakoski, T., Tasanen, K., Myllyla, R., Huhtala, M.-L., Koivu, J., and Kivirikko, K. I. (1987). Molecular cloning of the β-subunit of human prolyl 4-hydroxylase. This subunit and protein disulphide isomerase are products of the same gene. *EMBO J.* **6**, 643–649.

Pryor, A. (1987). The origin and structure of fungal disease resistance genes in plants. *Trends Genet.* **3**, 157–161.

Ralton, J. E., Howlett, B. J., Clarke, A. E., Irwin, J. A. G., and Imrie, B. (1988). Interaction of cowpea with *Phytophthora vignae*: Inheritance of resistance and production of phenylalanine ammonia-lyase as a resistance response. *Physiol. Mol. Plant Pathol.* **32**, 89–103.

Reisener, H. J., Tiburzy, R., Kogel, K. H., Moerschbacher, B., and Heck, B. (1986). Mechanism of resistance of wheat against stem rust in the Sr5/P5 interaction. *In* "Biology and Molecular Biology of Plant–Pathogen Interactions" (J. A. Bailey, ed.), pp. 141–148. Springer-Verlag, Berlin and New York.

Rigden, J., and Coutts, R. (1988). Pathogenesis-related proteins in plants. *Trends Genet.* **4**, 87–89.

Riggleman, R. C., Fristensky, B., and Hadwiger, L. A. (1985). The disease resistance response of pea is associated with increased levels of specific mRNAs. *Plant Mol. Biol.* **4**, 81–86.

Robbins, M. P., and Dixon, R. A. (1984). Induction of chalcone isomerase in elicitor-treated bean cells. Comparison of rates of synthesis and appearance of immunode-tectable enzyme. *Eur. J. Biochem.* **145**, 195–202.

Robbins, M. P., Bolwell, G. P., and Dixon, R. A. (1985). Metabolic changes in elicitor-treated bean cells: Selectivity of enzyme induction in relation to phytoalexin in-duction. *Eur. J. Biochem.* **148**, 563–569.

Roberts, W. K., and Selitrennikoff, C. P. (1988). Plant and bacterial chitinases differ in antifungal activity. *J. Gen. Microbiol.* **134**, 169–176.

Robertson, M. (1988). Homoeo boxes, POU proteins and the limits to promiscuity. *Nature (London)* **336**, 522–524.

Roby, D., and Gaynor, J. J. (1988). Regulation of chitinase gene expression in melon plants infected with *Colletotrichum lagenarium*. *Plant Physiol., Suppl.* **86**, 22.

Roby, D., Toppan, A., and Esquerré-Tugayé, M.-T. (1985). Cell surfaces in plant–microorganism interactions. V. Elicitors of fungal and of plant origin trigger the synthesis of ethylene and of cell wall hydroxyproline-rich glycoproteins in plants. *Plant Physiol.* **77**, 700–704.

Ryan, C. A. (1984). Systemic responses to wounding. *In* "Plant-Microbe Interactions. Molecular and Genetic Perspectives" (T. Kosuge and E. W. Nester, eds.), Vol. 1, pp. 307–320. Macmillan, New York.

Ryan, C. A. (1988). Oligosaccharides as recognition signals for the expression of defensive genes in plants. *Biochemistry* **27**, 8879–8883.

Ryan, C. A., and An, G. (1988). Molecular biology of wound-inducible proteinase inhibi-tors in plants. *Plant, Cell Environ.* **11**, 345–349.

Ryder, T. B., Cramer, C. L., Bell, J. N., Robbins, M. P., Dixon, R. A., and Lamb, C. J. (1984). Elicitor rapidly induces chalcone synthase mRNA in *Phaseolus vulgaris* cells at the onset of the phytoalexin response. *Proc. Natl. Acad. Sci. U.S.A.* **81**, 5724–5728.

Ryder, T. B., Hedrick, S. A., Bell, J. N., Liang, X., Clouse, S. D., and Lamb, C. J. (1987).

Organization and differential activation of a gene family encoding the plant defense enzyme chalcone synthase in *Phaseolus vulgaris*. *Mol. Gen. Genet.* **210**, 219–233.

Saluz, H. P., and Jost, J. P. (1989a). Genomic footprinting with Taq polymerase. *Nature (London)* **338**, 277.

Saluz, H. P., and Jost, J. P. (1989b). Genomic sequencing and *in vivo* footprinting. *Anal. Biochem.* **176**, 201–208.

Sauer, N., Corbin, D. R., Keller, B., and Lamb, C. J. (1990). Cloning and characterization of a wound-specific hydroxyproline-rich glycoprotein in *Phaseolus vulgaris*. *Plant Cell Environ.* **13**, 257–266.

Schaffner, W. (1989). How do different transcription factors binding the same DNA sequence sort out their jobs? *Trends Genet.* **5**, 37–38.

Scherer, G. F. E., Martiny-Baron, G., and Stoffel, B. (1988). A new set of regulatory molecules in plants: A plant phospholipid similar to platelet-activating factor stimulates protein kinase and proton-translocating ATPase in membrane vesicles. *Planta* **175**, 241–253.

Schlumbaum, A., Mauch, F., Vogeli, U., and Boller, T. (1986). Plant chitinases are potent inhibitors of fungal growth. *Nature (London)* **324**, 365–367.

Schmelzer, E., Borner, H., Grisebach, H., Ebel, J., and Hahlbrock, K. (1984). Phytoalexin synthesis in soybean (*Glycine max*). Similar time courses of mRNA induction in hypocotyls infected with a fungal pathogen and in cell cultures treated with fungal elicitor. *FEBS Lett.* **172**, 59–63.

Schmelzer, E., Jahnen, W., and Hahlbrock, K. (1988). *In situ* localization of light-induced chalcone synthase mRNA, chalcone synthase, and flavonoid end products in epidermal cells of parsley leaves. *Proc. Natl. Acad. Sci. U.S.A.* **85**, 2989–2993.

Schmid, J., Doerner, P. W., Clouse, S. D., Dixon, R. A., and Lamb, C. J. (1990). Developmental and environmental regulation of a bean chalcone synthase promoter in transgenic tobacco. *Plant Cell* (in press).

Schmidt, W. E., and Ebel, J. (1987). Specific binding of a fungal glucan phytoalexin elicitor to membrane fractions from soybean *Glycine max*. *Proc. Natl. Acad. Sci. U.S.A.* **84**, 4117–4121.

Schmidt, R. J., Burr, F. A., and Burr, B. (1989). Molecular analyses of *opaque-2*; a zein regulatory locus. *J. Cell. Biochem., Suppl.* **13d**, 237.

Schroder, G., Brown, J. W. S., and Schroder, J. (1988). Molecular analysis of resveratrol synthase. cDNA, genomic clones and relationship with chalcone synthase. *Eur. J. Biochem.* **172**, 161–169.

Schulze-Lefert, P., Dangl, J. L., Becker-André, M., Hahlbrock, K., and Schulz, W. (1989). Inducible *in vivo* DNA footprints define sequences necessary for UV-light activation of the parsley chalcone synthase gene. *EMBO J.* **8**, 651–657.

Sharp, J. K., Albersheim, P., Ossowski, P., Pilotti, A., Garegg, P., and Lindberg, P. (1984). Comparison of the structures and elicitor activities of a synthetic and a mycelial-wall-derived hexa-(β-D-glucopyranosyl)-D-glucitol isolated from the mycelial walls of *Phytophthora megasperma* f. sp. *glycinea*. *J. Biol. Chem.* **259**, 11341–11345.

Shields, R. (1989). Moving in on plant genes. *Nature (London)* **337**, 308.

Shinshi, H., Mohnen, D., and Meins, F., Jr. (1987). Regulation of a plant pathogenesis-related enzyme: Inhibition of chitinase and chitinase mRNA accumulation in cultured tobacco tissues by auxin and cytokinin. *Proc. Natl. Acad. Sci. U.S.A.* **84**, 89–93.

Shinshi, H., Wenzler, H., Neuhaus, J.-M., Felix, G., Hofsteenge, J., and Meins, F., Jr. (1988). Evidence for N- and C-terminal processing of a plant defense-related enzyme: Primary structure of tobacco prepro-β-1,3-glucanase. *Proc. Natl. Acad. Sci. U.S.A.* **85**, 5541–5545.

Showalter, A. M., Bell, J. N., Cramer, C. L., Bailey, J. A., Varner, J. E., and Lamb, C. J.

(1985). Accumulation of hydroxyproline-rich glycoprotein mRNAs in response to fungal elicitor and infection. *Proc. Natl. Acad. Sci. U.S.A.* **82**, 6551–6555.

Singh, H. Clerc, R. G., and Le Bowitz, J. H. (1989). Molecular cloning of sequence-specific DNA binding proteins using recognition site probes. *BioTechniques* **7**, 252–261.

Somssich, I. E., Schmelzer, E., Bollmann, J., and Hahlbrock, K. (1986). Rapid activation by fungal elicitor of genes encoding pathogenesis-related proteins in cultured parsley cells. *Proc. Natl. Acad. Sci. U.S.A.* **83**, 2427–2430.

Somssich, I. E., Schmelzer, E., Kawalleck, P., and Hahlbrock, K. (1988). Gene structure and *in situ* transcript localization of pathogenesis-related protein 1 in parsley. *Mol. Gen. Genet.* **213**, 93–98.

Stab, M. R., and Ebel, J. (1987). Effects of Ca^{2+} on phytoalexin induction by fungal elicitor in soybean cells. *Arch. Biochem. Biophys.* **257**, 416–423.

Stanford, A., Bevan, M., and Northcote, D. (1989). Differential expression within a family of novel wound-induced genes in potato. *Mol. Gen. Genet.* **215**, 200–208.

Staskawicz, B. J., Dahlbeck, D., and Keen, N. T. (1984). Cloned avirulence gene of *Pseudomonas syringae* pv. *glycinea* determines race-specific incompatibility on *Glycine max* (L.) Merr. *Proc. Natl. Acad. Sci. U.S.A.* **81**, 6024–6028.

Staskawicz, B., Dahlbeck, D., Keen, N., and Napoli, C. (1987). Molecular characterization of cloned avirulence genes from race 0 and race 1 of *Pseudomonas syringae* pv. *glycinea. J. Bacteriol.* **169**, 5789–5794.

Staskawicz, B. J., Bonas, U., Dahlbeck, D., Huynh, T., Kearney, B., Ronald, P., and Whalen, M. (1988). Molecular determinants of specificity in plant–bacterial interactions. *In* "Physiology and Biochemistry of Plant–Microbial Interactions" (N. T. Keen, T. Kosuge, and L. L. Walling, eds.), pp. 124–130. Am. Soc. Plant Physiol., Rockville, Maryland.

Stayton, M., Tamaki, S., Kobayashi, D., and Keen, N. T. (1989). A cultivar-specific elicitor of the hypersensitive response in soybean has been identified and may be the signal molecule that interacts with the plant disease resistance gene product to trigger host defense. *J. Cell. Biochem., Suppl.* **13D**, 326.

Stermer, B. A., and Hammerschmidt, R. (1987). Association of heat shock induced resistance to disease with increased accumulation of insoluble extensins and ethylene synthesis. *Physiol. Mol. Plant Pathol.* **31**, 453–461.

Stermer, B. A., Edwards, L. A., Edington, B. V., and Dixon, R. A. (1990a). Molecular cloning and analysis of 3-hydroxy-3-methylglutaryl coenzyme A reductase cDNAs from potato. *Plant Physiol.* (submitted).

Stermer, B. A., Schmid, J., Lamb, C. J., and Dixon, R. A. (1990b). Infection and stress activation of bean chalcone synthase promoters in transgenic tobacco. *Mol. Plant Microbe Int.* (submitted).

Stougaard, J., Sandal, N. N., Gron, A., Kühle, A., and Marckes, K. A. (1987). 5' Analysis of the soybean leghaemoglobin *1bc3* gene: Regulatory elements required for promoter activity and organ specificity. *EMBO J.* **6**, 3565–3569.

Strasser, H., Hoffmann, C., Grisebach, H., and Matern, U. (1986). Are polyphosphoinositides involved in signal transduction of elicitor-induced phytoalexin synthesis in cultured plant cells? *Z. Naturforsch., C: Biosci.* **41C**, 717–724.

Tabata, T., Takase, H., Takayama, S., Mikami, K., Nakatsuka, A., Kawata, T., Nakayama, T., and Iwabuchi, M. (1989). A protein that binds to a *cis*-acting element of wheat histone genes has a leucine zipper motif. *Science* **245**, 897–1020.

Tamaki, S., Dahlbeck, D., Staskawicz, B., and Keen, N. T. (1988). Characterization and expression of two avirulence genes cloned from *Pseudomonas syringae* pv. *glycinea. J. Bacteriol.* **170**, 4846–4854.

Tanaka, Y., Matsuoka, M., Yamamoto, N., Ohashi, Y., Kano-Murakami, Y., and Ozeki, Y. (1989). Structure and characterization of a cDNA clone for phenylalanine ammonia-lyase from cut-injured roots of sweet potato. *Plant Physiol.* **90**, 1403–1407.

Tate, B. F., Schaller, G. E., Sussman, M. R., and Crain, R. C. (1989). Characterization of a polyphosphoinositide phospholipase C from the plasma membrane of *Avena sativa*. *Plant Physiol.* **91**, 1275–1279.

Templeton, M. D., and Lamb, C. J. (1988). Elicitors and defense gene activation. *Plant, Cell Environ.* **11**, 395–401.

Tepper, C. S., and Anderson, A. J. (1986). Two cultivars of bean display a differential response to extracellular components from *Colletotrichum lindemuthianum*. *Physiol. Mol. Plant Pathol.* **29**, 411–420.

Tepper, C. S., Albert, F. G., and Anderson, A. J. (1989). Differential mRNA accumulation in three cultivars of bean in response to elicitors from *Colletotrichum lindemuthianum*. *Physiol. Mol. Plant Pathol.* **34**, 85–98.

Thornburg, R. W., An, G., Cleveland, T. E., Johnson, R., and Ryan, C. A. (1987). Wound-inducible expression of a potato inhibitor II—Chloramphenicol acetyltransferase gene fusion in transgenic tobacco plants. *Proc. Natl. Acad. Sci. U.S.A.* **84**, 744–748.

Tiemann, K., Hinderer, W., and Barz, W. (1987). Isolation of NADPH : isoflavone oxidoreductase, a new enzyme of pterocarpan phytoalexin biosynthesis in cell suspension cultures of *Cicer arietinum*. *FEBS Lett.* **213**, 324–328.

Ueda, T., Pichersky, E., Malik, V. S., and Cashmore, A. R. (1989). Level of expression of the tomato *rbcS-3A* gene is modulated by a far upstream promoter element in a developmentally regulated manner. *Plant Cell* **1**, 217–227.

van der Krol, A. R., Lenting, P. E., Veenstra, J., van der Meer, I. M., Koes, R. E., Gerats, A. G. M., Mol, J. N. M., and Stuitje, A. R. (1988a). An anti-sense chalcone synthase gene in transgenic plants inhibits flower pigmentation. *Nature (London)* **333**, 866–869.

van der Krol, A. R., Mol, J. N. M., and Stuitje, A. R. (1988b). Modulation of eukaryotic gene expression by complementary RNA or DNA sequences. *BioTechniques* **6**, 958–975.

van Tunen, A. J., Koes, R. E., Spelt, C. E., van der Krol, A. R., Stuitje, A. R., and Mol, J. N. M. (1988). Cloning of the two chalcone flavanone isomerase genes from *Petunia hybrida*: Coordinate, light-regulated and differential expression of flavonoid genes. *EMBO J.* **7**, 1257–1263.

van Tunen, A. J., Hartman, S. A., Mur, L. A., and Mol, J. N. M. (1989). Regulation of chalcone flavanone isomerase (CHI) gene expression in *Petunia hybrida*: The use of alternative promoters in corolla, anthers and pollen. *Plant Mol. Biol.* **12**, 539–551.

Vaslet, C. A., Strausberg, R. L., Sykes, A., Levy, A., and Filpula, C. (1988). cDNA and genomic cloning of yeast phenylalanine ammonia-lyase genes reveal genomic intron deletions. *Nucleic Acids Res.* **16**, 11382.

Vogeli, U., Meins, F., Jr., and Boller, T. (1988). Co-ordinated regulation of chitinase and β-1,3-glucanase in bean leaves. *Planta* **174**, 364–372.

Vogeli-Lange, R., Hansen-Gheri, A., Boller, T., and Meins, F., Jr. (1988). Induction of the defense-related glucanohydrolases, β-1,3-glucanase and chitinase, by tobacco mosaic virus infection of tobacco leaves. *Plant Sci.* **54**, 171–176.

Voisey, C. R., and Slusarenko, A. J. (1989). Chitinase mRNA and enzyme activity in *Phaseolus vulgaris* (L.) increase more rapidly in response to avirulent than to virulent cells of *Pseudomonas syringae* pv. *phaseolicola*. *Physiol. Mol. Plant Pathol.* **35**, 403–412.

Walker-Simmons, M., and Ryan, C. A. (1984). Proteinase inhibitor synthesis in tomato leaves. Induction by chitosan oligomers and chemically modified chitosan and chitin. *Plant Physiol.* **76,** 787–790.

Walter, M. H., Grima-Pettenati, J., Grand, C., Boudet, A. M., and Lamb, C. J. (1988). Cinnamyl alcohol dehydrogenase, a molecular marker specific for lignin synthesis: cDNA cloning and mRNA induction by fungal elicitor. *Proc. Natl. Acad. Sci. U.S.A.* **85,** 5546–5550.

Wang, J., Holden, D. W., and Leong, S. A. (1988). Gene transfer system for the phytopathogenic fungus *Ustilago maydis. Proc. Natl. Acad. Sci. U.S.A.* **85,** 865–869.

Wasylyk, B. (1988). Enhancers and transcription factors in the control of gene expression. *Biochim. Biophys. Acta* **951,** 17–35.

Weising, K., Schell, J., and Kahl, G. (1988). Foreign genes in plants: Transfer, structure, expression and applications. *Annu. Rev. Genet.* **22,** 421–477.

Welle, R., and Grisebach, H. (1988a). Induction of phytoalexin synthesis in soybean: Enzymatic cyclization of prenylated pterocarpans to glyceollin isomers. *Arch. Biochem. Biophys.* **263,** 191–198.

Welle, R., and Grisebach, H. (1988b). Isolation of a novel NADPH-dependent reductase which coacts with chalcone synthase in the biosynthesis of 6′-deoxychalcone. *FEBS Lett.* **236,** 221–225.

Weltring, K.-M., Turgeon, B. G., Yoder, O. C., and VanEtten, H. D. (1988). Isolation of a phytoalexin-detoxification gene from the plant pathogenic fungus *Nectria haematococca* by detecting its expression in *Aspergillus nidulans. Gene* **68,** 335–344.

Whalen, M. C., Stall, R. E., and Staskawicz, B. J. (1988). Characterization of a gene from a tomato pathogen determining hypersensitive resistance in non-host species and genetic analysis of this resistance in bean. *Proc. Natl. Acad. Sci. U.S.A.* **85,** 6743–6747.

Wingate, V. P. M., Lawton, M. A., and Lamb, C. J. (1988). Glutathione causes a massive and selective induction of plant defense genes. *Plant Physiol.* **87,** 206–210.

Young, D. H., and Pegg, G. F. (1981). Purification and characterization of 1,3-β-glucan hydrolases from healthy and *Verticillium albo-atrum*-infected tomato plants. *Physiol. Plant Pathol.* **19,** 391–417.

Young, N. D., Zamir, D., Ganal, M. W., and Tanksley, S. D. (1988). Use of isogenic lines and simultaneous probing to identify DNA markers tightly linked to the *Tm-2a* gene in tomato. *Genetics* **120,** 579–585.

MOLECULAR AND CELLULAR BIOLOGY OF THE HEAT-SHOCK RESPONSE

Ronald T. Nagao,* Janice A. Kimpel,† and Joe L. Key*,†

*Botany Department and †Office of the Vice President for Research,
University of Georgia, Athens, Georgia 30602

I. Introduction

The response of organisms to temperature stress has been a subject of study for many years, beginning with a significant observation by Ritossa in 1962 that several new puffs appeared in the polytene chromosomes of *Drosophila* upon a shift from 20 to 37°C. This observation led to many interesting studies over the years, including the important discovery by Tissiere's group that synthesis of a new set of proteins was induced following this shift in temperature. Related studies demonstrated that the puffs observed by Ritossa were the sites of vigorous RNA synthesis and that these RNAs were translated into the "heat-shock" proteins (HSPs) (see Ashburner and Bonner, 1979). Extensive reviews on the heat-shock (HS) response subsequently appeared (Craig, 1985; Pelham, 1985, 1986, 1987, 1988; Bienz, 1985; Kimpel and Key, 1985b; Lindquist, 1986; Nagao *et al.*, 1986; Schlesinger, 1986; Nover,

ADVANCES IN GENETICS, Vol. 28

1987; Neidhardt, 1987; Bienz and Pelham, 1987; Lindquist and Craig, 1988; Vierling, 1990; Nagao and Key, 1989).

There are several features of the HS response that make it an attractive system for study. The HS response is one of the most highly conserved biological responses known; it is universal, having been observed in organisms ranging from eubacteria to archebacteria, from lower eukaryotes to plants and man. [A species of *Hydra* appears to be the only organism studied to date that does not undergo a HS response (Bosch *et al.*, 1988).] The rapidity of induction of HS gene expression, the magnitude of the response in terms of the levels of heat-shock messenger RNAs (HS mRNAs) and HSPs that accumulate over a short period of time, and the ease of manipulation of the stimulus make the HS system an excellent model for studies on regulated gene expression. The HS response appears to be critical in the protection of cells from thermal damage caused by excessively high temperatures, in the repair and/or removal of structures damaged by heat stress, and in the maintenance of cellular structures during stress. These features presumably allow organisms to recover normal functions quickly following the stress.

In general terms, the HS response is characterized by (1) a very fast readout of polyribosomes with the release and accumulation of monoribosomes and a concomitant decrease in normal protein synthesis; (2) the induction of synthesis of a new set of mRNAs, the HS mRNAs, concomitant with reduced transcription of most preexisting mRNAs; (3) preservation of most preexisting mRNAs; (4) the reformation of polyribosomes engaged primarily in translation of HS mRNAs; (5) the accumulation of high levels of HSPs; (6) a gradual decline in HS mRNA synthesis during prolonged heat treatment; (7) the acquisition of thermotolerance to otherwise lethal temperatures; (8) the selective intracellular localization of many HSPs during the HS followed by delocalization during recovery; (9) a gradual decline in HSP synthesis and return to normal protein synthesis during prolonged heat treatments; and (10) its induction, in whole or in part, by other stresses (e.g., arsenite, cadmium, and amino acid analogs). Some HS genes are expressed under tight developmental controls, whereas other HS genes have counterparts that are constitutively expressed at non-HS temperatures (HS cognate genes) and that also show patterns of regulated expression during development.

This article concentrates on three major aspects of the HS response: (1) a comparative analysis of HS gene families and their expression, (2) regulation of HS gene transcription, and (3) functions of the HSPs.

II. Heat-Shock Protein Genes

A. HIGH-MOLECULAR-WEIGHT HEAT-SHOCK PROTEIN GENES

The most abundant HSP induced in many organisms has a molecular weight (MW) between 68K and 74K. This major protein family is generally referred to as HSP70, reflecting the molecular weight of the major *Drosophila* HSP for which the initial characterization was done. This heat-inducible protein has been highly conserved throughout evolution, and closely related proteins have been identified in a wide range of organisms, including prokaryotes (for reviews, see Lindquist and Craig, 1988; Nagao and Key, 1989). *Escherichia coli* has a single *HSP70*-related gene, *dnaK*, with 48% sequence identity to the *Drosophila HSP70* gene (Bardwell and Craig, 1984). The amino acid sequence identity of eukaryotic HSP70s (derived from the nucleotide sequence) ranges from about 50% to greater than 90%. In some cases, different members of a protein family from one species have greater sequence diversity than do representative members from widely different species. Plants also synthesize a family of HSP70 proteins, but generally the level of expression is not nearly as high as in other organisms, relative to other HSPs.

In all eukaryotic organisms that have been carefully examined, the *HSP70* genes consist of a family of closely related genes. The complexity and number of genes comprising the *HSP70* family differ among species. Despite the number of organisms that have been investigated, it is difficult to make generalizations about the composition of groups of HSPs from various species. The difficulty arises in part due to the existence of very closely related proteins, known as heat-shock cognates (HSCs), whose expression is not heat inducible. These genes may show constitutive expression or developmentally regulated expression, depending on the cell or tissue type. In addition to the HSCs, some HSPs also are expressed in response to developmental signals in the absence of heat stress.

Biochemical and genetic studies indicate that some HSPs or their cognates are present in organisms at normal growth temperatures and perform essential roles in normal cell function. The identification of some of these functions has been valuable in providing clues to the function of HSPs at high temperatures (see Lindquist and Craig, 1988; Vierling, 1990; see also below). Yeast (*Saccharomyces cerevisiae*) has at least nine different *HSP70-* like genes organized into five gene families (*SSA1* to *SSA4, SSB1* and *SSB2, SSC1, SSD1,* and *KAR2*) varying in sequence identity from 50 to 96%. *SSA3* and *SSA4* are heat-inducible genes (showing a low level of expression at 23°C), whereas *SSA1* and *SSA2* are constitutively expressed (and are thus considered HSCs). The *SSA1* and SSA2 genes have 80% sequence identity (Craig *et al.*, 1985). Some of the structural differences among these highly conserved proteins may be due in part to the different intracellular compartments in which the proteins are located. For example, a mammalian endoplasmic reticulum (ER)-localized protein, grp78 (glucose-regulated protein) [also found to be identical to the immunoglobulin heavy-chain binding protein (BiP)], has about 60% homology to HSP70.

The very highly conserved nature of the *HSP70* genes has been utilized to isolate cross-hybridizing homologs from a variety of plant species. For example, a *Drosophila HSP70* probe was used to isolate corresponding genes from maize (Rochester *et al.*, 1986), *Arabidopsis* (Wu *et al.*, 1988), and soybean (Roberts and Key, 1990). Two *HSP70* genes were isolated and sequenced from maize; one contains the entire coding region (predicting a 646-amino acid protein), and the other contains only part of the coding sequence and 270 base pairs (bp) of 5' upstream sequence (Rochester *et al.*, 1986). Both maize genes are heat inducible, and each contains an intron. The position of the introns is interesting from an evolutionary perspective because each is in the same position (interrupting a codon specifying aspartic acid at position 71) as in the HSC gene, *HSC1*, of *Drosophila* (Ingolia and Craig, 1982). The *HSC1* gene is, however, constitutively expressed and is not heat inducible. In *Drosophila* and yeast, the heat-inducible members of the *HSP70* gene family do not contain introns. While this is clearly not the case for plants, the relationship between thermal inducibility, cognates, and introns in plants is not yet clear. Winter *et al.* (1988) reported the isolation of a *HSP70* cDNA from petunia using a maize *HSP70* clone as probe. This cDNA is homologous to other members of the petunia *HSP70* gene family, but the corresponding gene to this cDNA appears to be present as a single copy. This gene is expressed constitutively; however, HS, arsenite, and heavy metals significantly enhance expression. The gene contains a 618-bp intron in the same position as the intron in the maize *HSP70* genes (Rochester *et al.*, 1986).

The *HSP70*-related gene family from *Arabidopsis* codes for at least 12 polypeptides, most of which are constitutively expressed (Wu *et al.*, 1988). Comparisons of *in vitro* translation products of mRNA isolated from HS and control leaves indicate that the amount of mRNA encoding four HSP70 polypeptides is increased strongly by HS. Three *HSP70*-related genes were isolated, and two of the three genes are arranged in direct orientation approximately 1.5 kilobases (kb) apart. Only one of these genes, *HSP70-1*, is expressed at significant levels in unstressed tissue, and this expression is enhanced approximately 4- or 5-fold after HS. Sequence analysis predicts a putative intron in all three *Arabidopsis* genes at the same position as that in the two maize *HSP70* genes and the petunia HSP70 gene mentioned above, as well as in the *Drosophila HSC70* gene (Ingolia and Craig, 1982) and a rat *HSC70* gene (Sorger and Pelham, 1987b). By analogy with results of similar studies from other systems, Wu and colleagues have concluded that these three *Arabidopsis* genes are members of a small family and are most closely related to *HSP70*-cognate genes found in the other species.

Two-dimensional polyacrylamide gel electrophoresis (PAGE) resolves soybean 70-kDa HSPs into at least 10 polypeptides. The p*I* of the various peptides ranges from 5.6 to 7.0. Although the number of HSP70 proteins varies slightly among different plant species, all appear to have two sets of isoforms, one having a p*I* around 5.6 and another with a p*I* around 7.0 (Mansfield and Key, 1987). The HSP70s characterized from animal systems generally have a p*I* around 5.6–6.1 (Welch *et al.*, 1985; Palter *et al.*, 1986); therefore, the HSP70 isoforms with a p*I* around 7.0 may be unique to plants.

The sequence of a near full-length cDNA and its corresponding genomic sequence for a soybean *HSP70* gene has recently been determined. This gene is not expressed (or expressed at very low levels) at control temperatures. DNA sequence characterization of the soybean *HSP70* gene showed high sequence identity (from 65 to 73%) to *HSP70* genes isolated from *Drosophila,* yeast, rat, maize, and *Petunia* (Roberts and Key, 1990). This *HSP70* gene does not contain an intron, in contrast to the results noted above for corn and petunia *HSP70* genes.

A second prominent set of high-molecular-weight heat-shock proteins (HMW HSPs) produced by all eukaryotic organisms is termed the *HSP90* family (Lindquist and Craig, 1988). The nomenclature convention of naming proteins based on mobilities on polyacrylamide gels has resulted in different size designations (e.g., *Drosophila* HSP83, yeast HSP90, and chicken HSP89), but evidence indicates that the *HSP90* genes represent a highly conserved family of HSPs. Members of the *HSP90* gene family have been cloned and sequenced from several evolutionarily diverse organisms, including *Drosophila,* yeast, chickens, mammals, trypanosomes, plants, and bacteria (see Lindquist and Craig, 1988; Nagao and Key, 1989). Sequence analysis of the cloned genes demonstrates that the proteins are very highly conserved; proteins from even the most distantly related eukaryotes have 50% amino acid sequence identity, and all have greater than 40% identity with the corresponding *Escherichia coli* protein (Lindquist and Craig, 1988).

In virtually all organisms, proteins of the HSP90 family are present at normal temperatures, but they are induced to higher levels (e.g., 5- to 10-fold) by HS. The number of genes within the *HSP90* family varies with the organism; for example, *Drosophila* has only one gene, but yeast has two genes (Lindquist and Craig, 1988). In yeast, one member of this gene family encodes a signal sequence to transport the protein across the ER. This ER protein is larger than the cytosolic protein, with apparent MW on sodium dodecyl sulfate (SDS) gels from 94K to 108K versus from 87K to 92K for the cytosolic form. The ER protein is also a glucose-regulated protein (grp94) induced by glucose starvation whereas the cytosolic version is induced by glucose restoration (Lindquist and Craig, 1988).

Conservation of the *HSP90* gene family sequence has been used to isolate the corresponding *HSP90*-related genes from soybean (Roberts and Key, 1985) and corn (Sinibaldi *et al.*, 1985) using a *Drosophila HSP83* gene fragment as probe. Two cDNA clones from this family have also been isolated from soybean (Roberts, 1989). These two cDNAs differ in their expression pattern; one cDNA is constitutively expressed and shows little enhancement with heat treatment, the other shows a very low level of constitutive expression with significant enhancement during HS (Roberts, 1989). Preliminary analysis of members of this gene family in plants indicates the presence of introns, but again the significance of this fact is unclear; the *Drosophila HSP83* contains one intron, yeast and *Trypanosoma cruzi* contain no introns, and the *HSP90*-related hen oviduct *HSP108* contains 17 introns.

B. LOW-MOLECULAR-WEIGHT HEAT-SHOCK PROTEIN GENES

Most organisms synthesize one or more 15- to 35-kDa HSPs, usually varying up to about 6–8 polypeptides, as in *Drosophila*. Plants, on the other hand, synthesize a large number (15–30 depending on the species) of low-molecular-weight heat-shock proteins (LMW HSPs) ranging in size from 15 to 30 kDa. In contrast to animal systems where the most abundant HSP family is HSP70, the LMW HSPs are the most abundant proteins induced by high temperature stress in most plant species (Nagao *et al.*, 1986; Mansfield and Key, 1987).

1. 15- to 18-kDa Families

Characterization of soybean HS cDNA and genomic clones has demonstrated that soybean LMW HSP genes represent several multigene families that share domains of homology with genes from evolutionarily distant organisms, including *Drosophila, Xenopus,* and *Caenorhabditis elegans* (see Nagao *et al.,* 1986; Lindquist and Craig, 1988; Nagao and Key, 1989). The area of highest conservation is in the carboxyl portion of these proteins. Based on hybrid-select translation and DNA sequence analyses, the largest soybean family (Class I) consists of 13 proteins, and DNA sequences of representative genomic clones of this family have been published (Schöffl *et al.,* 1984; Czarnecka *et al.,* 1985; Nagao *et al.,* 1985). Based on genes sequenced thus far, the MW range of this family is 17.3K–18.5K (Key *et al.,* 1987b). Comparative analysis of four Class I soybean HSP genes of 17.3–17.6 kDa showed greater than 90% amino acid homology, because approximately two-thirds of the nucleotide changes were silent substitutions (Nagao *et al.,* 1985). Comparison of an 18.5-kDa HSP sequence, *Gmhsp18.5-V,* with the four Class I sequences mentioned previously showed approximately 75% amino acid identity among all five sequences but greater than 90% identity when the *Gmhsp18.5-V* protein sequence is compared individually to each of the four 17-kDa proteins. Of the 11 amino acid changes unique to *Gmhsp18.5-V,* 9 are located in one region close to the amino-terminus of the protein. This is the same region in which a single amino acid deletion is postulated in two of the genes to maintain amino acid alignment (Key *et al.,* 1987b; Nagao and Key, 1989). This suggests that this region is less significant functionally and therefore less conserved evolutionarily. The sequence of a 17.9-kDa soybean gene demonstrates that this gene is a member of a different gene family, having only 37–44% homology with the Class I LMW HSPs discussed previously (Raschke *et al.,* 1988). This is expected, because the 13 Class I proteins identified by hybrid-select translation do not include all of the soybean HSPs in this MW range. Despite the divergence in sequence among eukaryotic LMW HSPs, they maintain a strong conservation in two characteristics: (1) hydropathic profiles (see Nagao *et al.,* 1986; Lindquist and Craig, 1988) and (2) the existence of a five-amino-acid sequence, within the carboxyl-terminus, that is also present in a lens crystal protein, α-crystallin, of the vertebrate compound eye. As discussed below, the LMW HSPs, like the lens crystal protein, form aggregates or granules during HS.

2. 21- to 24-kDa Families

Based on hybrid-select translation and DNA sequence analyses, families of soybean HSP genes encoding approximately 21- to 24-kDa proteins are represented by cDNA clones pFS2033, pEV1, pEV2, and pEV6 and by genomic clone *Gmhsp22-K*. Amino acid sequence alignment shows that homology variation among clones within this size class (e.g., *Gmhsp22-K* versus pEV2; 42% identity within a 193-amino-acid sequence) may be as much as the variation between these and the 17-kDa HSPs (e.g., *Gmhsp22-K* versus *Gmhsp17.6-L;* 47% identity within a 151-amino-acid sequence). The region of maximum conservation is located near the carboxyl-terminus of the protein. The lesser homology in the amino-terminal portion of the HSPs may represent functional divergence for such phenomena as organellar-specific localization (see Section IV).

3. 26- to 28-kDa Families

HS genes encoding 26- to 28-kDa proteins can be divided into at least two classes. One class, represented by the cDNA pCE54, represents a family of general stress proteins. This family of four to six genes is expressed constitutively at control temperatures, and synthesis is enhanced with HS as well as with numerous other stress agents, including arsenite, heavy metals, high salt concentration, anaerobiosis, water stress, high 2,4-dichlorophenoxy acetic acid (2,4-D) concentration, and abscisic acid (ABA) treatment (Czarnecka *et al.*, 1984). The sequence of the gene corresponding to pCE54, *Gmhsp26-A,* contains a single intron of 388 bp that occurs between codons 107 and 108 of an open reading frame of 225 codons, encoding a putative 26-kDa protein. Processing of the intron was preferentially inhibited by treatment of soybean seedlings with $CdCl_2$ and $CuSO_4$ but not by elevated temperature (Czarnecka *et al.*, 1988). Comparison of hydropathy plots of this deduced amino acid sequence and those of the smaller HSPs of soybean indicates a high degree of relatedness in the carboxyl half of the protein. However, the *Gmhsp26-A* sequence does not retain identity to the conserved Asn-Gly-Val-Leu-Thr sequence found in the highly conserved hydrophobic domain of the other LMW HSPs. This may be related to the fact that this protein does not localize to membranous fractions in the cell during HS as do other LMW HSPs; it remains a soluble protein during HS. While clearly related to HSPs, the lower amino acid sequence identity suggests that this protein is highly diverged and may therefore be specialized for general stress adaptation in soybean (Czarnecka *et al.*, 1988).

A second class of LMW HSPs is specifically localized in the chloroplasts of plants. These proteins are encoded by nuclear genes, synthesized in the cytoplasm, and transported into the chloroplasts. They range in size from 21 to 28 kDa, depending on the plant species. Vierling *et al.* (1988) have isolated and sequenced cDNA clones from soybean and pea that specify some members of this class. Nucleotide sequence comparison shows that the derived amino acid sequences of the mature pea and soybean proteins are 79% identical, and all but two changes represent conservative replacements. The soybean cDNA encodes 181 amino acids or 20.5 kDa of the 22-kDa mature protein. Comparison of this soybean cDNA to *Gmhsp22-K* (encoding a 22-kDa protein) showed 39% identity within a 94-amino-acid stretch, and comparison to *Gmhsp17.5-E* (a 17.5-kDa protein of Class I described above) showed 34% identity within a 102-amino-acid stretch. (These stretches were chosen by computer searches that determined the "best fit" for maximal homology between two sequences.) The stretches of homology between these proteins are located in the carboxyl-terminal portion of the polypeptides. While the amino-terminal portion of the *Gmhsp22-K* protein contains a putative transit peptide sequence that has some characteristics typical of transit sequences, it lacks two domains considered important for import and maturation of other chloroplast proteins.

Vierling and co-workers raised polyclonal antibodies to the carboxyl-terminal domain and the amino-terminal domain of the pea chloroplast HSP. The carboxyl-terminal antibodies cross-reacted with chloroplast HSP precursor proteins from *Arabidopsis,* petunia, and maize, but the amino-terminal antibodies effectively recognized only the pea precursor (Vierling *et al.,* 1989). Confirming the results of the sequence comparison between soybean HSP genes of different families, structural similarities among LMW HSPs occur in the carboxyl-terminal half of the proteins. Vierling *et al.* (1988) have proposed that the LMW HSPs may have evolved from a single nuclear gene; the less conserved amino-terminal region specifies cellular location and the more conserved carboxyl-terminal region specifies a functional domain(s) (Vierling *et al.,* 1988).

III. Regulation of the Heat-Shock Response

A. REGULATION OF HEAT-SHOCK GENE TRANSCRIPTION

Although the HS response is highly conserved, there are some marked differences in the regulation of HS gene expression in bacteria, lower eukaryotes, and some higher eukaryotes (see Bienz and Pelham, 1987; Lindquist and Craig, 1988). The response is controlled primarily at the level of transcription in *E. coli* and yeast, whereas regulation occurs both transcriptionally and translationally in *Drosophila,* mammalian cells, and plants (see Lindquist, 1986; Nagao *et al.,* 1986). These differences might be expected, given the half-lives of mRNAs (minutes in *E. coli* up to several hours in *Drosophila* and plants) and the requirement for a rapid qualitative shift in protein synthesis that occurs during HS.

1. Activation of Transcription of Heat-Shock Genes

In *E. coli,* HS results in the rapid and transient expression of at least 17 proteins as a result of enhanced gene transcription (Neidhardt *et al.,* 1984). The HS genes occur in unlinked operons and constitute a single regulatory unit (regulon) under the control of a positive-acting protein. This protein is an RNA polymerase sigma factor (σ^{32}), which is specifically required for the efficient transcription of the HS genes; these genes have promoter sequences distinguishable from other bacterial promoters. The level of σ^{32} appears to be rate limiting under normal growth temperatures and increases rapidly upon shift to HS temperatures. The control of HS gene expression is apparently accomplished through changes in σ^{32} concentration, by both increasing its rate of synthesis and decreasing its rate of degradation (Grossman *et al.,* 1987; Erickson *et al.,* 1987).

In eukaryotes, induction of HSP synthesis also involves transcriptional activation of the HS genes. The first eukaryotic HS promoters to be studied in detail were those of the *Drosophila HSP70* genes. Sequence analysis identified 355 nucleotides (nt) of conserved sequence 5' to the transcription unit of two *HSP70* genes (Karch *et al.,* 1981). Identification of sequences necessary for heat induction was initially accomplished by introduction of the *Drosophila* gene into cells of other organisms (Corces *et al.,* 1981; Mirault *et al.,* 1982; Pelham and Bienz, 1982; Voellmy and Rungger, 1982). The heat inducibility of the *HSP70* gene in heterologous systems demonstrated that the recognition signal for HS responsiveness is similar in widely divergent species. Comparison of the upstream regions of different HS genes identified a consensus sequence with dyad symmetry, CtnGAAnnTTCnAG (Pelham, 1982). Further comparison of HS gene promoters revealed that the minimum consensus element is the dyad C--GAA--TTC--G (Pelham, 1985; Bienz, 1985). This sequence is now recognized as a key, cis-acting regulatory element, the heat-shock element (HSE). An 8 out of 10 match to the core inverted repeat of the HSE (7 out of 8 for minimum consensus) is sufficient for heat inducibility, although some exceptions are noted (see Bienz, 1985; Bienz and Pelham, 1987). Even though *Drosophila* HSEs function to permit heat-inducible expression in heterologous systems, the temperature optimum for expression corresponds to that of the recipient cells and not of *Drosophila* (Corces *et al.*, 1981; Beinz and Pelham, 1982; McMahon *et al.*, 1984; Spena *et al.*, 1985).

The HSE is the cis-acting element that interacts with a trans-acting transcription factor, known as the heat-shock factor (HSF) (Wiederrecht *et al.*, 1987; Wu *et al.*, 1987; Bienz and Pelham, 1987). The HSF, originally referred to as HSTF (e.g., Parker and Topol, 1984) or HS activator protein (Wu, 1984), has not been as widely characterized as the HSE, but the *Drosophila* and yeast factors are very similar (Wiederrecht *et al.*, 1987; Wu *et al.*, 1987; Sorger and Pelham, 1987a). Discrepancies in the DNA-binding ability of the HSF before and after HS remain to be resolved (e.g., Zimarino and Wu, 1987; Bienz and Pelham, 1987). Unlike the σ^{32} "transcription factor" of *E. coli*, which increases markedly in concentration at the onset of HS, the HSF is present constitutively in eukaryotes, and it requires only "activation" upon the perception of HS to elicit the HS response (Wu *et al.*, 1987; Sorger *et al.*, 1987; Kingston *et al.*, 1987; Zimarino and Wu, 1987). In human and *Drosophila* cells, this activation results in the ability of HSF to bind to promoter regions of the HS genes. In contrast, the yeast HSF binds to the HSE in the absence of HS. In an attempt to understand the activation of HS gene transcription in yeast, Sorger and Pelham (1988) cloned and characterized the HSF gene. This work has led to several important findings: (1) this gene is required for growth at normal temperatures, (2) the transcriptional activity of a promoter containing a synthetic HSF-binding site increases proportionally over 200-fold as the growth temperature is increased from 15 to 39°C, and (3) the enhanced transcriptional activity is associated with changes in the phosphorylation state of HSF. These workers proposed that the expression of HS genes in yeast is modulated by phosphorylation of DNA-associated HSF, which allows a more efficient interaction of the factor with the transcription complex. They further suggested that the phosphorylation of clusters of amino acids on the HSF may create the equivalent of acidic activation sequences found in a number of different transcription factors. Progressive increases in HSF phosphorylation and transcriptional activity occur in parallel over a range of growth temperatures, supporting the idea that the ability to alter the acidity of such a cluster(s) by phosphorylation would permit the cell to vary HSF activity rapidly and continuously depending on the physiological state of the cell. Even though the HSFs of higher eukaryotes usually differ from yeast in not binding to HS gene promoters at normal growth temperatures, the activation and modulation of HS gene transcription during HS induction may use a similar phosphorylation mechanism, as is suggested by the work of Kingston *et al.* (1987).

Gilmour and Lis (1985) examined the distribution of RNA polymerase II on both uninduced and induced *HSP70* genes of *Drosophila* by protein–DNA cross-linking. The enzyme was present at very high density over the entire induced gene. However, the enzyme was also found associated with the promoter region (between -12 and $+65$ nt) of the uninduced gene (Gilmour and Lis, 1986), suggesting that the HS-induced activation of *HSP70* gene transcription involved a step subsequent to RNA polymerase binding at the promoter. More recent data show that the bound RNA polymerase is transcriptionally engaged in the uninduced cells, but transcriptional arrest occurs after the production of a nascent RNA chain about 25 nt in length (Rougvie and Lis, 1988). These results suggest that the HSF-mediated activation of this *HSP70* gene on HS is associated with the release of this postinitiation block in transcription. Rougvie and Lis (1988) offer several hypotheses for how the HSF, which in *Drosophila* binds to the promoter region following an activation step induced by HS (see above), may influence the arrested polymerase complex. These include an interaction between HSF and either (1) RNA polymerase II, (2) a hypothetical molecule that inhibits chain elongation at the point of arrest, (3) the constitutively bound TATA factor, or (4) the chromatin or DNA, causing changes that are propagated along the gene from the HSF-binding site to the site of the block.

In addition to the HSE proximal to the TATA sequence, other upstream sequences are often required for maximal HS-induced expression (Cohen and Meselson, 1984; Dudler and Travers, 1984; Morganelli *et al.*, 1985; Mestril *et al.*, 1986; Bienz and Pelham, 1987). The HSF shows cooperative binding to some HSEs (Topol *et al.*, 1985; Shuey and Parker, 1986), and most HS genes contain two or more HSEs (Pelham, 1985; Nagao *et al.*, 1986). Pauli *et al.*, (1986) reported that several hundred base pairs upstream of the coding region were required for expression of HSP23 and HSP26 in transformed flies; the sites of HSF interaction are separated by more than 200 bp in at least the HSP26 gene (Thomas and Elgin, 1988). Some features of this activation are conserved, because the introduction of a HS gene or a reporter gene under the control of a HS promoter sequence into a heterologous system results in expression of the gene in response to HS (see Nover, 1987). It is clear that HSF interacts with HSEs in many different DNA sequence environments to influence heat induction of HS genes, while other "elements" modulate the level, but not the inducibility, of transcription of HS genes (e.g., Hoffman and Corces, 1986; Bienz and Pelham, 1986; 1987; Gurley *et al.*, 1986, 1987; Wu *et al.*, 1986; Baumann *et al.*, 1987; Czarnecka *et al.*, 1989).

The presence of a single HSE in some DNA sequences is not always sufficient, however, to ensure significant heat inducibility of the gene (Sorger and Pelham, 1987a). A rat *HSC* gene promoter has a HSE that matches the sequence that was recently identified as the strongest HSE sequence for HS activation (Xiao and Lis, 1988; see below). However, this gene is only weakly heat inducible and strongly expressed at non-HS temperatures. Several other cis elements (e.g., CCAAT and Sp1 elements) are present in the vicinity of the HSEs. These elements apparently account for the constitutive expression of this gene and for the lack of heat inducibility as well. Additionally, when a *Drosophila HSP70* promoter was fused to a marker gene and introduced back into *Drosophila* cells, a single HSE was not sufficient for activity (Dudler and Travers, 1984; Amin *et al.*, 1985; Simon *et al.*, 1985). Additional 5' sequences, including a second HSE with only a 6 out of 8 match to the consensus, were required. Induction of *HSP70* expression in *Drosophila* apparently requires two HSEs, whose locations are somewhat flexible. This specific requirement, in contrast to the sufficiency of a single HSE in some eukaryotes, appears to represent a genuine species difference (Pelham, 1987).

The critical components of the HSE sequences have been called into question by the recent work of Xiao and Lis (1988) and Amin *et al.*, (1988), who have shown that it is actually the 10-bp sequence nTTCnnGAAn that functions as a cis element and binds HSF. Nonetheless, additional sequences to this 10-bp consensus are required to fully describe a functional HSE. For example, the strongest upstream binding site for HSF has a consensus sequence of nGAAnnTTCnnGAAn, which represents $1\frac{1}{2}$ copies of the 10-bp sequence. Xiao and Lis (1988) suggest that the functional HSE includes bases beyond the 14-bp HSE, and the actual regulatory element may better be described as a dimer of the 10-bp sequence nTTCnnGAAn. Interestingly, all of the sequenced genes of the Class I group of soybean LMW HSPs have the sequence TTCnnGAA as part of two overlapping pairs of HSEs proximal to the TATA motif, i.e., nTTCnnGAAnnT(T/A)Cn (Nagao and Key, 1989).

In soybean, deletion mutagenesis of two HS genes encoding 17- to 18-kDa HSPs indicates that additional sequences, upstream of the TATA-proximal HSEs, contribute to the level of thermal induction (Gurley *et al.*, 1986; Baumann *et al.*, 1987; Czarnecka *et al.*, 1989). One highly expressed HS gene of soybean, *Gmhsp17.5E*, has been characterized in considerable detail, and at least five regions 5' to the transcription start site play important regulatory roles in expression (Czarnecka *et al.*, 1989). These studies were carried out using sunflower tumors that were transformed with a vector containing the altered promoter construct as well as a reference gene. From 100 to 200 independently transformed tumors were used in each experiment to minimize differences attributable to the location of insertion in the plant chromosomes. These regions include the TATA motif, two regions in the −40 to −100 upstream sequence including the TATA-proximal HSE (site 1) and an upstream HSE (site 2), an AT-rich region at −120 to −150 that is required for full activity and that specifically binds a nuclear protein (Czarnecka *et al.*, 1990), and a region at −180 to −244 that contains a TATA and partial HSE dyad (TATAAAGAATTTC) that contributes about 65% to the promoter activity. This latter sequence is conserved in many *Drosophila* (Southgate *et al.*, 1983) and soybean HS genes (Nagao *et al.*, 1985). These observations, together with the general conservation of a TATA/dyad region in genes from other organisms, indicate that this may be an important regulatory region not previously recognized. Multiple HSEs located 5′ to this TATA/dyad motif have little influence on expression of this particular gene, in contrast to the results for some *Drosophila* genes discussed earlier.

Regions of the promoter required for expression of a closely related soybean gene, *HS6871* (*Gmhsp17.3-B*), were evaluated by *in vitro*-generated deletion mutations of the promoter element using transformed, regenerated tobacco plants (Baumann *et al.*, 1987). Large clonal variation (greater than several hundred percent in some cases) of independently isolated transgenic plants containing the same deletion of *HS6871* was observed. On average, the longest deletion construct tested, *HS6871*-439 [−336 from transcription start site (TSS)], gave levels of HS-induced expression in tobacco similar to the levels of mRNA for that gene in soybean. Deletion *HS6871*-407 (−304 from TSS) interrupted an imperfect dyad of 34 bp (centered at −407), and expression was reduced to 50% of the full-length construct. Deletion *HS6871*-298 (−195 from TSS), which removed a sequence of 14 adenine nucleotides between −357 and −371 (−254 and −268 from TSS) reduced heat-inducible expression to 10%. Further deletions could not be quantified due to resolution limits, but similar, very low HS-inducible activity was observed for both deletion *HS6871*-242 (−139 from TSS), which still contained two overlapping HSEs (the deletion removed 3 nt at the 5′ end of a third overlapping HSE), and deletion *HS6871*-181 (−78 from TSS), which removed these HSEs. The most dramatic effect on expression, however, resulted from the removal of the region from −439 (−336 from TSS) to −298 (−195 from TSS), which does not contain any HSEs. An inversion construction *HS6871*-181/-407 (an inversion of the sequence −78 to −304 from TSS) gave an intermediate level of expression, suggesting a bidirectional function for this region (Baumann *et al.*, 1987). Thus, despite the different experimental systems (transgenic tobacco plants versus transformed sunflower callus), the promoter regions of *Gmhsp17.5-E* (Gurley *et al.*, 1986; Czarnecka *et al.*, 1989) and *Gmhsp17.3-B* (Baumann *et al.*, 1987) appear very similar.

In addition to the preceding sequences, which seem to play the most important roles in expression of the *Gmhsp17.5-E* gene, other sequences within some of these deletions may have a regulatory role in HS-induced expression (Czarnecka *et al.*, 1989). One sequence of interest is a CCAAT-box-like sequence (-87 to -81) immediately 3' to the HSE. Studies in other systems have indicated that the CCAAT-binding protein may interact with HSE to enhance its binding to HSF, especially in the absence of HS (Bienz, 1986; Bienz and Pelham, 1986). Additionally, the sequences around one HSE with a minimum consensus include a SV40 enhancer-like sequence (11 of 14 bp identity) that overlaps the HSE. This motif is preceded by a stretch of alternating purine/pyrimidine residues; while it has not been evaluated experimentally to date, this GT motif raises the possibility that this promoter may be regulated by a variety of developmental and environmental signals (see Czarnecka *et al.*, 1989).

The importance of the HSE to heat-induced expression of genes in plants is also illustrated by the report of Strittmatter and Chua (1987) on chimeric promoter constructions in transgenic tobacco. They demonstrated that a construct containing a synthetic 36-bp oligonucleotide, based on the overlapping HSEs of the *Gmhsp17.3B* soybean gene discussed earlier, conferred heat inducibility to a light-inducible, leaf-specific promoter. This oligonucleotide, which contains the TTCnnGAA motif as part of the proximal overlapping HSE structure, was inserted into the 5' region of the pea *rbcS-3A* 5' flanking region (position -410 to $+15$). The element conferred heat inducibility whether it was inserted at position -410 or at position -49 proximal to the TATA box (the latter is the normal position for proximal HSEs). There was no expression from this promoter in the dark under HS conditions; these researchers hypothesized that this lack of expression may result from negative elements in the *rbcS-3A* promoter that normally suppress expression in the dark and in other organs (e.g., roots).

Given the high conservation in promoter structure, it is not surprising that a chimeric gene containing a *Drosophila HSP70* promoter is also expressed in a heat-inducible manner in transgenic tobacco tissues (Spena *et al.*, 1985; Spena and Schell, 1987). The plant's trans-acting factors presumably correctly recognize the cis-acting elements of the *Drosophila HSP70* promoter. The actual level of expression relative to a plant *HSP70* promoter was not assessed, and it may be significantly lower, because this was the case for expression of the *Drosophila HSP70* promoter in sea urchin (McMahon *et al.*, 1984). Two maize *HSP70* genes have been isolated and used to construct a hybrid gene that was transferred into *Petunia* cells (Rochester *et al.*, 1986). The hybrid gene had 1.1 kb of a 5'-flanking sequence that contained two HSEs, with an 8 out of 10 identity to the *Drosophila* consensus located 29 and 75 bp upstream of the TATA motif. The expression of this gene in a regenerated *Petunia* plant was thermally induced, and the transcript was initiated correctly; however, the level of expression appeared to be significantly less than the HS-induced level in maize shoots.

Overall, these results on HS promoters in plant genes are generally consistent with the more detailed studies of *Drosophila* HS gene expression, but they also illustrate differences among eukaryotes (see Bienz and Pelham, 1987), reflecting the complexity of transcriptional regulation of HS genes.

2. Self-Regulation of the Heat-Shock Response

The production of HSPs in *E. coli* is transient, increasing within 1 minute of the temperature shift and declining after 15–20 minutes. This decrease in synthesis appears to occur as a result of the enhanced accumulation of a particular HSP, the dnaK protein, during HS (see Neidhardt *et al.*, 1984; Craig, 1985). Cells carrying a *ts* mutation (*dnaK 756*) in the dnaK protein do not turn off the HS response for several hours, and thus they overproduce HSPs; revertants of this mutation regain the ability to limit the HS response. Conversely, strains of *E. coli* carrying the *dnaK* gene on an overproducing plasmid underproduce HSPs at the elevated temperature. These results suggest that the dnaK protein is a negative regulator of HSP synthesis. Because the dnaK protein is a HSP, the HS response can be viewed as a self-regulating or "autoregulated" response. It should also be noted, however, that the dnaK protein is present in non-HS *E. coli* cells, and indeed it is required for normal growth as well as for phage replication. The mechanism of action of the dnaK protein is not fully understood, but it apparently interferes directly or indirectly with σ^{32} activity. The dnaK protein has been found associated with purified preparations of RNA polymerases (which contain the appropriate σ factors) from both the unstressed cells ($E\sigma^{70}$) and the heat-shocked cells ($E\sigma^{32}$) (Skelly *et al.*, 1988); interestingly, σ^{32} antibody cross-reacts with the dnaK protein.

There is significant sequence conservation (45–48% sequence identity) between the prokaryotic dnaK HSP and the eukaryotic HSP70 family of proteins. Because the dnaK protein appears to have a self-regulatory role in the HS response, several investigators have studied self-regulation of the HS response in eukaryotes, focusing on the *HSP70* gene and its protein. In *Drosophila,* the Lindquist laboratory has demonstrated a strong correlation between the accumulation of HSP70 protein and the expression of self-regulation (DiDomenico *et al.,* 1982; Lindquist and DiDomenico, 1985; Lindquist, 1986). When HSPs are synthesized in the presence of amino acid analogs (e.g., ca-navanine), self-regulation is not observed; further, these HSPs are apparently nonfunctional, because they fail to undergo the normal, intracellular localization during HS (see the following discussion). It appears that a threshold value of functional HSP70 is critical for the repression of further transcription of HS genes; the level of HSP70 protein that accumulates is proportional to the severity of the heat treatment. The onset of transcription repression also is delayed with increasing severity of the HS. One note of caution, however, relates to the fact that the level of HSP70 proteins during HS never reaches the basal level of the major HSC70 proteins under non-HS conditions (Palter *et al.,* 1986). The deletion of two *HSC70* genes of yeast (*YG100* and *YG102,* now referred to as *SSA1* and *SSA2,* respectively) results in the constitutive synthesis of a HSP70 protein not normally expressed at control temperatures (Craig and Jacobsen, 1984; Werner-Washburne *et al.,* 1987). Taken together, the data provide strong correlative evidence for a role of HSP70 and HSP70-like proteins in the self-regulation of the HS response. As noted above, the proposed mechanism of dnaK action in regulating the HS response is an interaction with σ^{32}. An analogous mechanism may apply to eukaryotes; i.e., a HSP70 protein(s) might interact with transcription regulatory proteins (e.g., HSF) to elicit the observed responses. Evidence is accumulating that HSP70 and HSP70-like proteins, as well as other stress proteins, can bind specifically to certain cellular proteins (see Lindquist and Craig, 1988; Vierling, 1990).

Soybean seedlings also exhibit the properties of self-regulation during continuous HS. HS mRNAs accumulate to maximal levels within the first 1–2 hours of HS, and then the levels gradually decline (Nagao *et al.,* 1986; Key *et al.,* 1987a). HSPs synthesized from these mRNAs accumulate in the tissue and are stable for many hours. Nuclear runoff transcription analyses also demonstrate that transcriptional activity is maximal in the first few hours of the heat treatment and not detectable after 4 hours (Kimpel *et al.,* 1990). Together, these results indicate that self-regulation of HS mRNA synthesis becomes operational after about

2 hours. HSP synthesis gradually declines to undetectable levels over the next 3–4 hours. HS mRNAs made during continuous 40°C heat treatment gradually decline in relative abundance, but remain detectable even after a 12-hour exposure to 40°C. However, the HS mRNAs rapidly deplete to barely detectable levels within the first 3 hours at 30°C following a 2-hour HS at 40°C HS. These results reflect a severalfold decrease in HS mRNA stability at 30°C compared to 40°C. In summary, experiments measuring steady-state HS mRNA levels and relative transcription rates indicate that heat is the primary signal for initiating HS gene transcription; however, additional mechanisms determine the steady-state level of HS mRNAs in the tissue. Soybean seedlings also respond to a sequential series of HS treatments (e.g., cycles of 2 hours at 40°C followed by 4 hours at 30°C) by repeated induction of HS mRNA and HSP synthesis (Nagao *et al.*, 1986). The levels of accumulated HS mRNAs diminish somewhat with each successive round of HS. Apparently, the presence of previously synthesized HSPs in soybean tissue is not in itself sufficient to suppress the typical HS response in successive cycles. This may be reconciled with the putative regulatory role for HSPs in the HS response by several possibilities: (1) the loss of self-regulation during successive HS recovery cycles could be due to a change in functional state of a HSP(s) or other non-HS-induced regulatory protein(s), HSF for example, during the intervening 28°C treatment; (2) a specific regulatory HSP(s) could have a much shorter half-life than the bulk of the HSPs; or (3) HSPs may not directly determine the regulation of the response. It is noteworthy that a 10-minute 45°C HS followed by incubation for 2 hours at 28°C (a treatment sufficient to induce a complete HS response in soybean when given initially) does not induce a substantial HS response when it follows an initial cycle of a 40°C 2-hour HS and a 4-hour recovery period at 30°C. However, a 10-minute 50°C HS (a lethal initial treatment) does induce a substantial HS response when administered following an initial HS/recovery period. Thus, it appears that a new threshold must be exceeded before a thermotolerant tissue will undergo a significant HS response. As mechanism(s) involved in induction of HS gene transcription are formulated, these observations should prove useful in gaining an insight into the mechanism(s) operative in self-regulation of the HS response. They should also assist in the identification and isolation of any additional HS regulatory protein(s) and of any modifications to HSF or other transcription factors that influence HS gene transcription. Clearly, cells and tissues respond to elevated temperatures quite differently depending on whether the tissue has received an earlier HS treatment. Thus, the HS response is

not simply the result of cells having experienced an excessive temperature; it must also involve a complex network of regulatory events.

B. ABNORMAL PROTEINS AS A SIGNAL FOR INDUCTION OF THE HEAT-SHOCK RESPONSE

Much has been learned about the regulation of expression of HS genes in a variety of organisms, yet little is known about how HS causes the dramatic shifts in physiology and biochemistry. Several lines of evidence indicate that during HS, abnormal proteins (e.g., heat denatured, foreign, aberrant, or modified) rapidly accumulate, and it has been suggested that these abnormal proteins may actually trigger the HS response. In fact, abnormal proteins trigger a HS response in the absence of heat (Finley et al., 1984; Voellmy, 1984; Goff and Goldberg, 1985; Ananthan et al., 1986; Lee and Hahn, 1988; Grant et al., 1989). One HSP with a known function is ubiquitin (Bond and Schlesinger, 1985; Finley et al., 1987; Christensen and Quail, 1989). In normal cells, ubiquitin is involved in the ATP-mediated proteolysis of unstable or abnormal cellular proteins in eukaryotes (Finley and Varshavsky, 1985). The lon protease apparently plays a similar role in degrading unstable and abnormal proteins in an ATP-dependent manner in prokaryotes. During HS, synthesis of both of these proteins is enhanced (Neidhardt et al., 1984; Goff and Goldberg, 1985; Schlesinger, 1986; Finley et al., 1987; Christensen and Quail, 1989). In the case of ubiquitin genes in yeast (Özkaynak et al., 1987), the UBI4 gene is strongly inducible by HS, starvation, and other stresses; this gene contains an element similar to a HSE in its 5' upstream region. It appears that the UBI4 gene is specifically required for resistance of yeast cells to stress (e.g., HS, amino acid analogs, and starvation), and it is essential for sporulation. The UBI4 mutants are also hypersensitive to HS, amino acid analogs, and starvation, indicating that this gene is an essential component to the stress response system (Finley et al., 1987). A ts mutant of mouse mammary cells, ts85, is unable to form ubiquitin–protein complexes or to degrade abnormal polypeptides (Finley et al., 1984). These ts85 cells also synthesize HSPs at 39°C, a temperature that produces no HS response in wild-type parental cells. Drosophila cells carrying mutations in the actin III gene (Hiromi and Hotta, 1985) have been shown to undergo a HS response at normal temperatures; these cells express abnormal actin proteins (Okamoto et al., 1986). Similarly, the production of some foreign (abnormal) proteins in E. coli at normal growth temperatures enhances transcription of some HS genes (Goff and Goldberg, 1985). Ananthan *et al.* (1986) showed that the injection of denatured proteins (e.g., bovine serum albumin) into *Xenopus* oocytes induced HSP synthesis, whereas the injection of native proteins did not.

Amino acid analogs also induce HS gene transcription at non-HS temperatures (Kelley and Schlesinger, 1978; DiDomenico *et al.*, 1982; White and Hightower, 1984; Li and Laszlo, 1985; Lindquist, 1986; J. Y.-R. Lee, R. T. Nagao, C. Y. Lin, and J. L. Key, unpublished observations). At least some HSPs are probably rendered nonfunctional by virtue of amino acid analogs incorporation into the proteins. Again, these abnormal proteins may serve as the signal for triggering the HS response in the presence of amino acid analogs. Amino acid analog-treated cells have elevated levels of ubiquitinated proteins (see Schlesinger, 1986), consistent with the fact that analog-containing proteins are more rapidly degraded in animal cells than are normal cellular proteins (Mizzen and Welch, 1988). This suggests that these proteins may be recognized as abnormal proteins.

A recent report (Grant *et al.*, 1989) demonstrates that mistranslation induces the HS response in yeast. The antibiotic paromomycin induces errors in translation *in vivo* and *in vitro* as a consequence of loss in fidelity of codon–anticodon interactions, resulting in both missense and nonsense errors. The induction of the HS response by paromomycin appears to be the result of the accumulation of abnormal "proteins" and not due to inhibition of protein synthesis by nonlethal levels of this antibiotic, because the inhibition of protein synthesis in yeast by cycloheximide does not induce the HS response.

Research to date provides strong evidence that the accumulation of abnormal proteins in both eukaryotic and prokaryotic organisms is sufficient to induce a HS response. It is reasonable to assume that the presence of abnormal proteins could be a cellular signal that triggers the expression of proteases and protease-related proteins (e.g., ubiquitin) and other cellular repair functions in response to HS and other stresses. It has been further speculated that the rapid accumulation of abnormal proteins at HS temperatures is the actual trigger for induction of the HS response. Such a mechanism could account for the fact that yeast cells carrying mutations in the *SSA1* and *SSA2* genes (which encode constitutively expressed HSC70 proteins) synthesize HSPs at non-HS temperatures. The absence of the SSA1 and SSA2 proteins would undoubtedly lead to the accumulation of proteins in inappropriate cell compartments and/or in an abnormal structure, because these proteins are involved in maintaining cellular proteins in a conformation to permit transport across membranes (see Section IV on function).

IV. Functions of the Heat-Shock Proteins

Research efforts directed at elucidating the function of HSPs have progressed in at least two aspects: the association of HSPs with the development of thermal tolerance, and the biochemical activities of individual HSPs. As mentioned previously, HSC proteins that are very closely related at the sequence level to the HSPs are constitutively expressed and often developmentally regulated. A popular hypothesis is that they fulfill a role at non-HS temperatures similar to that of HSPs at HS temperatures, but the nature of that role has not been precisely identified in most cases. The following sections briefly summarize the current knowledge in these two areas of HSP functions.

A. THERMOTOLERANCE

There is considerable correlative evidence supporting a role for HSPs in the development of thermotolerance. As defined here, thermotolerance is the ability of an organism to withstand an otherwise lethal heat treatment once it has been pretreated with a nonlethal HS. Generally, HS conditions that lead to the development of thermotolerance are also conditions that optimally induce the synthesis of HSPs (Neidhardt et al., 1984; Lindquist, 1986; Nagao et al., 1986; Li and Laszlo, 1985). In mammalian cells, heat-resistant variants isolated from cell cultures synthesize HSP70 constitutively, leading Laszlo and Li (1985) to conclude that HSP70 is involved in conferring thermotolerance. In many organisms, stresses other than heat often induce the synthesis of HSPs, and these stresses can also confer thermotolerance (Plesset et al., 1982; Li and Werb, 1982; Lin et al., 1984; Neidhardt et al., 1984; Lee and Hahn, 1988). One major exception to these observations involves the induction of HSPs by amino acid analogs. In Chinese hamster fibroblast cells, treatment with canavanine or azetidine-2-carboxylic acid, while inducing the synthesis of HSPs, resulted in a sensitization of the cells to heat treatments, rather than development of any thermotolerance (Li and Laszlo, 1985). As discussed previously, HSPs made in the presence of amino acid analogs are probably recognized as abnormal proteins by the cells. As first suggested by Lindquist's laboratory following their studies on HSP70 synthesis in Drosophila, HSPs synthesized during exposure to amino acid analogs are probably non- or poorly functional, and thus they would be unable to fulfill a role in protecting cells from subsequent heat damage (DiDomenico et al., 1982). Li and Laszlo (1985) further supported this interpretation by showing that Chinese hamster fibroblast cells allowed to synthesize normal HSPs, prior to treatment with the amino acid analogs during a subsequent HS, were thermo-tolerant.

Typically, to prove the function of a gene product, mutant lines of the organism that no longer contain the gene (or in which gene expression is specifically inactivated) are developed or isolated. Such mutants have been identified in *E. coli*. These mutants, which lack a functional *htpR* (*rpoH*) gene, did not acquire thermotolerance during a sublethal HS (Neidhardt *et al.*, 1984). However, in a more recent report, these re-searchers reported results that question the role of HSPs in the acqui-sition of thermotolerance in prokaryotes (Van Bogelen *et al.*, 1987). They developed *E. coli* cells that contained multiple copies of the *htpR* gene under the control of an isopropyl β-D-thiogalactoside (IPTG)-inducible promoter. In the presence of IPTG, HSPs under the control of the HS regulon are synthesized. However, such IPTG-induced cells were not protected at 50°C, whereas *E. coli* cells that were previously heat shocked at 42°C (and which synthesized the HSPs of the HS regulon) were tolerant of an incubation at 50°C. These researchers concluded that "thermotolerance appears to develop by processes other than *htpR*-dependent induction of heat shock proteins." As discussed below, it is possible that there are several mechanisms for the develop-ment of thermotolerance. However, as pointed out by Lindquist and Craig (1988), it must also be noted that the levels of HSPs in these cells following IPTG induction were not the same as the levels induced by heat, and a few HSPs were not induced at all.

The mutant approach has been difficult to apply in eukaryotes, be-cause, as described above, most organisms contain several copies of most HS genes. Still, the results from limited genetic studies generally support a role for some HSPs in thermotolerance. In *Dictyostelium,* a mutant was isolated that did not develop thermotolerance to normally lethal temperatures when first given a heat pretreatment (Loomis and Wheeler, 1982). This mutant did not synthesize the 26K–32K LMW HSPs, and synthesis of the HMW HSPs was only mildly enhanced during HS relative to the wild-type response.

Disruption mutation studies in yeast show that the heat-inducible *SSA3* and *SSA4* genes are not essential (Werner-Washburne *et al.*, 1987). Mutations in both *SSA1* and *SSA2* (which are normally ex-pressed constitutively) lead to constitutive expression of other HSPs, including *SSA4;* when these strains are exposed to a severe HS, they are about as tolerant as wild-type strains that are first given a precon-ditioning HS (Craig and Jacobsen, 1984). These data indicate that the SSA4 protein is functionally equivalent to the SSA1 and SSA2 proteins, allowing growth at normal temperatures in the mutant and being correlated with thermotolerance. Yeast strains lacking *SSA1, SSA2,* and *SSA4* genes are inviable, but they can be rescued by a plasmid carrying *SSA3* (normally only heat inducible) behind a strong constitutive promoter. These experiments indicate that *SSA3* encodes a protein functionally similar to that encoded by *SSA1, SSA2,* and *SSA4*.

In contrast, Petko and Lindquist (1986) created deletion mutations in yeast involving the *hsp26* gene. This is a single-copy gene in the yeast chromosome, and the encoded HSP represents a unique size class of HSPs of yeast that does not have significant homology to other classes of HSPs. Still, removal of this HSP had no detectable effect on any growth parameters that were measured, including spore development, growth at higher temperatures, and the development of thermotolerance. As these researchers indicate, it is possible that members of other HS gene families could provide sufficient activities necessary for the expression of thermotolerance, even though they would be structurally distinct from the HSP26 protein. Such a "functional redundancy" may also be indicated by the overexpression of other HSPs in the *SSA1 SSA2* mutants described above. Although such an explanation may prove correct, many more experiments are required to explain the strong conservation of these divergent groups of HS proteins that may nevertheless compensate for each other in biological responses to heat stress.

Johnston and Kucey (1988) devised a novel method for modulating the level of the HS response. They constructed a plasmid that carried a dihydrofolate reductase (DHFR) gene with its normal promoter and polyadenylation signals and a fragment from the 5' control region of a *HSP70* gene from *Xenopus laevis*. This 5' fragment contained three HSE sequences. The plasmid was transfected into DHFR$^-$ CHO cells, and DHFR$^+$ colonies were isolated. These DHFR$^+$ colonies were grown in the presence of methotrexate, a drug that inhibits DHFR activity. The only cells that can survive exposure to methotrexate are those that overproduce DHFR, and mammalian cells accomplish this overproduction by gene amplification. With this particular plasmid, amplification of the transfected DHFR gene would be accompanied by coamplification of the *HSP70* 5' control region. Analysis of the HS response of these DHFR$^+$/*HSP70* 5' overproducing cell lines indicated that synthesis of the 72-kDa protein (the HSP70 equivalent in these mammalian cells) was reduced to 10% of the normal levels obtained upon HS. When examined for the ability to survive a severe HS, the amplified cell lines demonstrated a pronounced thermosensitivity relative to the DHFR$^-$ parent lines.

Riabowol *et al.* (1988) took another approach to directly assess the role of HSP70 in thermotolerance. They injected rat embryo fibroblasts with monoclonal antibodies specific to HSP70. Cells injected with these antibodies could not survive a 30-minute treatment at 45°C, but cells injected with either "control" antibodies (i.e., to actin, tubulin, FOS, or cAMP) or denatured HSP70 antibodies could survive such a HS. This lethal effect of injected HSP70 antibodies was a function of the concentration of antibodies injected.

A final piece of evidence in support of the role of HSPs in thermotolerance is the recent report by Bosch *et al.* (1988) on two species of *Hydra*. As mentioned previously, one species of *Hydra, Hydra oligactis,* does not synthesize a new set of proteins in response to a heat stress. This is in contrast to *Hydra attenuata,* which does synthesize HSPs in response to heat stress and acquires thermotolerance when preincubated at moderately elevated temperatures (30°C). *Hydra oligactis,* which shows particular sensitivity to thermal stress, did not develop any thermotolerance following preincubation at either 25 or 30°C. Thus, the lack of synthesis of HSPs is correlated with an inability to acquire thermotolerance. The discovery of a species that does not demonstrate a HS response is unexpected. Bosch *et al.* suggested that selection for retention of a strong HS response may only occur in organisms that live in relatively unstable habitats. In nature, *H. oligactis* is found in a restricted range of habitats (characterized by stable temperatures and no heavy metals), compared to *H. attenuata.*

On the other hand, there are experiments that question the involvement of HSPs in the expression of thermotolerance in eukaryotes. Hall (1983) observed thermotolerance in yeast in the absence of HSP synthesis, as did Landry and Chretien (1983) in rat hepatoma cells and Widelitz *et al.* (1986) in rat fibroblasts. Similarly, in contrast to the results of Laszlo and Li (1985) cited earlier, Anderson *et al.* (1986) did not find any changes in the constitutive level of HSPs in heat-resistant variants of rat B16 melanoma cells. In an attempt to resolve these seemingly contradictory results, Laszlo (1988) has recently proposed that there are (at least) two pathways for development of heat-induced thermotolerance, only one of which requires protein synthesis.

Recent work by Mizuno *et al.* (1989) may provide some explanation for the sets of contradictory results on the role of HSPs in thermotolerance. These researchers studied the synthesis of HSPs and the development of thermotolerance in the mutant mouse cell line, *ts85,* mentioned above. These cells are defective in the ubiquitin-dependent protein degradation process due to thermolability of the ubiquitin-activating enzyme. Mizuno and colleagues have found that ts85 cells, in contrast to the wild-type parental line, do not acquire thermotolerance during exposure to moderately elevated temperatures (e.g., 39.5°C).

Nonetheless, ts85 cells synthesize elevated levels of HSPs during exposure to 39.5°C, leading these authors to conclude that enhanced HSP synthesis is neither sufficient nor necessary for the acquisition of thermotolerance. In light of the known defect in the ts85 cell line, this conclusion may be somewhat expansive. As noted above, ubiquitin is a HSP. With its function impaired (as is the case in the ts85 cells), ubiquitin could not fulfill its role in the HS response, presumably to remove abnormal proteins. Under such conditions, the enhanced synthesis of other HSPs may not satisfy the requirements for development of thermotolerance. One may speculate that the HS response is a cascade of events, and disruption of early steps in this cascade could result in dramatic phenotypes that would appear to uncouple the presence of HSPs from the expression of thermotolerance.

Finally, it is important to recognize that studies on thermotolerance and HSPs do not always measure the same parameters. Laszlo (1988) has made a distinction between the correlation of the biological effects of heat with increased HSP synthesis versus increased HSP levels. Most studies of the correlation between HSPs and thermotolerance have referred to increased levels of HSPs. Anderson *et al.* (1988), in quantifying levels of HSP70 in Chinese hamster cells, saw no change in the level of HSP70 during a gradual temperature increase, but they did measure a doubling of the HSP70 protein levels in the 6 hours following a HS (45°C for 10 minutes). In all cases, increased synthesis of HSP70, as indicated by incorporation of radiolabeled amino acids, was detected. They conclude that thermotolerance can develop in some cases in the absence of elevated levels of HSP70. It is possible that the increased *synthesis* of HSP70 is the critical parameter that correlates with thermotolerance.

In plants, few studies have quantified the levels of HSPs, but inspection of Coomassie-stained two-dimensional protein gels from several heat-shocked plant species has revealed that HSPs accumulate to significant levels during the first few hours of a heat treatment. Lin *et al.* (1984) reported that these proteins are relatively stable, remaining in the cells for at least 21 hours following the stress, based on the persistence of the radioactivity (incorporated during HS) in the proteins. Vierling (1990), using antibodies to two LMW HSPs in pea, demonstrated that they accumulate dramatically and are still present at significant levels for 3 days following the stress.

In the field of plant breeding, thermal tolerance is an important and heritable agronomic trait. Breeders have often asked whether the thermal tolerance that they work with in cultivar development is related to the transient thermotolerance studied by researchers using HS in the laboratory. It has been firmly established that in the case of crop species growing in growth chambers or under field conditions, HS mRNAs and proteins are synthesized during conditions of heat stress (Kimpel and Key, 1985a; Burke *et al.*, 1985; Vierling *et al.*, 1988). However, Edelman and Key did not detect qualitative changes in the profile of HS mRNAs synthesized by soybean seedlings of cultivars with heritable differences in their tolerance to heat stress in the field (unpublished observations). In contrast, Krishnan *et al.* (1989), studying wheat, reported a positive correlation between genetic differences in thermal tolerance and the levels of LMW HSPs.

The availability of antibodies to HSPs should allow a more detailed study of these plant responses. It is quite possible that there are different mechanisms underlying responses to short (i.e., transient) fluctuations in temperature versus longer exposures to chronically elevated temperatures (for a discussion, see Kimpel and Key, 1985b).

B. BIOCHEMICAL FUNCTIONS OF THE HEAT-SHOCK PROTEINS

In the last few years, considerable progress has been made in assigning biochemical activities to individual HSPs. As described in detail below, the functions for HSP70 that have now been identified support an originally speculative hypothesis by Pelham (1986) that one purpose of the HS response is to repair the damage caused by the stress. Additionally, other HMW HSPs have biochemical activities that are consistent with a protective role. In the case of LMW HSPs, it is now recognized that during heat stress these abundant proteins form aggregates that contain mRNAs for many cellular proteins.

1. HSP70

Several laboratories have demonstrated that the 70-kDa family of HSPs localizes into nuclei during HS, concentrating in the nucleolus in a matrix form that is resistant to disruption by salt, suggesting a hydrophobic binding (Lindquist, 1986; Pelham, 1986; Neumann *et al.*, 1987). The binding is reversed by ATP but not by nonhydrolyzable ATP analogs; the HSP70-related proteins bind ATP tightly and in some cases hydrolyze it. As noted above, several of the HMW HSPs, including the HSP70-, HSP83-, and HSP90-related proteins, share homology with the glucose-regulated proteins (GRPs) of mammalian cells. This observation contributed to the proposal by Pelham (1986) that these proteins participate in protein folding, assembly, and/or disaggregation (e.g., of aggregated, denatured proteins) by virtue of their ability to bind exposed, hydrophobic regions. Similar suggestions have been presented by Finley *et al.* (1987) based on their studies on the role of ubiquitin in the HS response. More recently, Deshaies *et al.* (1988a) extended these earlier speculations to accommodate a function for HSP70 in relaxing the tertiary structure within a single polypeptide. These investigators demonstrated that conditional yeast mutants depleted of the SSA family of *HSP70* genes accumulated precursor forms of proteins that normally would have been imported into the ER and mitochondria. Chirico *et al.* (1988) further demonstrated that HSP70-related proteins are involved in the translocation of proteins across microsomal membranes in yeast. Pelham (1988) again summarized these and related results. The family of 70-kDa HSPs is involved in protein interactions that protect precursor proteins and/or maintain the correct unfolded structures until they reach the site of membrane insertion. Other examples of HSPs of plants localized in the ER include one HSP70 in maize (Cooper and Ho, 1987) and one in tomato (Neumann *et al.*, 1987). Other members of HSPs are transported into the mitochondria (e.g., in yeast; E. A. Craig, personal communication) and into the chloroplasts (Vierling, 1990).

Taken together, these observations indicate that, during HS, a major function of this family of HSPs may resemble the function of their counterparts in facilitating transport and subunit assembly in unstressed cells. Ellis (1987) has referred to this general function as that of a "molecular chaperone." He proposed this term to describe a class of proteins "whose function is to ensure that the folding of certain other polypeptide chains and their assembly into oligomeric structures occur correctly." Thus, in analogy to such a function in unstressed cells (perhaps fulfilled by HSC proteins), these HSPs may interact with other proteins to facilitate disaggregation of damaged protein aggregates, stabilization of cellular proteins or aggregates during the stress, and possibly reassembly of proper structures during recovery; they may also function in the assembly of membranes (e.g., Deshaies *et al.*, 1988b) and possibly in the stabilization of membranes.

2. *Other High-Molecular-Weight Heat-Shock Proteins*

In *E. coli* groEL is an approximately 60-kDa HSP that is required for cell viability (Fayet *et al.*, 1989). GroEL is another member of proteins referred to as "molecular chaperones." Analogous proteins to groEL have been identified as constitutive proteins in chloroplasts and mitochondria. In *Tetrahymena*, McMullin and Hallberg (1988) have shown that synthesis of the groEL-related protein is enhanced during HS. A nuclear gene has been isolated from yeast that encodes a 60-kDa HSP (Reading *et al.*, 1989). A HS treatment results in a 2- to 3-fold induction above the constitutive level. The nucleotide sequence of this gene shows strong homology to the groEL protein of *E. coli* and the *Rubisco*-binding protein of chloroplasts. In plants, there is a report of a mitochondrial 60-kDa HSP (Sinibaldi and Turpen, 1985), but another study has suggested that this may actually be a contaminating bacterial protein (Nieto-Sotelo and Ho, 1987). Data from the yeast and *Tetrahymena* systems suggest, however, that a HSP with similar activity might well be conserved for mitochondrial function in plants relating to assembly of multisubunit complexes. The *Rubisco*-binding protein, a constitutive, nuclear-encoded chloroplast protein, participates in the assembly of the ribulose bisphosphate carboxylase/oxygenase (Rubisco) holoenzyme, a "molecular chaperone" activity analogous to the groEL protein (Hemmingsen *et al.*, 1988). It is not known whether synthesis of this protein is enhanced during HS; it is possible that there are separate but closely related genes for such proteins that are heat inducible.

3. Low-Molecular-Weight Heat-Shock Proteins

As described above, the HS response in plants and other organisms differs in that the plant LMW HSPs are highly abundant and demonstrate much greater complexity in size and number. Due to their abundance, it is reasonable to presume that they have a structural function in the stressed cell (see Schlesinger, 1986). Earlier reports on localization indicated that the LMW HSPs were associated with several cellular compartments during HS (e.g., nucleus, ribosomes, ER, and mitochondria). During recovery from HS, these HSPs became soluble; that is, they were no longer found associated with specific cell fractions. The general conclusion from these studies was that these proteins reversibly assemble into large aggregates (and possibly associate with various cellular surfaces/structures). This conclusion was supported by the conservation in these LMW HSPs of a five-amino-acid sequence from the lens crystallin protein, a polypeptide that aggregates into large structures (see Section II).

More detailed studies in *Drosophila* (Arrigo, 1987; Leicht *et al.*, 1986), mammalian cells (Arrigo and Welch, 1987), and tomato cells (Nover *et al.*, 1983) have firmly established that these proteins form cytoplasmic aggregates during HS that concentrate in the perinuclear region. Nover *et al.* (1989) more fully characterized the composition of these aggregates, referred to as heat-shock granules, showing that they are tightly associated with mRNAs. The mRNAs in these granules represent the bulk of the normal cellular mRNAs (i.e., *not* the HS mRNAs), probably reflecting a protective function of these granules in conservation of most cellular mRNAs even though they are not translated during the HS.

In plants, these LMW HSPs are also found associated with other organelles, and it is probable that they are in an aggregate form (Key *et al.*, 1982; Lin *et al.*, 1984; Cooper and Ho, 1987; Mansfield and Key, 1988; P. R. LaFayette and R. L. Travis, personal communication). The experiments indicate that this association is with the membrane of the organelles, a structure that rapidly disintegrates during a lethal HS (Mansfield *et al.*, 1988). Again, one role of these organelle-associated HSPs is likely protective.

Lin *et al.* (1984) demonstrated selective localization of certain 21- to 24-kDa HSPs to a mitochondria-enriched fraction during HS. When the mitochondrial fraction was incubated with protease, the 21- to 24-kDa proteins were resistant to digestion, indicating that these proteins were protected within the membrane or transported into the mitochondria. These proteins did not delocalize from the mitochondria during recovery from HS, suggesting that they are within the membrane rather than attached to the outside. On the other hand, the 15- to 18-kDa HSPs do delocalize during the chase period, and they are protease sensitive.

Certain LMW HSPs are nuclear encoded, synthesized in the cytoplasm, and transported (with processing) into the chloroplasts (Kloppstech *et al.*, 1985; Vierling *et al.*, 1986; Süss and Yordanov, 1986). As described above, these chloroplast-directed HSPs share extensive homology with their cytoplasmic counterparts, suggesting that they provide a protective function analogous to those HSPs in the cytoplasm (Vierling *et al.*, 1988). Additionally, certain HSPs may have functions specific to the chloroplast. Schuster *et al.* (1988) have presented evidence of a role for the chloroplast-localized HSPs in preventing damage to the photosystem II reaction center during HS in the light.

V. Concluding Remarks

The highly conserved nature of the HS response among all classes of organisms has quickened the pace of progress in understanding the molecular mechanisms that control the response and the biological basis of the phenomenon. It is clear that the HSE sequence is a necessary feature of a heat-inducible promoter. Still, construction of a synthetic promoter using a single HSE would not, in most organisms, lead to strong, heat-inducible expression. Multiple and overlapping HSEs in combination with other elements that modulate expression are necessary to obtain the high-level expression noted for many of these genes during HS. Additionally, several research results indicate that there are multiple sequences in these promoters other than the HSE that also determine the level of expression. For example, the presence of sequences that regulate constitutive expression of some HS (or HS-related) genes may influence the effectiveness of the HSEs to elicit heat-inducible expression. Continued research is needed to understand the role of specific cis-acting regulatory elements of HS genes in (1) inducibility at specific stages of development in a tissue-specific manner and (2) inducibility by other stress agents. The mechanism(s) that transduce the HS signal to activate functional HSF binding to HSEs is poorly understood, but phosphorylation of the HSE upon the sensing of HS and dephosphorylation upon relief of the stress is involved in the response.

It is now established that the function of at least some HSPs, notably the family of HSP70s, is to repair the cellular damage caused by stress. Presumably HSP70 may also replace the function of HSC70s during HS in other protein–protein interactions essential to cell survival. There is good correlative evidence that some HSPs may provide protection by stabilizing membrane and protein structures and preserving synthetic machinery of the cell during the stress, permitting recovery when the stress is relieved. Perhaps such a protective role may be the sole purpose of the entire HS response.

There is striking conservation of some HS genes and proteins, yet the complexity in terms of copy number and sequence diversity among the families is not understood. In plants, there is a much higher complexity and abundance of LMW HSPs. Because the bulk of these LMW HSPs are associated with granules, other cellular matrices, and organelles during HS, their abundance may reflect the extensive organellar structure of plant cells (e.g., chloroplasts, peroxisomes, and vacuoles).

Though not reviewed in detail here, it is recognized that in many organisms, the HS response is also induced by other stresses such as ethanol, anoxia, amino acid analogs, heavy metal ion, or arsenite. The nature of the stresses that can induce the response and the extent of induction vary with the organism studied. Plants generally stand out in this observation, because they do not usually synthesize HSPs in response to as wide a variety of stresses as do other organisms. In plants, only arsenite and to a lesser extent cadmium treatments induce a response at normal temperature similar to a HS response. Instead, plants seem to express classes of proteins that are stress specific; for example, in response to anoxic conditions, a set of anaerobic proteins, including enzymes for alcohol metabolism such as alcohol dehydrogenase, is synthesized. None of these "anaerobiosis protein" genes shows induction by HS, and none of the cloned HS genes is induced to significant levels by anaerobiosis in plants. This apparent difference in the response of plants to stress may reflect the need for plants, as nonmotile organisms, to maintain a highly refined response to each of the variety of stresses to which they are routinely exposed.

At first reading, it may seem rather surprising that the role of HSPs in the development of thermotolerance remains ill-defined. However, it is important to note that experiments on this aspect of the HS response have been carried out in a variety of ways. Experimental protocols vary in both the physiological state of the organisms at the outset of the experiment, the length and timing of HS, and the radioactive labeling period for the HSPs; all too often the amount of radioactive amino acid incorporation is equated to rate and level of HSP synthesis. Absolute rates of synthesis and quantitative measurements of HSP levels are needed, but they are rarely assessed. Also, some researchers routinely label during the stress period, whereas others provide label during the recovery period. In addition, the relative level of stress that is applied varies tremendously. Some researchers use a stress that is not life threatening, while others study a level of stress that would be lethal if applied continuously. These differences in experimental design may particularly influence interpretation of data and hypotheses on the molecular bases of self-regulation and/or the acquisition of thermotolerance.

There is potential in plants to further understand and manipulate the regulatory mechanisms of the HS response through the use of genetic transformation. Promoter deletion analyses and the use of reporter genes under the control of HS promoters are allowing extensive dissection of the critical elements, including but not limited to the HSE, as well as the isolation and characterization of trans-acting transcription factors. In transformed plants, the HS promoter can be used as a convenient switch for introduced genes to study their activities. A HS-promoter cassette has been constructed using the soybean *Gmhsp17.5E* promoter region, which should allow the heat-inducible expression of any inserted coding sequence (Ainley and Key, 1990). This particular HS promoter is tightly regulated, having very high heat inducibility and very low constitutive expression. This HS-promoter cassette may be very useful both for the expression of selected genes [e.g., pathogen-related protein (PRP) genes] in transgenic crop species and as a research tool to evaluate the effects of the inducible overexpression or underexpression (using an antisense orientation) of a gene of interest. Many interesting questions could be pursued, such as the physiological effects of *in vivo* hormone production (both over- and underexpression) relative to translocation, apical dominance, cell differentiation, growth, and development. Preliminary experiments of this type are underway with cloning of a cytokinin biosynthetic gene (isopentenyl transferase) into the HS cassette, transformation into tobacco, and the demonstration of heat-inducible cytokinin gene expression (Ainley *et al.*, 1990).

Finally, it is important to consider the potential for manipulation of this response in plants, wherein cyclical heat stress is a daily experience for most prominent crop species. As described above, one characteristic of the HS response is that genes other than HS genes are not expressed, or are weakly expressed, during stress. For plants in the field, this could result in a restricted ability to respond to other stresses (anaerobiosis in water-logged soils, pathogens, etc.) for a substantial part of the day. For example, because pathogen-related proteins are not made during HS (Walter, 1989), it seems reasonable to suggest that additional copies of resistance genes under the control of HS promoters may improve the pathogen resistance of crop species during a growing season.

Thus, despite the strong conservation of several features of the HS response, there are clearly differences within and among classes of organisms. With a general consensus on the more conserved features of HS, perhaps the differences can serve as the focus for promising research efforts in the coming years.

Acknowledgments

The authors thank Dr. Laura Hoffman for helpful comments and Joyce Kochert for assistance in the preparation of this manuscript. Work from the laboratory of J. L. Key was supported by a contract from DOE (DE-FG09-86ER13602).

References

Ainley, W. M., and Key, J. L. (1990). Development of a heat shock inducible expression cassette for plants: Characterization of parameters for its use in transient expression assays and transgenic plants. *Plant Mol. Biol.* (in press).

Ainley, W. M., Simpson, R. B., Hill, J. W., and Key, J. L. (1990). Expression of a heat shock promoter-*tmr* chimeric gene in transgenic tobacco plants: Heat inducible shoots. *Plant Physiol.* (submitted).

Amin, J., Mestril, R., Lawson, R., Klapper, H., and Voellmy, R. (1985). The heat shock consensus sequence is not sufficient for hsp70 gene expression in *Drosophila melanogaster. Mol. Cell. Biol.* **5,** 197–203.

Amin, J., Ananthan, J., and Voellmy, R. (1988). Key features of heat shock regulatory elements. *Mol. Cell. Biol.* **8,** 3761–3769.

Ananthan, J., Goldberg, A. L., and Voellmy, R. (1986). Abnormal proteins serve as eukaryotic stress signals and trigger the activation of heat shock genes. *Science* **232,** 522–524.

Anderson, R. L., Tao, T. W., Betten, D. A., and Hahn, G. M. (1986). Heat shock protein levels are not elevated in heat-resistant B16 melanoma cells. *Radiat. Res.* **105,** 240–246.

Anderson, R. L., Herman, T. S., Van Kersen, I., and Hahn, G. M. (1988). Thermotolerance and heat shock protein induction by slow rates of heating. *Int. J. Radiat. Oncol., Biol. Phys.* **15,** 717–726.

Arrigo, A. P. (1987). Cellular localization of HSP23 during *Drosophila* development and following subsequent heat shock. *Dev. Biol.* **122,** 39–48.

Arrigo, A. P., and Welch, W. J. (1987). Characterization and purification of the small 28,000-dalton mammalian heat shock protein. *J. Biol. Chem.* **262,** 15359–15369.

Ashburner, M., and Bonner, J. J. (1979). The induction of gene activity in *Drosophila* by heat shock. *Cell* **17,** 241–254.

Bardwell, J. C. A., and Craig, E. A. (1984). Major heat shock gene of *Drosophila* and the *Escherichia coli* heat-inducible *dnaK* gene are homologous. *Proc. Natl. Acad. Sci. U.S.A.* **81,** 848–852.

Baumann, G., Raschke, E., Bevan, M., and Schöffl, F. (1987). Functional analysis of sequences required for transcriptional activation of a soybean heat shock gene in transgenic tobacco plants. *EMBO J.* **6,** 1161–1166.

Bienz, M. (1985). Transient and developmental activation of heat-shock genes. *Trends Biochem. Sci. (Pers. Ed.)* **10,** 157–161.

Bienz, M. (1986). A CCAAT box confers cell-type-specific regulation on the *Xenopus hsp70* gene in oocytes. *Cell* **46,** 1037–1042.

Bienz, M., and Pelham, H. R. B. (1982). Expression of a *Drosophila* heat-shock protein in *Xenopus* oocytes: Conserved and divergent regulatory signals. *EMBO J.* **1,** 1583–1588.

Bienz, M., and Pelham, H. R. B. (1986). Heat shock regulatory elements function as an inducible enhancer in the *Xenopus hsp70* gene and when linked to a heterologous promoter. *Cell* **45,** 753–760.

Bienz, M., and Pelham, H. R. B. (1987). Mechanisms of heat-shock gene activation in higher eukaryotes. *Adv. Genet.* **24,** 31–72.

Bond, U., and Schlesinger, M. J. (1985). Ubiquitin is a heat shock protein in chicken embryo fibroblasts. *Mol. Cell. Biol.* **5,** 949–956.

Bosch, T. C. G., Krylow, S. M., Bode, H. R., and Steele, R. E. (1988). Thermotolerance and synthesis of heat shock proteins: These responses are present in *Hydra attenuata* but absent in *Hydra oligactis. Proc. Natl. Acad. Sci. U.S.A.* **85,** 7927–7931.

Burke, J. J., Hatfield, J. L., Klein, R. R., and Mullet, J. E. (1985). Accumulation of heat shock proteins in field-grown cotton. *Plant Physiol.* **78,** 394–398.

Chirico, W. J., Waters, M. G., and Blobel, G. (1988). 70K heat shock related proteins stimulate protein translocation into microsomes. *Nature (London)* **332,** 805–810.

Christensen, A. H., and Quail, P. H. (1989). Sequence analysis and transcriptional regulation by heat shock of polyubiquitin transcripts from maize. *Plant Mol. Biol.* **12,** 619–632.

Cohen, R. S., and Meselson, M. (1984). Inducible transcription and puffing in *Drosophila melanogaster* transformed with *hsp70*-phage lambda hybrid heat shock genes. *Proc. Natl. Acad. Sci. U.S.A.* **81,** 5509–5513.

Cooper, P., and Ho, T. H. D. (1987). Intracellular localization of heat shock proteins in maize. *Plant Physiol.* **84,** 1197–1203.

Corces, V., Pellicer, A., Axel, R., and Meselson, M. (1981). Integration, transcription, and control of a *Drosophila* heat shock gene in mouse cells. *Proc. Natl. Acad. Sci. U.S.A.* **78,** 7038–7042.

Craig, E. A. (1985). The heat shock response. *CRC Crit. Rev. Biochem.* **18**, 239–280.

Craig, E. A., and Jacobsen, K. (1984). Mutations of the heat inducible 70 kilodalton genes of yeast confer temperature sensitive growth. *Cell* **38**, 841–849.

Craig, E. A., Slater, M. R., Boorstein, W. R., and Palter, K. (1985). Expression of *S. cerevisiae* hsp70 multigene family. *In* "Sequence Specificity in Transcription and Translation" (R. Calendar and L. Gold, eds.), pp. 659–667. Liss, New York.

Czarnecka, E., Edelman, L., Schöffl, F., and Key, J. L. (1984). Comparative analysis of physical stress responses in soybean seedlings using cloned heat shock cDNAs. *Plant Mol. Biol.* **3**, 45–58.

Czarnecka, E., Gurley, W. B., Nagao, R. T., Mosquera, L. A., and Key, J. L. (1985). DNA sequence and transcript mapping of a soybean gene encoding a small heat shock protein. *Proc. Natl. Acad. Sci. U.S.A.* **82**, 3726–3730.

Czarnecka, E., Nagao, R. T., Key, J. L., and Gurley, W. B. (1988). Characterization of *Gmhsp26-A*, a stress gene encoding a divergent heat shock protein of soybean: Heavy-metal-induced inhibition of intron processing. *Mol. Cell. Biol.* **8**, 1113–1122.

Czarnecka, E., Key, J. L., and Gurley, W. B. (1989). Regulatory domains of the *Gmhsp17.5-E* heat shock promoter of soybean: A mutational analysis. *Mol. Cell. Biol.* (in press).

Czarnecka, E., Fox, P. C., and Gurley, W. B. (1990). *In vitro* interaction of nuclear proteins with the promoter of soybean heat shock gene *Gmhsp 17.5E. Plant Physiol.* (in press).

Deshaies, R. J., Koch, B. D., Werner-Washburne, M., Craig, E. A., and Schekman, R. (1988a). A subfamily of stress proteins facilitates translocation of secretory and mitochondrial precursor polypeptides. *Nature (London)* **332**, 800–805.

Deshaies, R. J., Koch, B. D., and Schekman, R. (1988b). The role of stress proteins in membrane biogenesis. *Trends Biochem. Sci. (Pers. Ed.)* **13**, 384–388.

DiDomenico, B. J., Bugaisky, G. E., and Lindquist, S. (1982). Heat shock and recovery are mediated by different translational mechanisms. *Proc. Natl. Acad. Sci. U.S.A.* **79**, 6181–6185.

Dudler, R., and Travers, A. A. (1984). Upstream elements necessary for optimal function of the hsp70 promoter in transformed flies. *Cell* **38**, 391–398.

Ellis, J. (1987). Proteins as molecular chaperones. *Nature (London)* **328**, 378–379.

Erickson, J. W., Vaughn, V., Walter, W. A., Neidhardt, F. C., and Gross, C. A. (1987). Regulation of the promoters and transcripts of *rpoH*, the *Escherichia coli* heat shock regulatory gene. *Genes Dev.* **1**, 419–432.

Fayet, O., Ziegelhoffer, T., and Georgopoulos, C. (1989). The *groES* and *groEL* heat shock gene products of *Escherichia coli* are essential for bacterial growth at all temperatures. *J. Bacteriol.* **171**, 1379–1385.

Finley, D., and Varshavsky, A. (1985). The ubiquitin system: Functions and mechanisms. *Trends Biochem. Sci. (Pers. Ed.)* **10**, 343–347.

Finley, D., Ciechanover, A., and Varshavsky, A. (1984). Thermolability of ubiquitin-activating enzyme from the mammalian cell cycle mutant ts85. *Cell* **37**, 43–55.

Finley, D. Özkaynak, E., and Varshavsky, A. (1987). The yeast polyubiquitin gene is essential for resistance to high temperatures, starvation, and other stresses. *Cell* **48**, 1035–1046.

Gilmour, D. S., and Lis, J. T. (1985). *In vivo* interactions of RNA polymerase II with genes of *Drosophila melanogaster. Mol. Cell. Biol.* **5**, 2009–2018.

Gilmour, D. S., and Lis, J. T. (1986). RNA polymerase II interacts with the promoter region of the noninduced *hsp70* gene in *Drosophila melanogaster* cells. *Mol. Cell. Biol.* **6**, 3984–3989.

Goff, S. A., and Goldberg, A. L. (1985). Production of abnormal proteins in *E. coli* stimulates transcription of *lon* and other heat shock genes. *Cell* **41**, 587–595.

Grant, C. M., Firoozan, M., and Tuite, M. F. (1989). Mistranslation induces the heat-shock response in the yeast *Saccharomyces cerevisiae*. *Mol. Microbiol.* **3,** 215–220.

Grossman, A. D., Straus, D. B., Walter, W. A., and Gross, C. A. (1987). σ^{32} Synthesis can regulate the synthesis of heat shock proteins in *Escherichia coli*. *Genes Dev.* **1,** 179–184.

Gurley, W. B., Czarnecka, E., Nagao, R. T., and Key, J. L. (1986). Upstream sequences required for efficient expression of a soybean heat shock gene. *Mol. Cell. Biol.* **6,** 559–565.

Gurley, W. B., Bruce, W. B., Czarnecka, E., Bandyopadhyay, R., Nagao, R. T., and Key, J. L. (1987). *Cis*-regulatory elements in heat shock and T-DNA promoters. *In* "Plant Gene Systems and Their Biology" (J. L. Key and L. McIntosh, eds.), pp. 279–288. Liss, New York.

Hall, B. G. (1983). Yeast thermotolerance does not require protein synthesis. *J. Bacteriol.* **156,** 1363–1365.

Hemmingsen, S. M., Woolford, C., van der Vies, S. M., Tilly, K., Dennis, D. T., Georgopoulos, C. P., Hendrix, R. W., and Ellis R. J. (1988). Homologous plant and bacterial proteins chaperone oligomeric protein assembly. *Nature (London)* **333,** 330–334.

Hiromi, Y., and Hotta, Y. (1985). Actin gene mutations in *Drosophila*; heat shock activation in the indirect flight muscles. *EMBO J.* **4,** 1681–1687.

Hoffman, E., and Corces, V. (1986). Sequences involved in temperature and ecdysterone-induced transcription are located in separate regions of a *Drosophila melanogaster* heat shock gene. *Mol. Cell. Biol.* **6,** 663–673.

Ingolia, T. D., and Craig, E. A. (1982). *Drosophila* gene related to the major heat shock-induced gene is transcribed at normal temperatures and not induced by heat shock. *Proc. Natl. Acad. Sci. U.S.A.* **79,** 525–529.

Johnston, R. N., and Kucey, B. L. (1988). Competitive inhibition of *hsp70* gene expression causes thermosensitivity. *Science* **242,** 1151–1554.

Karch, F., Török, I., and Tissières, A. (1981). Extensive regions of homology in front of the two *hsp70* heat shock variant genes in *Drosophila melanogaster*. *J. Mol. Biol.* **148,** 219–230.

Kelley, P., and Schlesinger, M. J. (1978). The effect of amino-acid analogues and heat shock on gene expression in chicken embryo fibroblasts. *Cell* **15,** 1277–1286.

Key, J. L., Lin, C. Y., Ceglarz, E., and Schöffl, F. (1982). The heat-shock response in plants: Physiological considerations. *In* "Heat Shock from Bacteria to Man" (M. L. Schlesinger, M. Ashburner, and A. Tissières, eds.), pp. 329–336. Cold Spring Harbor Lab. Cold Spring Harbor, New York.

Key, J. L., Kimpel, J., and Nagao, R. T. (1987a). Heat shock gene families of soybean and the regulation of their expression. *In* "Plant Gene Systems and Their Biology" (J. L. Key and L. McIntosh, eds.), pp. 87–97. Liss, New York.

Key, J. L., Nagao, R. T., Czarnecka, E., and Gurley, W. B. (1987b). Heat stress: Expression and structure of heat shock protein genes. *In* "Plant Molecular Biology" (D. von Wettstein and N. H. Chua, eds.), pp. 385–397. Plenun, New York.

Kimpel, J. A., and Key, J. L. (1985a). Presence of heat shock mRNAs in field grown soybeans. *Plant Physiol.* **79,** 672–678.

Kimpel, J. A., and Key, J. L. (1985b). Heat shock in plants. *Trends Biochem. Sci. (Pers. Ed.)* **10,** 353–357.

Kimpel, J. A., Nagao, R. T., Goekjian, V., and Key, J. L. (1990). Regulation of the heat shock response in soybean seedlings. *Plant Physiol.* (in press).

Kingston, R. E., Schuetz, T. J., and Larin, Z. (1987). Heat-inducible human factor that binds to a human hsp70 promoter. *Mol. Cell. Biol.* **7,** 1530–1534.

Kloppstech, K., Meyer, G., Schuster, G., and Ohad, I. (1985). Synthesis, transport and localization of a nuclear coded 22-kd heat-shock protein in the choloroplast membranes of peas and *Chlamydomonas reinhardi. EMBO J.* **4**, 1901–1909.

Krishnan, M., Nguyen, H. T., and Burke, J. J. (1989). Heat shock protein synthesis and thermal tolerance in wheat. *Plant Physiol.* **90**, 140–145.

Landry, J., and Chretien, P. (1983). Relationship between hyperthermia-induced heat-shock proteins and thermotolerance in Morris hepatoma cells. *Can. J. Biochem.* **61**, 428–437.

Laszlo, A. (1988). The relationship of heat-shock proteins, thermotolerance, and protein systhesis. *Exp. Cell Res.* **178**, 401–414.

Laszlo, A., and Li, G. C. (1985). Heat-resistant variants of Chinese hamster fibroblasts altered in expression of heat shock protein. *Proc. Natl. Acad. Sci. U.S.A.* **82**, 8029–8033.

Lee, K. J., and Hahn, G. M. (1988). Abnormal proteins as the trigger for the induction of stress responses: Heat, diamide, and sodium arsenite. *J. Cell. Physiol.* **136**, 411–420.

Leicht, B. G., Biessmann, H., Palter, K. B., and Bonner, J. J. (1986). Small heat shock proteins of *Drosophila* associate with the cytoskeleton. *Proc. Natl. Acad. Sci. U.S.A.* **83**, 90–94.

Li, G. C., and Laszlo, A. (1985). Thermotolerance in mammalian cells: A possible role for heat shock proteins. *In* "Changes in Eukaryotic Gene Expression in Response to Environmental Stress" (B. G. Atkinson and D. B. Walden, eds.), pp. 227–254. Academic Press, Orlando, Florida.

Li, G. C., and Werb, Z. (1982). Correlation between synthesis of heat shock proteins and development of thermotolerance in Chinese hamster fibroblasts. *Proc. Natl. Acad. Sci. U.S.A.* **79**, 3219–3222.

Lin, C. Y., Roberts, J. K., and Key, J. L. (1984). Acquisition of thermotolerance in soybean seedlings: Synthesis and accumulation of heat shock proteins and their cellular localization. *Plant Physiol.* **74**, 152–160.

Lindquist, S. (1986). The heat-shock response. *Annu. Rev. Biochem.* **55**, 1151–1191.

Lindquist, S., and Craig, E. A. (1988). The heat shock proteins. *Annu. Rev. Genet.* **22**, 631–677.

Lindquist, S., and DiDomenico, B. (1985). Coordinate and noncoordinate gene expression during heat shock: A model for regulation. *In* "Changes in Eukaryotic Gene Expression in Response to Environmental Stress" (B. G. Atkinson and D. B. Walden, eds.), pp. 71–90. Academic Press, Orlando, Florida.

Loomis, W. F., and Wheeler, S. A. (1982). Chromatin-associated heat-shock proteins of *Dictyostelium. Dev. Biol.* **90**, 412–418.

Mansfield, M. A., and Key, J. L. (1987). Synthesis of the low molecular weight heat shock proteins in plants. *Plant Physiol.* **84**, 1007–1017.

Mansfield, M. A., and Key, J. L. (1988). Cytoplasmic distribution of heat shock proteins in soybean. *Plant Physiol.* **86**, 1240–1246.

Mansfield, M. A., Lingle, W. L., and Key, J. L. (1988). The effects of lethal heat shock on nonadapted and thermotolerant root cells of *Glycine max. J. Ultrastruct. Mol. Struct. Res.* **99**, 96–105.

McMahon, A. P., Novak, T. J., Britten, R. J., and Davidson, E. H. (1984). Inducible expression of a cloned heat shock fusion gene in sea urchin embryos. *Proc. Natl. Acad. Sci. U.S.A.* **81**, 7490–7494.

McMullin, T. W., and Hallberg, R. L. (1988). A highly evolutionarily conserved mitochondrial protein is structurally related to the protein encoded by the *Escherichia coli groEL* gene. *Mol. Cell. Biol.* **8**, 371–380.

Mestril, R., Schiller, P., Amin, J., Klapper, H., Ananthan, J., and Voellmy, R. (1986).

Heat-shock and ecdysterone activation of the *Drosophila* hsp23 gene: A sequence element implied in developmental regulation. *EMBO J.* **5**, 1667–1673.

Mirault, M. E., Southgate, R., and Delwart, E. (1982). Regulation of heat-shock genes: A DNA sequence upstream of *Drosophila hsp70* genes is essential for their induction in monkey cells. *EMBO J.* **1**, 1279–1285.

Mizuno, S., Ohkawara, A., and Suzuki, K. (1989). Defect in the development of thermo-tolerance and enhanced heat shock protein synthesis in the mouse temperature-sensitive mutant ts85 cells upon moderate hyperthermia. *Int. J. Hypertherm.* **5**, 105–113.

Mizzen, L. A., and Welch, W. J. (1988). Characterization of the thermotolerant cell. I. Effects on protein synthesis activity and the regulation of heat-shock protein 70 expression. *J. Cell Biol.* **106**, 1105–1116.

Morganelli, C. M., Berger, E. M., and Pelham, H. R. B. (1985). Transcription of *Drosophila* small *hsp–tk* hybrid genes is induced by heat shock and by ecdysterone in transfected *Drosophila* cells. *Proc. Natl. Acad. Sci. U.S.A.* **82**, 5865–5869.

Nagao, R. T., and Key, J. L. (1989). Heat shock protein genes of plants. *In* "Cell Culture and Somatic Cell Genetics of Plants" (I. K. Vasil and J. Schell, eds.), pp. 297–328. Academic Press, San Diego, California.

Nagao, R. T., Czarnecka, E., Gurley, W.B., Schöffl, F., and Key, J. L. (1985). Genes for low-molecular-weight heat shock proteins of soybeans: Sequence analysis of a multi-gene family. *Mol. Cell. Biol.* **5**, 3417–3428.

Nagao, R. T., Kimpel, J. A., Vierling, E., and Key, J. L. (1986). The heat shock response: A comparative analysis *Oxford Surv. Plant Mol. Cell Biol.* **3**, 384–438.

Neidhardt, F. C. (1987). What the bacteriologists have learned about heat shock. *Genes Dev.* **1**, 109–110.

Neidhardt, F. C., VanBogelen, R. A., and Vaughn, V. (1984). The genetics and regulation of heat-shock proteins. *Annu. Rev. Genet.* **18**, 295–329.

Neumann, D., Nieden, U. Z., Manteuffel, R., Walter, G., Scharf, K.D., and Nover, L. (1987). Intracellular localization of heat-shock proteins in tomato cell cultures. *Eur. J. Cell Biol.* **43**, 71–81.

Nieto-Sotelo, J., and Ho, T. H. D. (1987). Absence of heat shock protein synthesis in isolated mitochondria and plastids from maize. *J. Biol. Chem.* **25**, 12288–12292.

Nover, L. (1987). Expression of heat shock genes in homologous and heterologous sys-tems. *Enzyme Microb. Technol.* **9**, 130–144.

Nover, L., Scharf, K. D., and Neumann, D. (1983). Formation of cytoplasmic heat shock granules in tomato cell cultures and leaves. *Mol. Cell. Biol.* **3**, 1648–1655.

Nover, L., Scharf, K. D., and Neumann, D. (1989). Cytoplasmic heat shock granules are formed from precursor particles and are associated with a specific set of mRNAs. *Mol. Cell. Biol.* **9**, 1298–1308.

Okamoto, H., Hiromi, Y., Ishikawa, E., Yamada, T., Isoda, K., Maekawa, H., and Hotta, Y. (1986). Molecular characterization of mutant actin gene which induce heat-shock proteins in *Drosophila* flight muscles. *EMBO J.* **5**, 589–596.

Özkaynak, E., Finley, D., Solomon, M. J., and Varshavsky, A. (1987). The yeast ubiquitin genes: A family of natural gene fusions. *EMBO J.* **6**, 1429–1439.

Palter, K. B., Watanabe, M., Stinson, L., Mahowald, A. P., and Craig, E. A. (1986). Expression and localization of *Drosophila melanogaster* hsp70 cognate proteins. *Mol. Cell. Biol.* **6**, 1187–1203.

Parker, C. S., and Topol, J. (1984). A *Drosophila* RNA polymerase II transcription factor binds to the regulatory site of an hsp 70 gene. *Cell* **37**, 273–283.

Pauli, D., Spierer, A., and Tissières, A. (1986). Several hundred base pairs upstream of

Drosophila hsp23 and *26* genes are required for their heat induction in transformed flies. *EMBO J.* **5,** 755–761.

Pelham, H. R. B. (1982). A regulatory upstream promoter element in the *Drosophila HSP70* heat-shock gene. *Cell* **30,** 517–528.

Pelham, H. (1985). Activation of heat-shock genes in eukaryotes. *Trends Genet.* **1,** 31–35.

Pelham, H. R. B. (1986). Speculations on the functions of the major heat shock and glucose-regulated proteins. *Cell* **46,** 959–961.

Pelham, H. (1987). Properties and uses of heat shock promoters. *Genet. Eng.* **9,** 27–44.

Pelham, H. R. B. (1988). Evidence that luminal ER proteins are sorted from secreted proteins in a post-ER compartment. *EMBO J.* **7,** 913–918.

Pelham, H. R. B., and Bienz, M. (1982). A synthetic heat-shock promoter element confers heat-inducibility on the herpes simplex virus thymidine kinase gene. *EMBO J.* **1,** 1473–1477.

Petko, L., and Lindquist, S. (1986). Hsp26 is not required for growth at high temperatures, nor for thermotolerance, spore development, or germination. *Cell* **45,** 885–894.

Plesset, J., Palm, C., and McLaughlin, C. S. (1982). Induction of heat-shock proteins and thermotolerance by ethanol in *Saccharomyces cerevisiae*. *Biochem. Biophys. Res. Commun.* **108,** 1340–1345.

Raschke, E., Baumann, G., and Schöffl, F. (1988). Nucleotide sequence analysis of soybean small heat shock protein genes belonging to two different multigene families. *J. Mol. Biol.* **199,** 549–557.

Reading, D. S., Hallberg, R. L., and Myers, A. M. (1989). Characterization of the yeast *HSP60* gene coding for a mitochondrial assembly factor. *Nature (London)* **337,** 655–659.

Riabowol, K. T., Mizzen, L. A., and Welch, W. J. (1988). Heat shock is lethal to fibroblasts microinjected with antibodies against hsp70. *Science* **242,** 433–436.

Roberts, J. K. (1989). Characterization of high molecular weight from soybeans. Ph.D. dissertation, Univ. of Georgia, Athens.

Roberts, J. K., and Key, J. L. (1985). Characterization of the genes for the heat shock 70kD and 80kD proteins in soybean. *Proc. Int. Congr. Plant Mol. Biol., 1st* p. 137 (abstr.).

Roberts, J. K., and Key, J. L. (1990). Isolation and characterization of an hsp70 from soybean. *Plant Mol. Biol.* (submitted).

Rochester, D. E., Winter, J. A., and Shah, D. M. (1986). The structure and expression of maize genes encoding the major heat shock protein, hsp70. *EMBO J.* **5,** 451–458.

Rougvie, A. E., and Lis, J. T. (1988). The RNA polymerase II molecule at the 5′ end of the uninduced *hsp70* gene of *D. melanogaster* is transcriptionally engaged. *Cell* **54,** 795–804.

Schlesinger, M. J. (1986). Heat shock proteins: The search for functions. *J. Cell Biol.* **103,** 321–325.

Schöffl, F., Raschke, E., and Nagao, R. T. (1984). The DNA sequence analysis of soybean heat-shock genes and identification of possible regulatory promoter elements. *EMBO J.* **3,** 2491–2497.

Schuster, G., Even, D., Kloppstech, K., and Ohad, I. (1988). Evidence for protection by heat-shock proteins against photoinhibition during heat-shock. *EMBO J.* **7,** 1–6.

Shuey, D. J., and Parker, C. S. (1986). Binding of *Drosophila* heat-shock gene transcription factor to the hsp 70 promoter. *J. Biol. Chem.* **261,** 7934–7940.

Simon, J. A., Sutton, C. A., Lobell, R. B., Glaser, R. L., and Lis, J. T. (1985). Determinants of heat shock-induced chromosome puffing. *Cell* **40,** 805–817.

Sinibaldi, R. M., and Turpen, T. (1985). A heat shock protein is encoded within mitochondria of higher plants. *J. Biol. Chem.* **260**, 15382–15385.

Sinibaldi, R. M., Coldiron, P., and Dietrich, P. S. (1985). Expression of putative maize heat shock clones. *Proc. Int. Congr. Plant Mol. Biol., 1st* p. 34 (abstr.).

Skelly, S., Fu, C. F., Dalie, B., Redfield, B., Coleman, T., Brot, N., and Weissbach, H. (1988). Antibody to σ^{32} cross-reacts with DnaK: Association of DnaK protein with *Escherichia coli* RNA polymerase. *Proc. Natl. Acad. Sci U.S.A.* **85**, 5497–5501.

Sorger, P. K., and Pelham, H. R. B. (1987a). Purification and characterization of a heat-shock element binding protein from yeast. *EMBO J.* **6**, 3035–3041.

Sorger, P. K., and Pelham, H. R. B. (1987b). Cloning and expression of a gene encoding hsc73, the major hsp70-like protein in unstressed rat cells. *EMBO J.* **6**, 993–998.

Sorger, P. K., and Pelham, H. R. B. (1988). Yeast heat shock factor is an essential DNA-binding protein that exhibits temperature-dependent phosphorylation. *Cell* **54**, 855–864.

Sorger, P. K., Lewis, M. J., and Pelham, H. R. B. (1987). Heat shock factor is regulated differently in yeast and HeLa cells. *Nature (London)* **329**, 81–85.

Southgate, R., Ayme, A., and Voellmy, R. (1983). Nucleotide sequence analysis of the *Drosophila* small heat shock gene cluster at locus 67B. *J. Mol. Biol.* **165**, 35–57.

Spena, A., and Schell, J. (1987). The expression of a heat-inducible chimeric gene in transgenic tobacco plants. *Mol. Gen. Genet.* **206**, 436–440.

Spena, A., Hain, R., Ziervogel, U., Saedler, H., and Schell, J. (1985). Construction of a heat-inducible gene for plants. Demonstration of heat-inducible activity of the *Drosophila hsp70* promoter in plants. *EMBO J.* **4**, 2739–2743.

Strittmatter, G., and Chua, N. H. (1987). Artificial combination of two cis-regulatory elements generates a unique pattern of expression in transgenic plants. *Proc. Natl. Acad. Sci. U.S.A.* **84**, 8986–8990.

Süss, K. H., and Yordanov, I. T. (1986). Biosynthetic cause of *in vivo* acquired thermotolerance of photosynthetic light reactions and metabolic responses of chloroplasts to heat stress. *Plant Physiol.* **81**, 192–199.

Thomas, G. H., and Elgin, S. C. R. (1988). Protein/DNA architecture of the DNase I hypersensitive region of the *Drosophila* hsp26 promoter. *EMBO J.* **7**, 2191–2201.

Topol, J., Ruden, D. M., and Parker, C. S. (1985). Sequences required for *in vitro* transcriptional activation of a *Drosophila hsp70* gene. *Cell* **42**, 527–537.

VanBogelen, R. A., Acton, M. A., and Neidhardt, F. C. (1987). Induction of the heat shock regulon does not produce thermotolerance in *Escherichia coli*. *Genes Dev.* **1**, 525–531.

Vierling, E. (1990). Heat shock protein function and expression in plants. *In* "Stress Responses in Plants: Adaptation Mechanisms" (R. Alscher, ed.). Liss, New York, in press.

Vierling, E., Mishkind, M. L., Schmidt, G. W., and Key, J. L. (1986). Specific heat shock proteins are transported into chloroplasts. *Proc. Natl. Acad. Sci. U.S.A.* **83**, 361–365.

Vierling, E., Nagao, R. T., DeRocher, A. E., and Harris, L. M. (1988). A heat shock protein localized to chloroplasts is a member of a eukaryotic superfamily of heat shock proteins. *EMBO J.* **7**, 575–581.

Vierling, E., Harris, L. M., and Chen, Q. (1989). The major low-molecular-weight heat shock protein in chloroplasts shows antigenic conservation among diverse higher plant species. *Mol. Cell. Biol.* **9**, 461–468.

Voellmy, R. (1984). The heat shock genes: A family of highly conserved genes with a superbly complex expression pattern. *BioEssays* **1**, 213–217.

Voellmy, R., and Rungger, D. (1982). Transcription of a *Drosophila* heat shock gene is heat-induced in *Xenopus* oocytes. *Proc. Natl. Acad. Sci. U.S.A.* **79**, 1776–1780.

Walter, M. H. (1989). The induction of phenylpropanoid biosynthetic enzymes by ultraviolet light or fungal elicitor in cultured parsley cells is overridden by a heat-shock treatment. *Planta* **177**, 1–8.

Welch, W. J., Feramisco, J. R., and Blose, S. H. (1985). The mammalian stress response and the cytoskeleton: Alterations in intermediate filaments. *In* "Intermediate Filaments" (E. Wang, D. Fischman, R. K. H. Liem, and T. T. Sun, eds.), pp. 57–67. N. Y. Acad. Sci., New York.

Werner-Washburne, M., Stone, D. E., and Craig, E. A. (1987). Complex interactions among members of an essential subfamily of *hsp70* genes in *Saccharomyces cerevisiae. Mol. Cell. Biol.* **7**, 2568–2577.

White, C. N., and Hightower, L. E. (1984). Stress mRNA metabolism in canavanine-treated chicken embryo cells. *Mol. Cell. Biol.* **4**, 1534–1541.

Widelitz, R. B., Magun, B. E., and Gerner, E. W. (1986). Effects of cycloheximide on thermotolerance expression, heat shock protein synthesis, and heat shock protein mRNA accumulation in rat fibroblasts. *Mol. Cell. Biol.* **6**, 1088–1094.

Wiederrecht, G., Shuey, D. J., Kibbe, W. A., and Parker, C. S. (1987). The *Saccharomyces* and *Drosophila* heat shock transcription factors are identical in size and DNA binding properties. *Cell* **48**, 507–515.

Winter, J., Wright, R., Duck, N., Gasser, C., Fraley, R., and Shah, D. (1988). The inhibition of petunia hsp70 mRNA processing during $CdCl_2$ stress. *Mol. Gen. Genet.* **211**, 315–319.

Wu, C. (1984). Activating protein factors binds *in vitro* to upstream control sequences in heat shock gene chromatin, *Nature (London)* **311**, 81–84.

Wu, B. J., Kingston, R. E., and Morimoto, R. I. (1986). Human *HSP70* promoter contains at least two distinct regulatory domains. *Proc. Natl. Acad. Sci. U.S.A.* **83**, 629–633.

EFFECTS OF HEAT AND CHEMICAL STRESS ON DEVELOPMENT

Nancy S. Petersen

Department of Molecular Biology, University of Wyoming,
Laramie, Wyoming 82071

I. Introduction

Both heat and chemicals can alter developmental programs with dramatic consequences. Some birth defects are caused by environmental agents, such as thalidomide, and others are clearly due to inherited molecular defects, such as sickle cell anemia and cystic fibrosis. However, the most common human birth defects, anencephaly, spina bifida aperta, cleft lip and palate, and heart defects, are due to the combined effects of inheritance and environment (Inouye and Nishimura, 1976). The biochemical bases for many genetic defects, including

275

sickle cell anemia, phenylketonuria, and glycogen storage diseases, are well understood. In contrast, the molecular bases for the most common types of morphological defects are still a matter of debate. Even the molecular basis for the severe limb defects caused by thalidomide is not well understood in spite of a great deal of research on this subject during the last 30 years (Newman, 1985; Stephens, 1988).

The theme of this article is a discussion of the common features of environmentally induced morphological defects during mammalian and *Drosophila* development. Cellular response to heat is the prototype system for this analysis. Heat is easy to apply and withdraw, and the effects of heat on cells have been extensively studied in a wide variety of experimental systems. Evidence that several teratogenic chemicals cause defects by a mechanism similar to heat is presented. The reader is referred to several reviews on teratology and on the heat-shock response for more general treatments of these subjects (see Table 1).

II. Developmental Defects Induced in Vertebrates by Environmental Stress

A. Heat-Induced Defects

Short-term exposure of pregnant guinea pigs, hamsters, mice, or rats to temperatures above 42°C induces a variety of developmental defects depending on the stage of development at the time of heating. The

TABLE 1
Reviews on Heat-Shock Proteins and Teratology

Topic	Reference(s)
Heat-shock proteins—general	Schlesinger *et al.* (1982), Atkinson and Walden (1985), Craig (1985), Lindquist (1986), Nover (1984)
Heat-shock proteins—vertebrate	Hightower *et al.* (1985), Subjeck and Shy (1986)
Heat-shock proteins—*Escherichia coli* and insects	Neidhardt *et al.* (1984), Petersen and Mitchell (1985)
Heat-shock proteins—functions	Pelham (1986, 1988), Welch (1987), Lindquist and Craig (1988), Tomasovic (1989), Bond and Schlesinger (1987)
Thermotolerance	Gerner (1983), Landry (1986), Tomasovic (1989)
Teratology	Warkany (1986), Bennett *et al.* (1990a), Fraser (1970), Inouye and Nishimura (1976)

pioneering work of Edwards (1967, 1969) defined time periods for induction of microcephaly and various limb malformations in guinea pigs. Cockroft and New (1978) showed that similar defects were induced in rat embryos cultured above 41°C. Subsequent studies have shown that severe cases of cleft palate, cleft lip, and exencephaly can be induced by heating pregnant mice to 43°C for short periods (Webster and Edwards, 1984; Finnell *et al.*, 1986). The percentage of the animals affected varies with genetic background, indicating a contribution of genes as well as environment to the induction of the defects.

In humans, retrospective studies have linked anencephaly, mental deficiency, facial dysmorphogenesis, and altered muscle tone to maternal exposure to heat either from prolonged high fever or excessive sauna or hot tub use (Pleet *et al.*, 1981; Fisher and Smith, 1981; Miller *et al.*, 1978). However, other studies contradict these results (Rapola *et al.*, 1978; Saxen *et al.*, 1982). Although it seems certain from studies on other mammals that heat is a potential teratogen, the actual contribution of heat to the incidence of human birth defects remains in question (Warkany, 1986; Edwards, 1979).

Whether or not heat is an important teratogen in humans, it is a good model teratogen for animal studies because both temperature and the length of exposure can be rigorously controlled. Heat shock has many of the characteristics of classical chemical teratogens. It is stage specific, induces similar types of defects, and interacts with the genetic background in a similar way to other teratogens.

B. NEURAL TUBE AND HEART DEFECTS

Neural tube defects are among the most common human malformations (Nakano, 1973; Lemire *et al.*, 1978). They include anencephaly, encepholoceles, meningoceles, and spina bifida. In experimental animal systems, neural tube defects can be induced by a variety of environmental teratogens. Cadmium, valproic acid, insulin, retinoic acid, and ethanol, as well as heat treatment, have been shown to induce neural tube defects in mice (Table 2). No matter which teratogen is used, the sensitive period during which neural tube defects can be induced is the same. For mice this is between the eighth and ninth day of development. The defects induced by different chemicals are morphologically indistinguishable. These observations are consistent with the hypothesis that these chemicals act via a common molecular mechanism. Heat and cadmium have also been shown to induce neural tube defects in hamsters and rats (Table 2). As with mice, the sensitive period for induction is critical, and the types of defects induced by the different treatments are indistinguishable.

TABLE 2
Inducers of Developmental Defects in Vertebrates

Defect	Animal	Inducer	Reference(s)
Cleft palate	Mice	Cortisone	Fraser *et al.* (1954)
Neural tube	Mice	Heat	Webster and Edwards (1984), Finnell *et al.* (1986)
Neural tube	Mice	Cadmium	Layton and Ferm (1980)
Neural tube	Mice	Ethanol	Daft *et al.* (1986)
Neural tube	Mice	Insulin	Cole and Trasler (1980)
Neural tube	Mice	Retinoic acid	Tibbles and Wiley (1988)
Neural tube	Mice	Valproic acid	Finnell *et al.* (1988)
Neural tube	Rats	Heat	Cockroft and New (1978), Webster *et al.* (1984)
Neural tube	Hamsters	Heat	Kilham and Ferm (1976)
Neural tube	Hamsters	Cadmium	Ferm and Layton (1979)
Neural tube	Chickens	Diazepam	Nagele *et al.* (1989)
Neural tube	Chickens	Verapamil	Lee and Nagele (1986)
Heart	Chickens	Heat	Nilson (1984)
Heart	Chickens	Ethanol	Fang *et al.* (1987)
Heart	Chickens	Arsenite	Johnston *et al.* (1980)
Heart	Chickens	Phenobarbital	Nishikawa *et al.* (1986)
Heart	Chickens	Valproic acid	H. J. Bruyere, Jr. (personal communication)

The idea that several apparently unrelated chemicals in addition to heat may cause congenital defects by a common mechanism is not new. In 1958, Landauer suggested that the molecular basis for chemically induced defects in chickens and heat-induced defects in flies was the same. However, there is new evidence that environmental insults from heat and chemicals do have a common effect on cells. Heat, steroid hormones, cadmium, and ethanol have all been shown to induce the synthesis of a group of proteins called heat-shock proteins (Table 3). No matter what the function of heat-shock proteins in the cell may be, the fact that all of these treatments induce their synthesis suggests that they do have a common effect on the cell.

TABLE 3
Induction of Heat-Shock Protein Synthesis by Teratogens

Teratogen	Animal/system	Heat-shock protein	Reference
Ethanol	CHO cells	All	Li (1983)
Cadmium	*Drosophila*	All	Courgeon *et al.* (1984)
Cadmium	CHO cells	All	Li and Laszlo (1985)
Arsenite	Rabbit embryos	HSP70	Heikkila and Schultz (1984)
Arsenite	*Drosophila*	All	Bournias-Vardiabasis and Buzin (1986)
Steroids	*Drosophila*	HSP22	Bournias-Vardiabasis and Buzin (1986)
Phenobarbital	*Drosophila*	HSP22	Bournia-Vardiabasis and Buzin (1986)

C. Heart Defects

Heart defects are also common human birth defects that are thought to be caused by the combined effects of genetic and environmental factors (Nora and Nora, 1976). Heart defects have been studied extensively in chick embryos. As with the neural tube defects, heart defects are induced by a wide variety of treatments, including heat, ethanol, and phenobarbital (Table 2). The type of defect depends on the time in development when the treatment is administered, rather than on the chemical used. When chicken eggs are incubated at 37°C, ventricular septal defects are induced if chemicals are administered on days 3 and 4, while aortic arch anomalies are induced on days 5 and 6 (Bruyere *et al.*, 1983; Hodach *et al.*, 1974). After day 7 of incubation, no defects are induced, even at the highest doses. Here again, the timing of the environmental insult is the critical factor rather than the specific type of environmental treatment.

Induction of heart defects by ethanol has also been reported in mice (Chernoff, 1980) and humans (Clarren and Smith, 1978; Streissguth *et al.*, 1985). In clinical studies as many as 30% of children with fetal alcohol syndrome also have heart defects (Jones *et al.*, 1973; Hanson *et al.*, 1976). It appears that ethanol does not act in a species-specific manner.

The same group of chemicals that induces neural tube defects can also induce heart defects. The type of defect induced depends critically on the timing of the environmental insult (Schardein, 1985). The similarities in the inducing treatments for different developmental defects suggest that these agents are affecting some very basic cellular process or processes that are important to both heart and nervous system development.

D. Development of Tolerance

Heat-induced developmental defects can be prevented if the embryos are allowed to adapt to the higher temperature in a stepwise manner. This phenomenon was first reported in *Drosophila* in which raising the temperature from 25 to 35°C for 30 minutes before a 40.5°C heat shock prevents the induction of developmental defects (Mitchell *et al.,* 1979; Petersen and Mitchell, 1982, 1985). A similar effect has been observed in mouse and in rat embryos cultured *in vitro* (Walsh *et al.,* 1987; Mirkes, 1987). The same treatments that prevent heat-induced developmental defects also enhance survival following heat shock at other stages of development (Mitchell *et al.,* 1979; Lindquist, 1986). The short-term adaptation to survive heat stress is called "acquisition of thermotolerance." A brief discussion of thermotolerance and the role of heat-shock proteins in its development is presented in Section III.

Pretreatment with lower doses of chemicals that induce developmental defects can also lower the incidence of the developmental defect. This is true for induction of cleft palate by cortisone. Pretreatment with cortisone lowers the frequency of defects from 100 to 17% (Fraser *et al.,* 1954). Cadmium pretreatment also can prevent cadmium-induced neural tube defects in hamsters by a maternally mediated mechanism (Layton and Ferm, 1980; Ferm and Hanlon, 1987). It is interesting that both steroid hormones and cadmium have been reported to induce the synthesis of heat-shock proteins in several systems (Table 3). This suggests the possibility that chemically induced tolerance may operate by the same mechanism as heat-induced thermotolerance.

The difficulty in determining whether heat is an important teratogen in humans is, in part, due to the development of thermotolerance. Under most natural conditions, a developing embryo is not exposed to a sudden heat shock, but rather the temperature rises gradually, allowing the development of thermotolerance. The teratogenic effects of heat may, therefore, often be prevented by the development of thermotolerance (Walsh *et al.,* 1987; Mirkes, 1987).

E. Interaction of Genetic and Environmental Factors

The importance of both genetics and environment in the induction of developmental defects has been known for many years (Fraser, 1976; Nora *et al.,* 1970). Several different genes may be involved in determining susceptibility to birth defects, but environmental conditions can make the critical difference. Congenital heart defects represent such a situation. Striking evidence of the importance of both environment and genetic constitution in susceptibility to heart defects comes from twin studies. In these studies, monozygotic twins have a very high (23%) incidence of the same defect (Ando et al., 1976). However, the fact that most of the monozygotic twins do not share the same defect argues also for an important role for prenatal environment. The incidence of the most common human birth defects, anencephaly, spina bifida aperta, cleft lip and palate, and heart defects, is related to both genetic and environmental factors. Cortisone-induced cleft palate in mice has been studied as a model of this type of interaction (Fraser et al., 1954; Fraser, 1976). The severity of these defects is known to depend on both the maternally administered dosage of cortisone and the genetic background of the mice.

More recently, the genetic component of susceptibility to heat-induced neural tube defects has been studied in mice. Strains of mice more susceptible to cortisone-induced defects are also more sensitive to heat-induced defects. Finnell et al. (1986) have shown that four strains of mice have susceptibilities to heat induction of neural tube defects ranging from 14 to 44%. The susceptibility of the most sensitive stain, SWV, maps as a single locus and is therefore thought to be due to a single gene or at most a group of closely linked genes. Mice of these strains were then tested for sensitivity to valproic acid induction of neural tube defects (Finnell et al., 1988). The same gradient of sensitivity was observed when valproic acid was used as an inducer. The fact that the same gene(s) confers sensitivity to both heat and valproic acid suggests that they induce neural tube defects by affecting a common pathway involving this gene.

The striking facts are that the same group of environmental agents induces a variety of different developmental defects. The type of defect induced depends on the timing of the treatment. Furthermore, the same types of agents induce defects in several different species, indicating that the effects are not due to species-specific metabolic events.

III. Environmentally Induced Developmental Defects in *Drosophila* as a Model System

A. HISTORICAL BACKGROUND

Heat has been known to act as a teratogen in insects for more than a century. In 1854, 2 years before Mendel published his treatise on the genetics of peas, and 5 years after the publication of Darwin's *Origin of Species,* an Austrian entomologist, George Dorfmeister, reported that exposure to extreme temperatures during a specific period in pupal life could change the pattern on the wings of butterflies. In the 1930s, Goldschmidt studied the effects of heat on *Drosophila* pupal development. He showed that heating during specific sensitive periods of development induced defects that resembled mutant phenotypes, which he called phenocopies (Goldschmidt, 1935, 1949). Goldschmidt showed that the type of defect induced depended on the time of heating, each defect having its own specific sensitive period for induction. He also showed that even though they resembled mutant phenotypes, heat-induced defects were not inherited and, therefore, were not due to permanent changes in the genes. Later, Milkman showed that the severity of crossveinless phenocopy in *Drosophila* depends on the genetic background of the fly (Milkman, 1966). Also in the 1960s, Mitchell studied induction of the blond phenocopy in wild-type strains of flies. The blond phenocopy is due to failure to activate phenol oxidase in the bristles on the back of the thorax and results in yellow rather than black bristles. The blond phenocopy resembles the phenotypes of the *blond* and *straw* mutations in which phenol oxidase fails to be activated. In the normal pupa, phenol oxidase is activated by a cascade of reactions occurring late in pupal development. A striking feature concerning the sensitive period for induction of the blond phenocopy is that it is 2–10 hours earlier than the time of normal activation of phenol oxidase. This observation led Mitchell to propose that heat affects the expression of a gene that codes for a protein involved in the phenol oxidase activation cascade (Mitchell, 1966; Seybold *et al.,* 1975).

B. HEAT-INDUCED DEVELOPMENTAL DEFECTS IN *Drosophila*

Heat can induce a wide variety of developmental defects in both embryos and pupae of *Drosophila*. During embryonic development, phenocopies of bithorax can be induced by heating during a 10-minute sensitive period at blastoderm stage (Gloor, 1947; Santamaria, 1979). As in mammals it is the timing of the environmental insult that is most critical. A variety of other developmental defects that resemble the phenotypes of homeotic mutants can also be induced during this time period (Eberlein, 1986). With embryos it is difficult to achieve a frequency of 100% for a single type of defect. This is most likely due to the very short window of sensitivity and difficulties in accurate staging so that embryos are actually treated at slightly different times in development. Fortunately, in *Drosophila,* there is a later developmental stage, the pupal stage, during which the sensitive periods for induction of developmental defects are longer. During the pupal period it is relatively simple to produce 100% induction of a specific developmental defect. Pupae also have the advantage that they contain large numbers of morphologically undifferentiated cells. Most epithelial cell division takes place during the larval stage, so that during pupal development there is an adult complement of cells whose fate has already been determined, but which are morphologically undifferentiated. This has a further advantage in that effects on differentiation can be separated from effects on cell division.

More than 20 different developmental defects can be induced by heat in *Drosophila* pupae (Goldschmidt, 1935; Milkman, 1966; Mitchell and Petersen, 1982). Each of these defects can only be induced during a specific developmental stage. As first noted by Goldschmidt, many environmentally induced defects closely resemble mutant defects. In fact, the same type of defect may be induced at different times on different parts of the animal. This is exemplified by the branched hair phenocopy that is induced when heating is done at 38 hours on the wing, 42 hours on the thorax, and at 45 hours on the abdomen (Mitchell and Petersen, 1981). The timing of phenocopy-sensitive periods is correlated with specific patterns of protein synthesis. In differentiating epithelial tissues such as the wing, gene expression changes rapidly and different stages of development can be recognized by characteristic patterns of proteins synthesized and separated on one- or two-dimensional gels (Mitchell and Petersen, 1981; Petersen *et al.*, 1985). The developmental changes seen in protein synthesis are similar for different tissues, but the timing of these protein synthesis patterns is different. The sensitive period for induction of the branched hair phenocopy corresponds to a particular pattern of protein synthesis that occurs at 38 hours in wings, but not until 42 hours in the thorax. Thus, the sensitive periods for induction of the branched hair defect are different in different tissues because the timing of the developmental program is different. The resemblance of phenocopies to mutants and the correlation with a specific stage in gene expression in different tissues both suggest that the heat-induced defects are caused by failure to synthesize specific gene products when they are needed during development.

The effects of heat on gene expression have been studied during the sensitive period for induction of the branched hair phenocopy in wings. The sensitive period for this defect occurs at a time of rapid change in mRNA levels and protein synthesis. A phenocopy -inducing heat shock (30–40 minutes at 40–40.1°C) interrupts this program by inhibiting both RNA and protein synthesis and causes a developmental delay of

15–20 hours. Protein synthesis is completely inhibited for a period of 8–10 hours, but then recovers gradually. During recovery heat-shock proteins are synthesized first, followed by a gradual recovery of the pattern of protein synthesis characteristic of the time in development when the heat shock was given (Petersen and Mitchell, 1982). Following heat shock, RNA synthesis is also inhibited but concentrations of message stay relatively constant because mRNA decay is inhibited as well. Synthesis of heat-shock mRNA appears to resume first, followed by the synthesis and decay of messages involved in the developmental program (Petersen and Mitchell, 1982). For the branched hair phenocopy, the delay in protein synthesis interrupts the order of the hair construction process by delaying the synthesis of cuticulin on the surface of the wings (Mitchell et al., 1983). The branches on wing hairs do not appear until about 15 hours after the heat shock, when internal pressure involved in the next stage of development pushes membrane through the holes in the incomplete cuticulin. Thus the branched hair phenocopy is due to failure to complete one process in development (cuticulin synthesis) before the next process begins (Petersen and Mitchell, 1982).

C. PHENOCOPY PREVENTION BY A THERMOTOLERANCE-INDUCING TREATMENT

Mild heat shocks do not induce developmental defects in *Drosophila;* they induce thermotolerance. Heat shocks sufficient to induce developmental defects are severe enough to compromise gene expression, including RNA synthesis and processing, as well as protein synthesis. Phenocopies can be prevented by a short pretreatment at 35°C, which induces synthesis of heat-shock proteins prior to the higher temperature heat shock. This has been shown for many different phenocopies, including the branched hair phenocopy previously described (Mitchell *et al.,* 1979; Petersen and Mitchell, 1982, 1985). The effect of the preshock is not simply to alter the timing of the sensitive period, because the pretreatment is equally effective at the beginning and end of the sensitive period (Mitchell *et al.,* 1979; Petersen and Mitchell, 1989). Furthermore, the high-temperature treatment still inhibits both RNA and protein synthesis. The primary difference under thermotolerant conditions is that recovery of RNA and protein synthesis is much more rapid. Protein synthesis recovers with a delay of only 4–5 hours, followed by recovery of the normal program of mRNA synthesis and decay (Petersen and Mitchell, 1981, 1982). These experiments suggest that the primary effect of the pretreatment is on the recovery of protein synthesis. Effects on mRNA synthesis and decay seem to follow the recovery of protein synthesis.

D. PHENOCOPIES INDUCED IN RECESSIVE MUTANT HETEROZYGOTES

A clearer picture of the relationship between mutants and phenocopies comes from studying pupae that are heterozygotes with one recessive mutant gene and one wild-type gene. In these pupae, the mutant phenotype can be induced by heating at a specific sensitive period. These unusual phenocopies can only be induced in pupae heterozygous for the recessive gene in question and not in wild-type pupae. As is the case for other phenocopies, heterozygote phenocopies can only be induced during specific sensitive periods. The sensitive periods for the induction of three such mutant phenotypes in mutant heterozygotes have been identified; these are multiple wing hair, forked, and singed (Mitchell and Petersen, 1985; Petersen and Mitchell, 1987). In each case, the phenocopy strongly resembles the mutant phenotype and can only be induced in mutant heterozygotes.

Heterozygote phenocopies behave very much like phenocopies induced in wild-type flies. They have different sensitive periods for induction on different parts of the animal, and they are prevented by thermotolerance-inducing treatments of the same type that prevent wild-type phenocopies (Petersen and Mitchell, 1987, 1989). The meaning of the phenomenon of heterozygote phenocopies becomes clearer when it is considered in the light of a failure of gene expression. In a recessive mutant heterozygote, it is expected that only half the normal level of gene product will be made. By definition, recessive means that this amount of gene product is enough to ensure normal development. If, however, an environmental insult in the form of a heat shock occurs at or just before the normal time of gene expression, the level of gene expression during this sensitive period is reduced even more, below the critical threshold level, and a morphological defect is induced. Of the herterozygote phenocopies studied so far, the time of expression of the normal gene product is known only for *forked*. The sensitive period for induction of the forked phenocopy coincides with the normal time of synthesis of the forked mRNA (Parkhurst and Corces, 1985; Mitchell and Petersen, 1985). Thus, it appears that the forked phenocopy is due to heat-induced changes in the level or timing of expression of the forked gene product. Because phenocopies appear to be caused by failure of gene expression, thermotolerance must act to prevent phenocopies by enhancing gene expression so that levels of the mutant gene product are restored to normal during the critical period.

E. Similarities to Mammalian Systems

The similarities between the induction of defects in *Drosophila* and mammals are striking.

1. The type of defect induced depends on the time of treatment.
2. The induction of a defect depends on both the genetic background and the environmental treatment.
3. Heat-induced defects can be prevented by a thermotolerance-inducing treatment before the environmental insult.

On a superficial level, it appears that the induction of defects in mammals and *Drosophila* are the same. At the cellular level, many known teratogens in mammals have been shown to induce the synthesis of heat-shock proteins and to affect the differentiation of nerve and muscle cells in *Drosophila* embryonic cells in culture (Bournias-Vardiabasis *et al.*, 1983). All of these facts reinforce the hypothesis that there may be a general cellular mechanism for both the induction of developmental defects and their prevention in thermotolerant animals. Based on the *Drosophila* model, the mechanism of induction appears to involve changes in the levels of critical gene products required for the developmental process.

IV. Effects of Heat Stress on Gene Expression during Development

A. Overview

Because heat and several chemical teratogens appear to act via a common mechanism, it seems worthwhile to ask what is known about the effects of heat on cells and especially what is known about how heat affects gene expression. A basic observation is that, as the temperature is raised above the normal basal temperature, cells respond by increasing the synthesis of a group of proteins called heat-shock proteins and by gradual reduction of synthesis of most cellular proteins (Tissières *et al.*, 1974; Craig, 1985; Lindquist, 1986). This is true in many varieties of cells, including bacteria, animal cells, and plant cells. As the temperature is raised even more, the synthesis of all proteins ceases, including synthesis of heat-shock proteins. In addition to protein synthesis, many other cellular functions are affected by heat—DNA synthesis is inhibited, RNA synthesis and processing are inhibited, ribosomal assembly is inhibited, the cytoskeleton collapses, and nuclear morphology changes. Whether a cell recovers from such an insult depends on the length and severity of the treatment as well as on whether the temperature is raised slowly or abruptly. In the former case cells can develop a surprising degree of thermotolerance. When Chinese hamster ovarian (CHO) cells are heated abruptly to 45°C, 99.99% of the cells are killed; however, if thermotolerant cells are given the same treatment, only 10–20% are killed (Li and Hahn, 1987). The development of thermotolerance is closely correlated with the synthesis of heat-shock proteins. In most cases, all aspects of cellular damage caused by heat shock are repaired much more rapidly in cells that have been induced to synthesize heat-shock proteins before the high-temperature treatment.

B. HEAT-SHOCK PROTEINS AND THERMOTOLERANCE

Proteins that are selectively synthesized at high temperatures are called heat-shock proteins. The most abundant and conserved of these proteins belong to three families of proteins, the HSP90 family, the HSP70 family, and the HSP60 or groEL family (Table 1). These proteins are among the most conserved known; the protein whose synthesis increases most with heat shock, HSP70, is 50% conserved among humans and *Escherichia coli* (Bardwell and Craig, 1984).

Heat-shock protein synthesis can be induced in experimental organisms by chemical agents as well as by heat shock. Inducers of heat-shock protein synthesis include such diverse agents as canavanine, cadmium, arsenite, and ethanol (Table 3). All of these chemicals can also induce (survival) thermotolerance (Li, 1983; Lindquist, 1986). Because one thing they are known to have in common is the induction of heat-shock proteins, it has been suggested that heat-shock protein synthesis is involved in the development of thermotolerance. Recently, the essential role of HSP70 in the development of thermotolerance has been demonstrated in mammalian cells. Mammalian cells in which HSP70 is not made or is inactivated by antibody binding are not capable of developing thermotolerance (Riabowol *et al.*, 1988; Johnston and Kucey, 1988).

The development of thermotolerance in the absence of heat-shock protein synthesis in cycloheximide-treated cells has also been reported (Hallberg *et al.*, 1985; Widelitz *et al.*, 1986). In *Tetrahymena*, cycloheximide treatment induces thermotolerance to a 43°C heat shock, whereas tolerance to higher temperatures (46°C) requires conditions in which heat-shock proteins are synthesized. These observations indicate that the phenomenon of thermotolerance is a more complicated process than simply the synthesis of several protective proteins.

The actual mechanism for the development of thermotolerance has been difficult to determine. Many possibilities have been suggested. Thermotolerance has been proposed to result from heat-shock protein prevention of mRNA decay or mRNA processing or on the refolding or degradation of denatured proteins (Mitchell *et al.*, 1979; Yost and Lindquist, 1986, 1988; Pelham, 1986, 1988). There are also several other heat-shock proteins that are synthesized in smaller quantities than are the major heat-shock proteins and it is still not clear how or whether these proteins may play a role in the development of thermotolerance. It is important to remember as well that there may be aspects of thermotolerance other than the synthesis of heat-shock proteins that remain to be discovered.

Although the roles of the heat-shock proteins in the development of thermotolerance are still in question, more progress has been made in understanding the roles of non-heat-induced members of the HSP70 and HSP90 families (reviewed by Lindquist and Craig, 1988). One of the most provocative recent developments is that a non-heat-inducible homolog of HSP70 (HSC70) has been shown to enhance protein transport across the endoplasmic reticulum in yeast (Chirico *et al.*, 1988; Deshaies *et al.*, 1988). Members of the HSP70 family of proteins have also been reported to be involved in uncoating of clathrin vesicles and in the transport of proteins into lysosomes (Chappell *et al.*, 1986; Chiang *et al.*, 1989). Another major group of heat-shock proteins, the HSP90-related proteins, has been shown to be associated with steroid hormone receptors in the absence of hormone binding (Catelli *et al.*, 1985). The ATP-dependent unfolding of proteins is the proposed mechanism by which HSC70 facilitates the transport across membranes. This ATP-dependent unfolding of proteins has also been proposed to interfere with hormone receptor binding to DNA when HSP90 is bound (Picard *et al.*, 1988). A similar role has been proposed for these proteins following heat shock in the ATP-dependent refolding of proteins (Pelham, 1986, 1988).

C. SYNTHESIS OF HEAT-SHOCK PROTEINS IN EARLY EMBRYOS

If heat-shock proteins are thought to play a role in the prevention of developmental defects, it is important to know when they are expressed in normal development and whether their synthesis can be induced in early embryos. The major heat-shock protein, HSP70, or a closely related HSC70 is normally present in large amounts very early in development in both mammalian embryos and in *Drosophila*. In the two-cell mouse embryo, HSP70 is among the first zygotic genes to be expressed (Bensaude *et al.*, 1983).

In *Drosophila,* HSC70 is the most abundant protein in oocytes and preblastoderm embryos (Palter *et al.*, 1986). The high level of expression of members of the HSP70 family in embryonic tissue even extends to plants in which HSP70 and its mRNA are stored in dried seeds (Abernethy *et al.*, 1988). A decrease in synthesis of HSP70 is an early event when teratocarcinoma cells or erythroleukemia cells are induced to differentiate in culture (Bensaude and Morange, 1983; Imperiale *et al.*, 1984; Hensold and Housman, 1988; Witting *et al.*, 1983). In embryos as well, the level of HSP70 synthesis decreases with development, and in general, it is only possible to induce thermotolerance after the levels of HSP70 have decreased somewhat. In rabbit embryos, the expression of HSP70 is inducible by heat and arsenite as early as the 6-day-old blastocyst stage (Heikkila and Schultz, 1984). The ability to induce heat-shock protein synthesis in early embryos indicates that they could play a role in the development of thermotolerance in embryos. The fact that members of the HSP70 family are normally expressed in large amounts at early stages of development suggests a role for the HSP70 family in the process of activating gene expression as a cell goes from dormancy to a state of active growth.

D. MOLECULAR MODELS FOR THE INDUCTION OF DEVELOPMENTAL DEFECTS

Several molecular models have been suggested to explain heat-induced developmental defects. The first theory is that heat shock destabilizes mRNA, and heat-shock proteins prevent phenocopies by stabilizing critical mRNAs (Mitchell *et al.*,1979; Haass *et al.*, 1989). This suggestion is attractive because it would explain why heat induces different developmental defects at different stages in development. A major argument against this model is that RNA in general is stabilized following heat shock whether or not thermotolerance is induced (Farrell-Towt and Sanders, 1984; Petersen and Mitchell, 1981, 1982). In fact, specific pupal mRNAs decay more rapidly in thermotolerant animals than in animals receiving a phenocopy inducing heat shock (Petersen and Mitchell, 1982; Petersen and Young, 1991).

Another theory for the induction of phenocopies holds that phenocopies are the result of a failure in mRNA processing following heat shock. In *Drosophila* cell lines, messages whose introns have not been removed can be transported into the cytoplasm and translated into abnormal proteins (Yost and Lindquist, 1986, 1988). According to this hypothesis, the abnormal proteins cause phenocopies, and heat-shock proteins prevent phenocopy induction by enhancing the recovery of the splicing apparatus in the nucleus. This interesting model may explain some of the phenocopies of homeotic mutants that can be induced during *Drosophila* embryonic development. This type of model would, however, be expected to produce a dominant effect and does not explain why the heterozygote phenocopies induced in *Drosophila* pupae are not induced in wild-type flies.

Another theory proposes that heat-shock proteins in mammals are the villains, rather than the rescuers. In the embryonic stress hypothesis, the takeover of the translational machinery by heat-shock mRNA prevents the synthesis of normal proteins, resulting in the induction of developmental defects (German, 1984). This theory is improbable because heat-shock protein synthesis occurs under conditions in which the animal is protected from defect induction as well as following defect-inducing heat shocks. In fact, the total amount of heat-shock protein synthesis in both cases is similar and is not correlated to the severity of the defect (Walsh *et al.*, 1987; Bennett *et al.*, 1990b; Velazquez and Lindquist, 1984).

Another hypothesis for the induction of defects by heat in mammalian systems states that the heat-induced defects are due to effects on cell division, especially the death of the more heat-sensitive cells undergoing mitosis (Edwards *et al.*, 1974). This model may explain some defects in size such as microcephaly, but it does not explain why specific stages in development are especially heat sensitive. Cell division goes on during a long period in the development of most organs. It also does not explain why nontoxic doses of some chemicals induce neural tube and heart defects. Finally, it does not explain the defects that are induced in *Drosophila* pupae, because, in most cases, these defects are induced after cell division is completed.

In order to determine which of these hypotheses may be true, it will be necessary to compare the level of expression of a specific gene during normal development with the levels of expression following heat shocks that induce the phenocopy. The heterozygote phenocopies in *Drosophila,* in which the gene whose expression is affected by heat shock can be identified, appear to be a good system in which to do this.

V. Summary

Similarities in the means by which developmental defects are induced in vertebrates and *Drosophila* suggest that some kinds of defects may be induced by similar mechanisms. The similarities include the fact that heat and a group of chemicals that induce synthesis of heat-shock proteins induce defects in mammals, chickens, and flies. Different kinds of defects are even produced in one type of animal, depending on the precise timing of the environmental insult. The effectiveness of the environmental treatment in inducing defects depends on the genetic background of the animal as well as on past exposure to chemicals and heat. Developmental defects induced by heat in mice, rats, and flies can all be prevented by thermotolerance-inducing treatments. The basis for these effects has been studied at the molecular level in *Drosophila,* and the evidence indicates that these teratogens and the thermotolerance-inducing treatments affect the level or timing of expression of specific genes during critical periods in the developmental program.

ACKNOWLEDGMENTS

I would like to thank Harold Bruyere, Richard Finnell, and Herschel Mitchell for reading the manuscript and making many helpful suggestions.

REFERENCES

Abernethy, R., Thiel, D. S., Petersen, N. S., and Helm, K. (1988). Thermotolerance is developmentally dependent in germinating wheat seed. *Plant Physiol.* **89,** 576–596.

Alsop, F. M. (1919). The effect of abnormal temperatures on the developing nervous system in chick embryos. *Anat. Rec.* **15,** 307.

Ando, M., Takao, A., and Mori, K. (1976). *In* "Gene Environment Interactions in Common Diseases" (E. Inouye and H. Nishimura, eds.), pp. 71–88. University Park Press, Baltimore.

Atkinson, B. G., and Walden, D. B., eds. (1985). "Changes in Eukaryotic Gene Expression in Response to Environmental Stress." Academic Press, Orlando, Florida.

Bardwell, J. C., and Craig, E. A. (1984). Major heat shock gene of *Drosophila* and the heat inducible gene *dnaK* are homologous. *Proc. Natl. Acad. Sci. U.S.A.* **81,** 848–851.

Bennett, G. D., Mohl, V. K., van Waes, M., and Finnell, R. H. (1990a). The murine heat shock response: Implications for the embryonic stress hypothesis. *Teratogen., Carcinogen. Mutagen.* (submitted).

Bennett, G. D., Mohl, V. K., and Finnell, R. H. (1990b). Embryonic and maternal heat shock responses to a teratogenic hyperthermic insult. *Reprod. Toxicol.* (in press).

Bensaude, O., and Morange, M. (1983). Spontaneous high expression of heat-shock proteins in mouse embryonal carcinoma cells and ectoderm from day 8 mouse embryo. *EMBO J.* **2,** 173–177.

Bensaude, O., Babinet, C., Morange, M., and Jacob, F. (1983). Heat shock proteins, the first major products of zygotic gene activity in mouse embryos. *Nature (London)* **305,** 331–332.

Bond, U., and Schlesinger, M. J. (1987). Heat-shock proteins and development. *Adv. Genet.* **24,** 1–29.

Bournias-Vardiabasis, N., and Buzin, C. H. (1986). Altered differentiation and induction of heat shock proteins in *Drosophila* embryonic cells associated with teratogen treatment. *In* "Banbury Report 26: Developmental Toxicology: Mechanisms and Risk" (J. Scandalios, ed.). Cold Spring Harbor Lab, Cold Spring Harbor, New York.

Bournias-Vardiabasis, N., Teplitz, R. L., Chernoff, G. F., and Seecof, R. L. (1983). Detection of teratogens in the *Drosophila* embryonic cell culture test: Assay of 100 chemicals. *Teratology* **28**, 109–122.

Bruyere, H. J., Jr., Matsuoka, R., Carlsson, E., Cheung, M. O., Dean, R., and Gilbert, E. F. (1983). Cardiovascular malformations associated with administration of prenalterol to young chick embryos. *Teratology* **28**, 75–82.

Catelli, M. G., Binart, N., Jung-Testas, I., Renoir, J. M., Baulieu, E.-E., Feramisco, J. M., and Welch, W. J. (1985). The common 90kD protein component of non-transformed '8S' steroid receptors is a heat-shock protein. *EMBO J.* **4**, 3131–3135.

Chappell, T. G., Welch, W. J., Schlossman, D. M., Palter, K. B., Schlesinger, M. J., and Rothman, J. E. (1986). Uncoating ATPase is a member of the 70 kilodalton family of stress proteins. *Cell* **45**, 3–13.

Chernoff, G. F. (1980). The fetal alcohol syndrome in mice: Maternal variables. *Teratology* **22**, 71–75.

Chiang, H.-L., Terlecky, S. R., Plant, C. P., and Dice, J. F. (1989). A role for the 70-kilodalton heat shock protein in lysosomal degradation of intracellular proteins. *Science* **246**, 382–384.

Chirico, W. J., Waters, M. G., and Blobel, G. (1988). 70K heat shock related proteins stimulate protein translocation into microsomes. *Nature (London)* **332**, 805–810.

Clarren, S. K., and Smith, D. W. (1978). The fetal alcohol syndrome. *N. Engl. J. Med.* **298**, 1063–1067.

Cockroft, D. L., and New, D. A. T. (1978). Abnormalities induced in cultured rat embryos by hyperthermia. *Teratology* **17**, 277–284.

Cole, W., and Trasler, D. G. (1980). Gene–teratogen interaction and insulin-induced mouse exencephaly. *Teratology* **22**, 125–139.

Courgeon, A. M., Maisonhaute, C. Best-Belpomme, M. (1984). Heat shock proteins are induced by cadmium in *Drosophila* cells. *Exp. Cell Res.* **153**, 515–521.

Craig, E. A. (1985). The heat shock response. *CRC Crit. Rev. Biochem.* **18**, 239–280.

Daft, P. A., Johnston, M. C., and Sulik, K. K. (1986). Abnormal heart and great vessel development following acute ethanol exposure in mice. *Teratology* **33**, 93–104.

Deshaies, R. J., Koch, B. D., Werner-Washburne, M., Craig, E. A., and Schekman, R. A. (1988). A subfamily of stress proteins facilitates translocation of secretory and mitochondrial precursor peptides. *Nature (London)* **332**, 800–805.

Eberlein, S. (1986). Stage specific embryonic defects following heat shock in *Drosophila*. *Dev. Genet.* **5**, 179–197.

Edwards, M. J. (1967). Congenital defects in guinea pigs. *Arch. Pathol.* **84**, 42–48.

Edwards, M. J. (1969). Congenital defects in guinea pigs: Fetal reabsorptions, abortions, and malformations following induced hyperthermia during early gestation. *Teratology* **2**, 313–328.

Edwards, M. (1979). Is hyperthermia a teratogen? *Am. Heart J.* **98**, 277–280.

Edwards, M. J., Mulley, R., Ring, S., and Wanner, R. A. (1974). Mitotic cell death and delay of mitotic activity in guinea-pig embryos following brief maternal hyperthermia. *J. Exp. Morphol.* **32**, 593–602.

Fang, T., Bruyere, H. J., Kargas, S. A., Nishikawa, T., Takagi, Y., and Gilbert, E. F. (1987). Ethyl alcohol-induced cardiovascular malformations in the chick embryo. *Teratology* **35**, 95–103.

Farrell-Towt, J., and Sanders, M. (1984). Noncoordinate histone synthesis in heat-shocked *Drosophila* cells is regulated at multiple levels. *Mol. Cell. Biol.* **4**, 2676–2685.

Ferm, V. H., and Hanlon, D. P. (1987). Inhibition of cadmium teratogenesis by a mercaptoacrylic acid. *Experientia* **43**, 208–210.

Ferm, V. H., and Layton, W. M. (1979). Reduction in cadmium teratogenesis by prior cadmium exposure. *Environ. Res.* **18**, 347–350.

Finnell, R. H., Moon, S. P., Abbott, L. C., Golden, J. A., and Chernoff, G. F. (1986). Strain differences in heat-induced neural tube defects in mice. *Teratology* **33**, 247–252.

Finnell, R. H., Bennett, G. D., Karras, S. B. and Mohl, V. K. (1988). Common hierarchies of susceptibility to the induction of neural tube defects by valproic acid and its 4-propyl-4-pentenoic acid metabolite. *Teratology* **38**, 313–320.

Fisher, N. L., and Smith, D. W. (1981). Occipital encephalocele and early gestational hyperthermia. *Pediatrics* **68**, 480–483.

Fraser, F. C. (1976). The multifactorial/threshold concept—Uses and misuses. *Teratology* **14**, 267–280.

Fraser, F. C., Kalter, H., Walker, B. E., and Fainstat, T. D. (1954). The experimental production of cleft palate with cortisone and other hormones. *J. Cell. Comp. Physiol.* **43**, 237–259.

German, J. (1984). Embryonic stress hypothesis of teratogenesis. *Am. J. Med.* **76**, 293–301.

Gerner, E. W. (1983). Thermotolerance. *In* "Hyperthermia and Cancer Therapy" (F. K. Storm, ed.), pp. 141–162. Hall Med. Publ., Boston, Massachusetts.

Gloor, H. (1947). Phanokapie-Versuche mit Aether an *Drosophila*. *Rev. Suissue Zool.* **54**, 637–712.

Goldschmidt, R. B. (1935). Gen und ausseneigenschaft (untersuchungan *Drosophila*). *Z. Indukt. Abstamm. Vererbungsl.* **69**, 38–131.

Goldschmidt, R. B. (1949). Phenocopies. *Sci. Am.* **181**, 46–49.

Haass, C., Falkenburg, P. E., and Kloetzel, P.-M. (1989). The molecular organization of the small heat shock proteins in *Drosophila*. *In* "Stress Induced Proteins" (M. L. Pardue, J. R. Feramisco, and S. Lindquist, eds.), pp. 175–185. Liss, New York.

Hallberg, R. L., Kraus, K. W., and Hallberg, E. M. (1985). Induction of acquired thermotolerance in *Tetrahymena thermophila* can be achieved without the prior synthesis of heat shock proteins. *Mol. Cell. Biol.* **5**, 2061–2070.

Hanson, J. W., Jones, K. L., and Smith, D. W. (1976). Fetal alcohol syndrome: Experience with 41 patients. *JAMA, J. Am. Med. Assoc.* **235**, 1458–1460.

Heikkila, J. J., and Schultz, G. A. (1984). Different environmental stresses can activate the expression of a heat shock gene in rabbit blastocysts. *Gamete Res.* **10**, 45–56.

Hensold, J. O., and Housman, D. E. (1988). Decreased expression of the stress protein hsp70 is an early event in murine erythroleukemic cell differentiation. *Mol. Cell. Biol.* **8**, 2219–2223.

Hightower, L. E., Guidon, P. T., Whelan, S. A., and White, C. N. (1985). Stress responses in avian and mammalian cells. *In* "Changes in Eukaryotic Gene Expression in Response to Environmental Stress" (B. G. Atkinson and D. B. Walden, eds.), pp. 197–210. Academic Press, Orlando, Florida.

Hodach, R. J., Gilbert, E. F., and Fallon, J. F. (1974). Aortic arch anomalies associated with the administration of epinephrine in chick embryos. *Teratology* **9**, 203–209.

Imperiale, M. J., Kao, H. T., Feldman, L. T., Nevins, J. R., and Strickland, S. (1984). Common control of the heat shock gene and early adenovirus genes: Evidence for a cellular E1A-like activity. *Mol. Cell. Biol.* **4**, 867–874.

Inouye, E., and Nishimura, H. (1976). "Gene–Environment Interaction in Common Diseases." Univ. Park Press, Baltimore, Maryland.

Johnston, R. N., and Kucey, B. L. (1988). Competitive inhibition of *hsp70* gene expression causes thermosensitivity. *Science* **242**, 1551–1554.

Johnston, D., Oppermann, H., Jackson, J., and Levinson, W. (1980). Induction of four proteins in chick embryo cells by sodium arsenite. *J. Biol. Chem.* **255**, 6875–6980.

Jones, D. L., Smith, D. W., Ulleland, C. N., and Streissguth, A. P. (1973). Pattern of malformation in offspring of chronic alcoholic women. *Lancet* **1**, 1267–1271.

Kilham, L., and Ferm, V. H. (1976). Exencephaly in fetal hamsters following exposure to hyperthermia. *Teratology* **14**, 323–326.

Landauer, W. (1958). On phenocopies, their developmental physiology and genetic meaning. *Am. Nat.* **92**, 201–213.

Landry, J. (1986). Heat shock proteins and cell thermotolerance. *In* "Hyperthermia and Cancer Treatment" (L. J. Angiheri and J. Robert, eds.), Vol. 1, pp. 37–58. CRC Press, Boca Raton, Florida.

Layton, W. M., and Ferm, V. H. (1980). Protection against cadmium-induced limb malformations by pretreatment with cadmium or mercury. *Teratology* **21**, 357–360.

Lee, H., and Nagele, R. G. (1986). Toxic and teratogenic effects of verapamil on early chick embryos: Evidence for the involvement of calcium in neural tube closure. *Teratology* **33**, 203–211.

Lemire, R. J., Beckwith, J. B., and Warkany, J. (1978). "Anencephaly," Raven, New York.

Li, G. C. (1983). Induction of thermotolerance and enhanced heat shock protein synthesis in Chinese hamster fibroblasts by sodium arsenite and ethanol. *J. Cell. Physiol.* **115**, 116–122.

Li, G. C., and Hahn, G. M. (1987). Influence of temperature on the development and decay of thermotolerance and heat shock proteins. *Radiat. Res.* **112**, 517–524.

Li, G. C., and Laszlo, A. (1985). Thermotolerance in mammalian cells: A possible role for heat shock proteins. *In* "Changes in Eukaryotic Gene Expression in Response to Environmental Stress" (B. G. Atkinson and D. B. Walden, eds.), pp. 227–254. Academic Press, Orlando, Florida.

Lindquist, S. (1986). The heat-shock response. *Annu. Rev. Biochem.* **55**, 1151–1191.

Lindquist, S., and Craig, E. A. (1988). The heat-shock proteins. *Annu. Rev. Genet.* **22**, 631–677.

Milkman, R. (1966). Analyses of some temperature effects on *Drosophila* pupae. *Biol. Bull. (Woods Hole, Mass.)* **131**, 331–345.

Miller, P., Smith, D. W., and Shepard, T. (1978). Maternal hyperthermia as a possible cause of anencephaly. *Lancet* **1**, 519.

Mirkes, P. E. (1987). Hyperthermia-induced heat shock response and thermotolerance in post-implantation rat embryos. *Dev. Biol.* **119**, 115–122.

Mitchell, H. K. (1966). Phenol oxidases and *Drosophila* development. *Insect Physiol.* **12**, 755–765.

Mitchell, H. K., and Petersen, N. S. (1981). Rapid changes in gene expression in differentiating tissues of *Drosophila*. *Dev. Biol.* **85**, 233–242.

Mitchell, H. K., and Petersen, N. S. (1982). Developmental abnormalities induced by heat shock. *Dev. Genet.* **3**, 91–102.

Mitchell, H. K., and Petersen, N. S. (1985). The recessive phenotype of *forked* can be uncovered by heat shock in *Drosophila*. *Dev. Genet.* **6**, 93–100.

Mitchell, H. K., Moller, G., Petersen, N. S., and Lipps-Sarmiento, L. (1979). Specific protection from phenocopy induction by heat shock. *Dev. Genet.* **1**, 181–192.

Mitchell, H. K., Roach, J., and Petersen, N. S. (1983). Morphogenesis of cell hairs in *Drosophila*. *Dev. Biol.* **95**, 387–398.

Nagele, R. G., Bush, K. T., Hunter, E. T., Kosciuk, M. C., and Lee, H. (1989). Biomedical basis of diazepam-induced neural tube defects in early chick embryos: A morphometric study. *Teratology* **40**, 29–36.

Nakano, K. K. (1973). Anencephaly: A review. *Dev. Med. Child Neurol.* **15**, 383–400.

Neidhardt, F. C., VanBogelen, R. A., and Vaughn, V. (1984). The genetics and regulation of heat-shock proteins. *Annu. Rev. Genet.* **18**, 295–329.

Newman, C. G. H. (1985). Teratogen update: Clinical aspects of thalidomide embryopathy—A continuing preoccupation. *Teratology* **32**, 133–144.

Nilson, N. O. (1984). Vascular abnormalities due to hyperthermia in chick embryos. *Teratology* **30**, 237–251.

Nishikawa, T., Bruyere, H. J., Jr., Takagi, Y., Gilbert, E. F., and Matsuoka, R. (1986). The teratogenic effect of phenobarbital on the embryonic chick heart. *J. Appl. Toxicol.* **6**, 91–94.

Nora, J. J., and Nora, A. H. (1976). Genetic and environmental factors in the etiology of congenital heart diseases. *South. Med. J.* **7**, 919–926.

Nora, J. J., Sommerville, R. J., and Fraser, F. C. (1970). Homologies for congenital heart diseases: Murine models, influenced by dextroamphetamine. *Teratology* **1**, 413–416.

Nover, L. (1984). "Heat Shock Response of Eukaryotic Cells." Springer-Verlag, Berlin and New York.

Palter, K. B., Wantanabe, M., Stinson, L., Mahowald, A. P., and Craig, E. A. (1986). Expression and localization of *Drosophila melanogaster* hsp70 cognate proteins. *Mol. Cell. Biol.* **6**, 1187–1203.

Parkhurst, S. M., and Corces, V. G. (1985). *Forked*, gypsys, and suppressors in *Drosophila*. *Cell* **41**, 429–437.

Pelham, H. R. B. (1986). Speculations on the functions of the major heat shock and glucose-regulated proteins. *Cell* **46**, 959–961.

Pelham, H. (1988). Coming in from the cold. *Nature (London)* **332**, 776–777.

Petersen, N. S., and Mitchell, H. K. (1981). Recovery of protein synthesis following heat shock: Prior heat treatment affects the ability of cells to translate mRNA. *Proc. Natl. Acad. Sci. U.S.A.* **78**, 1708–1711.

Petersen, N. S., and Mitchell, H. K. (1982). Effects of heat shock on gene expression during development: Induction and prevention of the multihair phenocopy in *Drosophila*. *In* "Heat Shock from Bacteria to Man" (M. J. Schlesinger, M. Ashburner, and A. Tissières, eds.), pp. 345–352. Cold Spring Harbor Lab., Cold Spring Harbor, New York.

Petersen, N. S., and Mitchell, H. K. (1985). Heat shock proteins. *In* "Comprehensive Insect Physiology, Biochemistry, and Pharmacology" (G. A. Kerkut and L. I. Gilbert, eds.), Vol. 10, pp. 347–365. Pergamon, Oxford.

Petersen, N. S., and Mitchell, H. K. (1987). The induction of a multiple wing hair phenocopy by heat shock in mutant heterozygotes. *Dev. Biol.* **121**, 335–341.

Petersen, N. S., and Mitchell, H. K. (1989). The forked phenocopy is prevented in thermotolerant pupae. *UCLA Symp. Moll. Cell. Biol., New Ser.* **96**, 235–244.

Petersen, N. S., and Young, P. (1991). Actin mRNA is stabilized by heat shock during *Drosophila* development. *Cell Differ. Dev.* (submitted).

Petersen, N. S., Bond, B., Mitchell, H. K., and Davidson, N. (1985). Stage specific regulation of actin genes in *Drosophila* wing cells. *Dev. Genet.* **5**, 219–225.

Picard, D., Salser, S. J., and Yamamoto, K. R. (1988). A movable and regulatable inactivation function within the steroid binding domain of the glucocorticoid receptor. *Cell* **54**, 1073–1080.

Pleet, H., Graham, J. M., and Smith, D. W. (1981). Central nervous system and facial defects associated with maternal hyperthermia at four weeks gestation. *Pediatrics* **67**, 785–789.

Rapola, J., Saxen, L., and Granoth, G. (1978). Anencephaly and the sauna. *Lancet* **1**, 1162–1166.

Riabowol, K. T., Mizzen, L. A., and Welch, W. J. (1988). Heat shock is lethal to fibroblasts injected with antibodies to hsp70. *Science* **242**, 433–436.

Santamaria, P. (1979). Heat shock induced phenocopies of dominant mutants of the bithorax complex in *Drosophila melanogaster*. *Mol. Gen. Genet.* **172**, 161–163.

Saxen, L., Holmberg, P. C., Nurminen, M., and Kuosma, E. (1982). Sauna and congenital defects. *Teratology* **25**, 309–313.

Schardein, J. L. (1985). "Chemically Induced Birth Defects." Dekker, New York.

Schlesinger, M. J., Ashburner, M., and Tissières, A., eds. (1982). "Heat Shock from Bacteria to Man." Cold Spring Harbor Lab., Cold Spring Harbor, New York.

Seybold, W. D., Meltzer, P. S., and Mitchell, H. K. (1975). Phenol oxidase activation in *Drosophila*: A cascade of reactions. *Biochem. Genet.* **13**, 85–108.

Stephens, T. D. (1988). Proposed mechanisms of action in thalidomide embryopathy. *Teratology* **38**, 229–239.

Streissguth, A. P., Clarren, S. K., and Jones, K. L. (1985). Natural history of the fetal alcohol syndrome: A 15 year follow-up of eleven patients. *Lancet* **1**, 85–91.

Subjek, J. R., and Shy, T.-T. (1986). Stress protein systems of mammalian cells. *Am. J. Physiol.* **250**, C1–C17.

Tibbles, L., and Wiley, M. J. (1988). A comparative study of the effects of retinoic acid given during the critical period for inducing spina bifida in mice and hamsters. *Teratology* **37**, 113–125.

Tissières, A., Mitchell, H. K., and Tracy, U. M. (1974). Protein synthesis in salivary glands of *D. melanogaster* cells. *J. Mol. Biol.* **84**, 389–398.

Tomasovic, S. P. (1989). Functional aspects of the mammalian heat-stress proteins response. *Life Chem. Rep.* **7**, 33–63.

Velazquez, J. M., and Lindquist, S. (1984). hsp70: Nuclear concentration during environmental stress and cytoplasmic storage during recovery. *Cell* **36**, 655–662.

Walsh, D. A., Klein, N. W., Hightower, L. E., and Edwards, M. J. (1987). Heat shock and thermotolerance during early rat embryo development. *Teratology* **36**, 181–191.

Warkany, J. (1986). Teratogen update: Hyperthermia. *Teratology* **33**, 365–371.

Webster, W. S., and Edwards, M. J. (1984). Hyperthermia and induction of neural tube defects in mice. *Teratology* **29**, 417–425.

Webster, W. S., Germain, M. A., and Edwards, M. J. (1984). The induction of microthalmia, encephalocele and other head defects following hyperthermia during the gastrulation process in the rat. *Teratology* **31**, 73–82.

Welch, W. J. (1987). Mammalian stress proteins. *In* "Advances in Experimental Medicine and Biology" (M. Z. Atassi, ed.), pp. 287–304. Plenum, New York.

Widelitz, R. B., Magun, B. E., and Gerner, E. W. (1986). Effects of cycloheximide on thermotolerance expression, heat shock protein synthesis, and heat shock protein mRNA accumulation in rat fibroblasts. *Mol. Cell. Biol.* **6**, 1088–1094.

Witting, S., Hensse, S., Keitel, C. V., Elsner, C., and Wittig, B. (1983). Heat shock gene expression is regulated during teratocarcinoma cell differentiation and early embryonic development. *Dev. Biol.* **96**, 507–514.

Yost, H. J., and Lindquist, S. (1986). RNA splicing is interrupted by heat shock and is rescued by heat shock protein synthesis. *Cell* **45**, 185–193.

Yost, H. J., and Lindquist, S. (1988). Translation of unspliced transcripts after heat shock. *Science* **242**, 1544–1548.

INDEX